My God, He Plays Dice!

How Albert Einstein Invented Most Of Quantum Mechanics

Bob Doyle

The Information Philosopher

"beyond logic and language"

Is it possible that the most famous critic of quantum mechanics actually invented most of its fundamentally important concepts?

In his 1905 Brownian motion paper, Einstein quantized matter, proving the existence of atoms. His light quantum hypothesis showed that energy itself comes in particles (photons). He showed energy and matter are interchangeable, $E = mc^2$. In 1905 Einstein was first to see nonlocality and instantaneous action-at-a-distance. In 1907 he saw quantum "jumps" between energy levels in matter, six years before Bohr postulated them in his atomic model. Einstein saw wave-particle duality and the "collapse" of the wave in 1909. And in 1916 his transition probabilities for emission and absorption processes introduced ontological chance when matter and radiation interact, making quantum mechanics statistical. He discovered the indistinguishability and odd quantum statistics of elementary particles in 1925 and in 1935 speculated about the nonseparability of interacting identical particles.

It took physicists over twenty years to accept Einstein's light-quantum. He explained the relation of particles to waves fifteen years before Heisenberg matrices and Schrödinger wave functions. He saw indeterminism ten years before the uncertainty principle. And he saw nonlocality as early as 1905, presenting it formally in 1927, but was ignored. In the 1935 Einstein-Podolsky-Rosen paper, he explored nonseparability, which was dubbed "entanglement" by Schrödinger. The EPR paper has gone from being irrelevant to Einstein's most cited work and the basis for today's "second revolution in quantum mechanics."

In a radical revision of the history of quantum physics, Bob Doyle develops Einstein's idea of objective reality to resolve several of today's most puzzling quantum mysteries, including the two-slit experiment, quantum entanglement, and microscopic irreversibility.

Schrödinger's Cat

Did Albert Einstein Invent

This book on the web
informationphilosopher.com/einstein/

chanics

ynia
tio

eci

ste

erp

Viol

lement

Special
Relativity

My God,
He Plays Dice!

How Albert Einstein
Invented Most Of
Quantum Mechanics

Are

eal

Particles or Fields?

These Great Physical Concepts?

Bob Doyle
The Information Philosopher
"beyond logic and language"

First edition, 2019
© 2019, Bob Doyle, The Information Philosopher

Publisher's Cataloging-In-Publication Data
(Prepared by The Donohue Group, Inc.)

Names: Doyle, Bob, 1936- author.

Title: My God, he plays dice! : how Albert Einstein invented most of quantum mechanics / Bob Doyle, the Information Philosopher.

Description: First edition. | Cambridge, MA, USA : I-Phi Press, 2019. | Includes bibliographical references and index.

Identifiers: ISBN 9780983580249 | ISBN 9780983580256 (ePub)

Subjects: LCSH: Einstein, Albert, 1879-1955--Influence. | Quantum theory. | Science--History.

Classification: LCC QC174.13 .D69 2019 (print) | LCC QC174.13 (ebook) | DDC 530/.12--dc23

I-Phi Press
77 Huron Avenue
Cambridge, MA, USA

Dedication

This book is dedicated to a handful of scholars who noticed that Albert Einstein's early work on quantum mechanics had been largely ignored by the great "founders" of quantum theory, overshadowed by his phenomenal creations of special and general relativity and by his dissatisfaction with "quantum reality."

Most notably I want to thank Leslie Ballentine, Frederik. J. Belinfante, David Cassidy, Carlo Cercigniani, Max Dresden, Gerald Holton, Don Howard, Max Jammer, Martin J. Klein, Thomas S. Kuhn, Cornelius Lanczos, Jagdish Mehra, Abraham Pais, Helmut Rechenberg, John C. Slater, John Stachel, A. Douglas Stone, and B. L. van der Waerden.

I also want to thank the many editors and translators of the *Collected Papers of Albert Einstein*, as well as the Hebrew University of Jerusalem and the Princeton University Press for making Einstein's work available online for scholars everywhere.

I have purchased all the volumes of CPAE over the years for my own library, but I am delighted that all these critical documents are now available online for free.

Information philosophy builds on the intersection of computers and communications. These two technologies will facilitate the sharing of knowledge around the world in the very near future, when almost everyone will have a smartphone and affordable access to the Internet and the World Wide Web.

Information is like love. Giving it to others does not reduce it. It is not a scarce economic good. Sharing it increases the total information in human minds, the *Sum* of human knowledge.

Information wants to be free.

Bob Doyle (bobdoyle@informationphilosopher.com)
Cambridge, MA
March, 2019

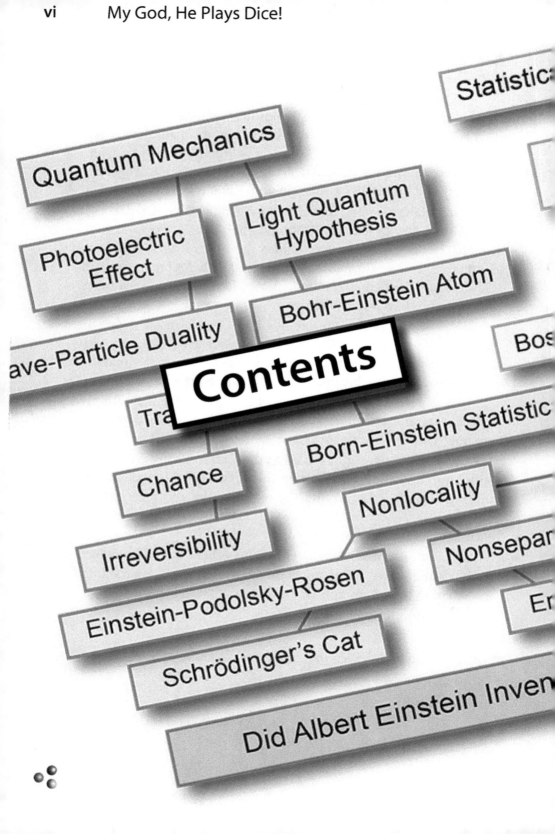

Statistic

Quantum Mechanics

Light Quantum
Hypothesis

Photoelectric
Effect

Bohr-Einstein Atom

ave-Particle Duality

Bos

Contents

Tra

Born-Einstein Statistic

Chance

Nonlocality

Irreversibility

Nonsepar

Einstein-Podolsky-Rosen

Er

Schrödinger's Cat

Did Albert Einstein Inven

Table of Contents

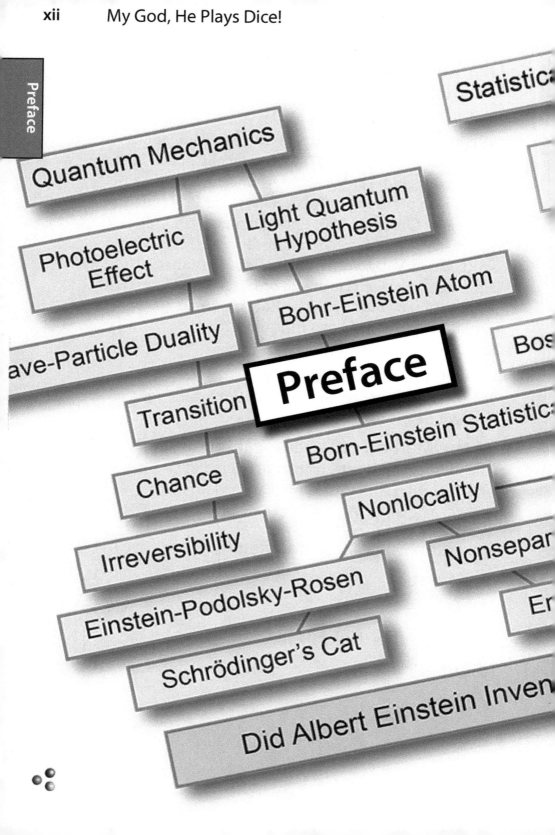

Preface

For well over a century, ALBERT EINSTEIN's many original contributions to quantum mechanics have been doubted by his colleagues. Some of those contributions have been credited to others, perhaps for the understandable reason that Einstein himself severely criticized his most revolutionary ideas.

MAX PLANCK is often cited today as discovering the photon. NIELS BOHR's discrete energy levels in atomic matter were first seen by Einstein in 1906 as explaining the anomalous specific heat of certain atoms. MAX BORN's 1926 statistical interpretation of the wave function was based on Einstein's 1909 insight that the light wave gives us probabilities of finding light particles. DAVID BOHM's particle mechanics with continuous paths and properties is an attempt to achieve Einstein's "objective reality." And JOHN BELL's claim that the "Einstein program fails" is based on a model of "hidden variables" that is physically unrealistic.

THE NEW YORK TIMES **INTERNATIONAL** THURSDAY, OCTOBER 22, 2015

Sorry, Einstein, but 'Spooky Action' Seems Real

By JOHN MARKOFF

In a landmark study, scientists at Delft University of Technology in the Netherlands reported that they had conducted an experiment that they say proved one of the most fundamental claims of quantum theory — that objects separated by great distance can instantaneously affect each other's behavior.

The finding is another blow to one of the bedrock principles of standard physics known as "locality," which states that an object is directly influenced only by its immediate surroundings. The Delft study, published Wednesday in the journal Nature, lends further credence to an idea that Einstein famously rejected. He said quantum theory necessitated "spooky action at a distance," and he refused to accept the no-

tion that the universe could behave in such a strange and apparently random fashion.

In particular, Einstein derided the idea that separate particles could be "entangled" so completely that measuring one particle would instantaneously influence the other, regardless of the distance separating them.

Einstein was deeply unhappy with the uncertainty introduced

Continued on Page A15

The New York Times in 2015 loudly proclaimed on its front page Einstein's mistake in doubting that measuring one particle can instantaneously influence another at an arbitrary distance. [1]

They did not mention it was Einstein who first saw "nonlocality" in 1905, reported it in 1927, and in his EPR paper of 1935 introduced it as "nonseparability," which he attacked. But without Einstein, it is likely no one ever would have seen "entanglement."

---

1   *The New York Times*, October 22, 2015, p.1

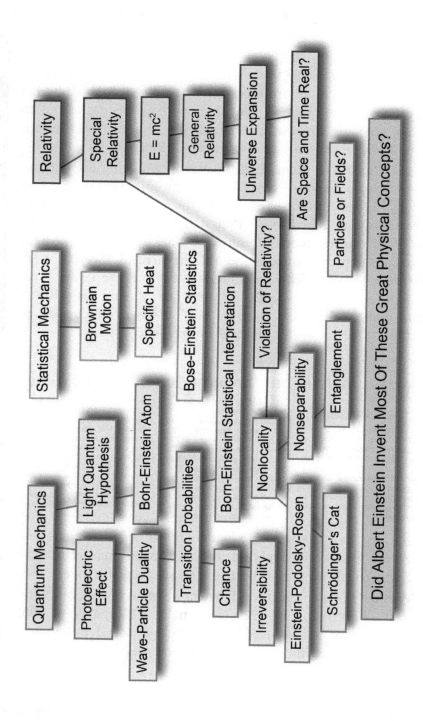

Thirty years ago, the *Economist* magazine described the "queerness of quanta." Quantum mechanics appears to say some rather odd things about the universe, they reported,

- There are no such things as "things". Objects are ghostly, with no definite properties (such as position or mass) until they are measured. The properties exist in a twilight state of "superposition" until then.

- All particles are waves, and waves are particles, appearing as one or the other depending on what sort of measurement is being performed.

- A particle moving between two points travels all possible paths between them simultaneously.

- Particles that are millions of miles apart can affect each other instantaneously. [2]

They also reported RICHARD FEYNMAN's critical analysis of the two-slit experiment. "The conclusion is inescapable. The photons somehow pass through both slits at once." [3]

All of these "queer" aspects of quantum mechanics were challenged by Einstein, even those that he was first to see as (perhaps unacceptable) possibilities. This led to his popular reputation as a critic of quantum mechanics. He was a critic, but he also accepted most of quantum mechanics!

> the reader should be convinced that I fully recognize the very important progress which the statistical quantum theory has brought to theoretical physics... This theory is until now the only one which unites the corpuscular and undulatory dual character of matter in a logically satisfactory fashion.. The formal relations which are given in this theory — i.e., its entire mathematical formalism — will probably have to be contained, in the form of logical inferences, in every useful future theory.

> What does not satisfy me in that theory, from the standpoint of principle, is its attitude towards that which appears to me to be the programmatic aim of all physics: the complete description of any (individual) real situation (as it supposedly exists irrespective of any act of observation or substantiation). [4]

---

2   *The Economist,* January 7, 1989, p.71
3   *ibid.,* p.72
4   "Reply to Critics," in Schilpp, 1949, p.666

This book is based on ALBERT EINSTEIN's web page on our Information Philosopher website, [5] which we started writing in 2007. We began the book in 2015 with our primary goal to review and correct the history of Einstein's contributions to quantum mechanics, which have been distorted for decades by the unfortunately biased accounts of the so-called "founders" of quantum mechanics, notably NIELS BOHR, WERNER HEISENBERG, and MAX BORN.

Besides hypothesizing light particles (1905) and seeing their interchangeability with matter, $E = mc^2$, Einstein was first to see many of the most fundamental aspects of quantum physics - the quantal derivation of Planck's blackbody radiation law, nonlocality and instantaneous action-at-a-distance (1905), the internal structure of atoms (1906), wave-particle duality and the "collapse" of the wave aspect (1909), transition probabilities for emission and absorption processes that introduce indeterminism whenever matter and radiation interact, making quantum mechanics a statistical theory (1916-17), the indistinguishability of elementary particles with their strange quantum statistics (1925), and the nonseparability and entanglement of interacting identical particles (1935).

It took the physics community eighteen years to accept Einstein's "very revolutionary" light-quantum hypothesis. He saw wave-particle duality at least ten years before LOUIS DE BROGLIE, ERWIN SCHRÖDINGER, Heisenberg, and Bohr. He saw indeterminism a decade before the Heisenberg uncertainty principle. He saw nonlocality as early as 1905, presenting it formally in 1927, but he was misunderstood and ignored. In the 1935 Einstein-Podolsky-Rosen paper, he examined nonseparability, which was dubbed "entanglement" by Schrödinger.

Our secondary goal is to show how a revised understanding of Einstein's contributions and his deep desire to describe an "objective reality" can lead to plausible solutions for some *unsolved* problems in statistical mechanics and quantum physics.

These problems or "mysteries" include:
- The 19th-century problem of microscopic irreversibility
- Nonlocality, first seen by Einstein in 1905
- Wave and particle "duality" (1909)
- The metaphysical question of ontological chance (1916)

- Nonlocality and action-at-a-distance (1927)
- The "mystery" of the two-slit experiment (1927)
- The measurement problem (1930)
- The role of a "conscious observer" (1930)
- Entanglement and "spooky" action-at-a-distance (1935)
- Schrödinger's Cat - dead and alive?
- No "hidden variables," but hidden constants
- Conflict between relativity and quantum mechanics?
- Is the universe deterministic or indeterministic?

A third ambitious goal is at once physically, metaphysically, and philosophically very deep, and yet we hope to explain it in such a simple way that it can be understood by almost everyone.

This goal is to answer a question that Einstein considered throughout his life. *Is nature continuous or discrete?*

Einstein's work on matter and light appears to show that the physical world is made up of nothing but *discrete discontinuous particles*. Continuous *fields* with well-defined values at all places and times may be simply abstract theoretical constructs, "free creations of the human mind" he called them, only "observable" as *averages* over very large numbers of discrete particles.

A year before his death, Einstein wrote to an old friend,

"I consider it quite possible that physics cannot be based on the field concept, i.e:, on continuous structures. In that case, nothing remains of my entire castle in the air, gravitation theory included, [and of] the rest of modern physics." [6]

No one did more than Einstein to establish the *reality* of particles of matter and energy. His study of Brownian motion proved that atoms are real. His analysis of the photoelectric effect proved that localized quanta of light are real. But Einstein wrestled all his life with the apparently continuous wave aspects of light and matter.

Einstein could not accept most of his quantum discoveries because their *discreteness* conflicted with his basic idea that nature is best described by a *continuous* field theory using differential equations that are functions of "local" variables, primarily the spacetime four-vector of his general relativistic theory.

---

6   Pais, 1982, p.467

Fields are "free creations of the human mind."

Einstein's description of wave-particle duality is as good as anything written today. He saw the relation between the wave and the particle as the relation between *probability* and the realization of one *possibility* as an *actual* event. He saw the *continuous* light wave spreading out in space as a mathematical construct giving us the probable number of *discrete* light particles in different locations.

But if light waves are carrying energy, Einstein feared their instantaneous "collapse" in the photoelectric effect might violate his special theory of relativity. He was mistaken.

Nonlocality is the idea that some interactions are transferring something, matter, energy, or minimally abstract information, faster than the speed of light. Einstein originated this idea, but this book will show that his hope for an "objective" *local* reality can be applied to deny the popular instances of nonlocal "action-at-a-distance," providing us a new insight into the mystery of "entanglement," the so-called "second revolution" in quantum mechanics.

DAVID BOHM thought "hidden variables" might be needed to communicate information between entangled particles. We shall show that most information is transported by "hidden" constants of the motion, but at speeds equal to or below the speed of light.

Nonlocality is only the *appearance* of faster-than-light action

Two particles travel away from the center in what quantum mechanics describes as a *superposition* of two possible states. Either particle has either spin down or spin up. The two-particle wave function is

$$\psi = (1/\sqrt{2}) \, (| + - > - | - + >).$$

In "objective reality," a specific pair starts off in just one of these states, say $| + - >$, as explained by PAUL DIRAC. See chapter 19.

A few moments later they are traveling apart in a $|+ - >$ state, with the left electron having spin $+1/2$ and the right $-1/2$. But neither has a definite spatial spin component in a given direction such as z+.

A *directionless* spin state is symmetric and isotropic, the same in all directions. It is rotationally invariant. Spin values of + and - are traveling with the particles from their entanglement in the center.

Because they are entangled, the + spin in the left-moving electron is always perfectly opposite that of the - spin electron moving right..

While there might not be Bohmian "hidden variables," the conserved spin quantities might be called "hidden constants" ("hidden in plain sight) that explain the *appearance* of nonlocal, nonseparable behavior.

But when the two particles are measured, they project spatial components of the two directionless spins, the two projections are occurring simultaneously in a spacelike separation. Einstein's special theory of relativity maintains such simultaneity is impossible.

Although nonlocality and nonseparability are only appearances, "objectively real" entanglement is all that is needed for quantum information, computing, encryption, teleportation, etc.

Information about probabilities and possibilities in the wave function is *immaterial*, not material. But this abstract information has real causal powers. The wave's *interference* with itself predicts *null points* where no particles will be found. And experiments confirm that no particles are found at those locations.

But how can mere probability influence the particle paths?

This is the one deep mystery in quantum mechanics.

Information philosophy sees this *immaterial* information as a kind of modern "spirit." Einstein himself described a wave as a "ghostly field" (*Gespensterfeld*) and as a "guiding field" (*Führungsfeld*). This idea was taken up later by LOUIS DE BROGLIE as "pilot waves," by ERWIN SCHRÖDINGER, who developed the famous equation that describes how his wave function moves through space *continuously* and *deterministically, and* by MAX BORN in his "statistical interpretation" (actually based on a suggestion by Einstein!).

Schrödinger objected his whole life to Born's idea that his *deterministic* wave function was describing the *indeterministic* behavior of particles. That quantum mechanics is statistical was of course the original idea of Einstein. But Born put it succinctly,

The motion of the particle follows the laws of probability, but the probability itself propagates in accord with causal laws. [7]

---

7   Born. 1926, p. 803.

Einstein believed that quantum mechanics, as good as it is, is "incomplete." Although the "founders" denied it, quantum theory is in fact incomplete. Its *statistical* predictions (phenomenally accurate in the limit of large numbers of identical experiments), tell us nothing but "probabilities" for *individual* systems.

Einstein's idea of an "objective reality" is that particles have paths and other properties independent of our measurements. He asked whether a particle has a position before we measure it and whether the moon only exists when we are looking at it? The fact that it is impossible *to know* the path or properties of a particle without measuring them does not mean that they do not exist.

Einstein's idea of a "local" reality is one where "action-at-a-distance" is limited to causal effects that propagate at or below the speed of light, according to his theory of relativity. This *apparent* conflict between quantum theory and relativity can be resolved using an explanation of nonlocality and nonseparability as merely "knowledge-at-a-distance," or "information-at-a-distance."

Einstein felt that his ideas of a local and objective reality were challenged by an *entangled* two-particle system which *appears* to produce instantaneous correlations between events in a space-like separation. He mistakenly thought this violated his theory of special relativity. This was the heart of his famous Einstein-Podolsky-Rosen paradox paper in 1935. But we shall show that Einstein had been concerned about faster-than-light transfer of energy or information from his very first paper on quantum theory in 1905.

In most general histories, and in the brief histories included in modern quantum mechanics textbooks, the problems raised by Einstein are usually presented as arising *after* the "founders" of quantum mechanics and their "Copenhagen Interpretation" in the late 1920's. Modern attention to Einstein's work on quantum physics often starts with the EPR paper of 1935, when his mysteries about nonlocality, nonseparability, and entanglement were not yet even vaguely understood as a problem by his colleagues.

Even today, when entanglement is advertised as the "second revolution" in quantum mechanics," few physicists understand it.

We will see that entanglement challenged Einstein's idea that his special theory of relativity shows the "impossibility of simultaneity."

Most physics students are taught that quantum mechanics *begins* with the 1925 Heisenberg (matrix/particle) formulation, the 1926 Schrödinger (wave) formulation, Born's statistical interpretation of the wave function in 1926, Heisenberg's uncertainty (indeterminacy) principle in 1927, then Dirac's transformation theory and von Neumann's measurement problem in 1930.

The popular image of Einstein post-EPR is either in the role of critic trying to expose fundamental flaws in the "new" quantum mechanics or as an old man who simply didn't understand the new quantum theory.

Both these images of Einstein are seriously flawed, as we shall see. It was actually the "founders" who did not understand Einstein's concerns, especially nonlocality. When physicists began to appreciate them between the 1960's and 1980's, they labeled them "quantum mysteries" that dominate popular discussions today.

Einstein and Schrödinger wanted to *visualize* quantum reality. Bohr and Heisenberg's Copenhagen Interpretation says don't even try to look for an underlying "quantum reality." But Einstein's ability to visualize quantum reality was unparalleled, despite errors that continue to mislead quantum physicists today.

While almost none of Einstein's contemporaries knew what his "spooky action-at-a-distance" was talking about, today "entanglement" is at the height of popularity and at the heart of quantum computing and encryption.

Einstein's best known biographer, Abraham Pais, said of the EPR paper, "It simply concludes that objective reality is incompatible with the assumption that quantum mechanics is complete. This conclusion has not affected subsequent developments in physics, and it is doubtful that it ever will."[8] Today, the EPR paper is the most cited of all Einstein's work, and perhaps of all physics!

We will focus on restoring Einstein's reputation as a *creator*, rather than a destructive critic of quantum mechanics. It is astonishing how many things that he was first to see have become central to quantum theory today. A close reading of Einstein recognizes him as the *originator* of both great theories of 20th-century physics, both relativity and quantum mechanics.

---

8   Pais, 1982, p. 456

Questions to Consider

As you read through this book, please keep in mind the following questions that we will explore throughout. Some of these issues Einstein was best known for denying, but he was first to see them and he considered them as very serious possibilities.

1) Are the fundamental constituents of the universe discrete discontinuous localized particles, and not continuous fields?

Nuclear, electromagnetic, and gravitational fields are theoretical constructs predicting the *forces* that would be felt by a test particle located at a given position in space.

Quantum mechanical fields, squares of the probability amplitudes $|\psi^2|$, predict the *probabilities* of finding particles at that position.

Probability amplitudes are calculated by solving the Schrödinger equation for eigenvalues consistent with the distribution of matter, the local "boundary conditions." Thus, probability amplitudes are different when one or two slits are open, independent of the presence of any test particle.

Can particles be successfully represented as singularities in continuous fields that carry substance? Can they be described as localized "wave packets," made from superimposed waves of different frequencies? Probably not.

2) Does ontological chance exist, or as Einstein might have put it, "Does God play dice"?

Einstein was the discoverer of ontological chance in his 1916 derivation of the Planck radiation law and the transition probabilities for emission and absorption needed to maintain thermal equilibrium. This led to his seeing the statistical nature of quantum mechanics.

Chance underlies indeterminacy and irreversibility. Without it there are no alternative possible futures and no free will.

3) Was Einstein right about an "objective reality"?

Can particles have continuous paths even though individual paths cannot be observed without disturbing them?

Just because we cannot continuously observe particles does not mean they are free to change their properties in ways that violate conservation principles.

Just because paths are not "observables" and we don't know them does not mean that those paths do not exist, as mistakenly insisted by the Copenhagen Interpretation, which claims that particle positions only come into existence when a measurement is made.

Regarding such extreme anthropomorphism, JOHN BELL quipped, does the experimenter need a Ph.D.?

Can "objective reality" give us a picture of particles moving along unobservable paths that conserve all the particle properties, so that when they are observed, properties like electron and photon spins are perfectly correlated with the values they were created with.

These "constants of the motion" would appear to be *communicating*, when they are actually just *carrying* information along their paths. We call them "hidden constants."

Measurements of electron spin spatial components by Alice and Bob are an exception, since they create the values.

6) Did Einstein see space and time as mathematical constructs?

We project continuous coordinates onto space to describe the changing relations between discrete discontinuous particles.

Are space and time just mathematical fictions, mere ideas invented by scientists? Two great nineteenth-century mathematicians were a great inspiration for Einstein.

One, LEOPOLD KRONECKER, said "God created the integers. All else is the work of man." The other, RICHARD DEDEKIND, said mathematical theories are "free creations of the human mind," a favorite phrase of Einstein, who called theories "fictions," however amazing they are in predicting phenomena.

7) Does the "expansion of space," which Einstein saw first, just mean that some particles are separating from one another?

Many visible objects, galaxies, stars, planets are not participating in the expansion. Their gravitational binding energy exceeds their kinetic energy, partly thanks to invisible dark matter.

Between large clusters of galaxies, the creation of more phase-space cells allows for new arrangements of particles into low-entropy information structures. New information created since the origin of the universe led first to the creation of elementary particles and atoms, then the galaxies, stars, and planets. The "negative entropy" radiating from the Sun supported the evolution of life.

Plausible, If Radical, Answers to Quantum Questions

- *On "spooky" action-at-a-distance.* Two entangled particles yield perfectly correlated properties at enormous distances, as long as they have not interacted with their environment. Have they somehow communicated with one another faster than light? Or do they simply conserve the same properties they had when first created, as the conservation laws suggest? Einstein showed that particles fired off in opposite directions, with equal and opposite momenta, can tell us the position of the second by measuring the first. Einstein used the conservation of momentum to reach this conclusion, which is still valid. But when David Bohm in 1952 changed the EPR experiment to include electron spins, the measurements by Alice and Bob of spin or polarization in spatial coordinates introduced a different kind of nonlocality. Alice's and Bob's values of spin components z+ and z- are *created* by her measurement. They are *nonlocal*, appearing simultaneously at a spacelike separation. But there is no action by one particle on the other! This nonlocality is only "knowledge-at-a-distance." See chapters 29 and 34.

- *On "hidden variables" and entanglement.* There are no hidden variables, local or nonlocal. But there are "hidden *constants*." Hidden in plain sight, they are the "constants of the motion," conserved quantities like energy, momentum, angular momentum, and spin, both electron and photon. These hidden constants explain why entangled particles retain their perfect correlation as they travel apart to arbitrary distances. The Copenhagen Interpretation says there are no properties until Alice's measurement, but this is wrong. The particles' objectively real properties are local and constant from their moment of entanglement, as long as they are not decohered by interactions with the environment. These + and - spins are *directionless*. Alice's measurement creates the *nonlocal* directional spin components z+ and z-. See chapters 30 to 32.

- *On the "one mystery" in the two-slit experiment.* RICHARD FEYNMAN made the two-slit experiment the defining mystery of quantum mechanics. How can a particle interfere with itself if it does not go through both slits? Einstein's "objective reality" imagines a continuous particle path, so it goes through one slit. But the wave function, determined by the solution of the Schrödinger equation given the surrounding boundary conditions, is different when two slits are open. Incoming particles show the two-slit interference pattern whichever slit they come through. See chapter 33.

- *On microscopic irreversibility.* Collisions between atoms and molecules are irreversible whenever radiation is emitted or absorbed. Einstein showed that an emitted photon goes off in a random direction, introducing the "molecular disorder" LUDWIG BOLTZMANN wanted. See chapter 12.

- *On nonlocality.* In his photoelectric effect explanation, Einstein wondered how the light wave going off in all directions could suddenly gather together and deposit all its energy at one location. No matter, energy, or information moves at greater than light speed when correlated information appears after a two-particle wave function collapse. See chapter 23.

- *On the conflict between relativity and quantum mechanics.* Einstein thought nonlocality - simultaneous events at space-like separations - cause a conflict between special relativity and quantum mechanics. He was wrong. We think there is a conflict between general relativity and quantum mechanics. The conflict disappears if gravity consists of discrete particles, whose separations are limited by inter-particle forces. Einstein suggested quantum mechanics and gravitation should be treated by discrete algebraic equations, not continuous differential equations with their unrealistic singularities.

- *On the "measurement problem."* Copenhageners think particles have no properties until they are measured. Indeed they say that those properties do not exist until they reach the mind of a "conscious observer." Einstein responded, "Look, I don't believe that when I am not in my bedroom my bed spreads out all over the room, and whenever I open the door and come in, it jumps into the corner." Conservation laws prevent the particles from moving erratically. See chapter 42.

- *On Schrödinger's Cat.* The cat was a challenge to the idea that a quantum system, actually the system's wave function Ψ, can be in a linear combination or *superposition* of states. It led to the absurd idea that a quantum cat can be both dead and alive, or that a particle can be in two places at the same time, or go through both slits in the two-slit experiment. Recall Einstein's view that the wave function is a "ghost field" guiding the particle, and is not "objectively real." See chapter 28.

- *On indeterminism.* Standard "orthodox" quantum mechanics accepts indeterminism and acausality. Einstein initially rejected indeterminism. "God does not play dice," he said repeatedly. But he came to accept that quantum physics is the most perfect theory we have at the moment, including its indeterminism. He thought nothing within the theory could change that fact. Only a much deeper theory might be found, he hoped, out of which the current theory might emerge, But quantum processes are statistical, introducing creative new possibilities, not pre-determined by past events. Indeterminism is the source of all creativity, physical, biological, and intellectual, "free creations of the human mind.".

- *On chance.* When Einstein explained the rates of "quantum jumps" between energy levels in the Bohr Atom, he found that a light particle had to be emitted in a random direction and at a random time, in order to maintain the equilibrium between radiation and matter, so they could both have the same temperature. This Einstein called "chance," and a "weakness in the theory." Einstein's chance is *ontological*. Heisenberg's uncertainty principle is *epistemological*. See chapter 11.

- *On the "collapse" of the wave function.* The Copenhagen Interpretation and standard quantum physics describe the "collapse" as the "reduction of the wave packet" from a linear combination or "superposition" of many quantum states into a single quantum state. WERNER HEISENBERG described the collapse as acausal, uncertain, indeterministic, and dependent on the "free choice" of the experimenter as to what to measure. This is correct, but he did not connect it to Einstein's *ontological* "chance." See chapter 24.

- *On waves and particles.* When Einstein showed that matter is made of *discrete* particles and hypothesized that light is also particles, he described the light waves as "ghost" fields, insubstantial but somehow governing the paths and ultimate positions of the substantial particles, so also "guiding" fields. The wave is only a mathematical device for calculating probabilities of finding photons. Only the light particles are "objectively real." Einstein pointed out that fields are convenient "fictions" that allow us to make amazingly accurate, though *statistical*, predictions. See chapter 9.

- *Why particles are more "objectively real" than fields* One of Einstein's earliest accomplishments was to reject the idea of a universal ether, a field which was the medium in which light could be the vibrations. James Clerk Maxwell's electric and magnetic fields have replaced the ether. Now quantum theory sees the electromagnetic field as only the average behavior of large numbers of Einstein's light quanta or photons. Particles are physical. Fields, especially continuous fields, are metaphysical.

- *On the incompleteness of quantum mechanics.* Einstein finally caught the attention of physicists and the general public with his claim in 1935 that quantum mechanics is "incomplete," that it is a *statistical* theory saying nothing certain about individual particles. Niels Bohr responded that the new quantum mechanics is complete, based on his philosophical idea of *complementarity*. But he offered no proof. Einstein was right. Quantum theory is incomplete. See chapters 26 to 29.

- *Is quantum mechanics epistemological or ontological?* Does quantum mechanics provide only the words and language we use to talk about the world, or does it access what philosophers call the "things in themselves.? Einstein's hopes for seeing an "objective reality" were dashed by almost all his physicist colleagues in the 1920's. We must give full credit to the "founders of quantum mechanics" who at that time gave us the extraordinary mathematical apparatus - and not just language - that allows us to predict the behavior of the physical world, albeit only statistically as Einstein was first to discover. But we hope to show that many of the concepts underlying their mathematics were discovered or invented by Einstein. Niels Bohr ignored or attacked those concepts for many years, especially light as a particle. Bohr was a positivist influenced by linguistic philosophers who think talk about an objectively real world is "metaphysics." He was unequivocal.

"There is no quantum world. There is only an abstract quantum physical description. It is wrong to think that the task of physics is to find out how nature is. Physics concerns what we can say about nature. [9]

My goal is to change Einstein's reputation from "the best known critic of quantum mechanics" [10] to the "inventor of most of the basic concepts in quantum mechanics," including his objective reality.

Bob Doyle
bobdoyle@informationphilosopher.com
Cambridge, MA
December, 2018

9   *Bulletin of the Atomic Scientists.* Sep 1963, Vol. 19 Issue 7, p.12
10   Nielsen and Chuang, 2010, p.2

How To Use This Book With The I-Phi Website

The content of this book comes primarily from the quantum section of the informationphilosopher.com website and from the individual web pages for Einstein, Planck, Heisenberg, Bohr, and other quantum physicists. You will find multiple entry points into the I-Phi site from this book, with URLs for the chapters and in many of the footnotes. I hope that you agree that the combination of a printed book and an online knowledge-base website is a powerful way to do philosophy in the twenty-first century.

The Quantum web page has a right-hand navigation menu with links to the many philosophers and scientists who have contributed to the development of quantum physics.

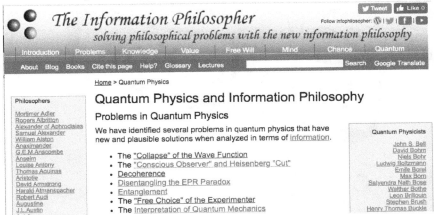

Figures in the text often link to full-color *animated* images on the I-Phi website. All images are original works or come from open-source websites.

Names in SMALL CAPS indicate philosophers and scientists with their own web pages on the I-Phi website.

It is not easy to navigate any website, and I-Phi is no exception. Find things of interest quickly with the Search box on every page. Once on a page, a "Cite this page" function generates a citation with the URL and the date you retrieved the page, in standard APA format that you can copy and paste into your work.

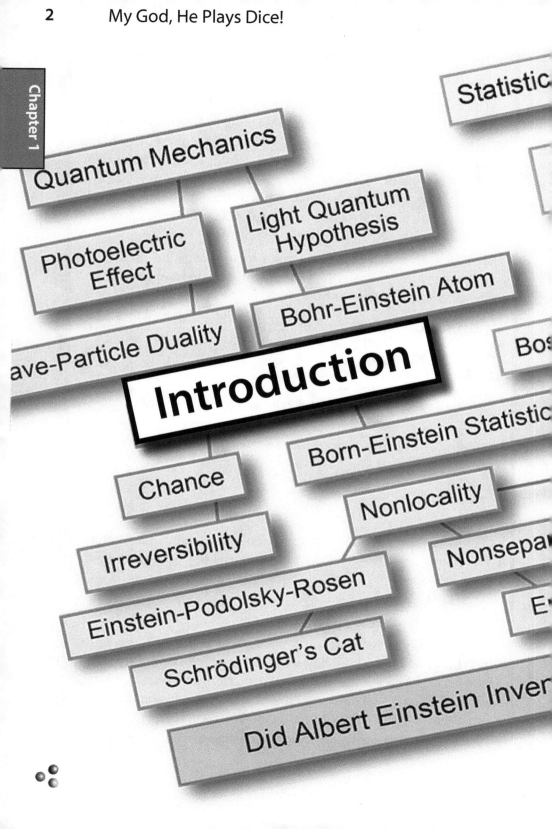

Statistic

Quantum Mechanics

Light Quantum
Hypothesis

Photoelectric
Effect

Bohr-Einstein Atom

ave-Particle Duality

Bos

Introduction

Born-Einstein Statistic

Chance

Nonlocality

Irreversibility

Nonsepa

Einstein-Podolsky-Rosen

E

Schrödinger's Cat

Did Albert Einstein Inven

Introduction

This book is the story of how ALBERT EINSTEIN analyzed what goes on when *light* interacts with *matter* and how he discovered ontological *chance* in the process. We can show that Einstein's chance explains the metaphysical possibilities underlying the creation of all of the *information structures* in the universe.

But the story begins with a deck of cards, a pair of dice, and the multiple flips of a coin.

Around 1700, ABRAHAM DE MOIVRE, a French Huguenot, emigrated to England to escape religious persecution. A brilliant mathematician, he worked with ISAAC NEWTON and other great English scientists, but he could never get an academic post, despite their excellent recommendations. To support himself, de Moivre wrote a handbook for gamblers called *The Doctrine of Chances*.

This was not the first book that calculated the odds for different hands of cards or rolls of the dice. But when de Moivre considered the flipping of a fair coin (with 50-50 odds of coming up heads and tails) he showed that as the number of flips gets large, the *discrete* binomial distribution of outcomes approaches a *continuous* curve we call the Gaussian distribution (after the great mathematician CARL FRIEDRICH GAUSS), the "normal" distribution, or just the "bell curve," from its familiar shape.

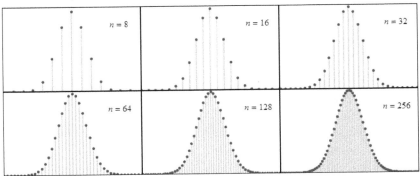

Figure 1-1. De Moivre's discovery of the continuous bell curve as a limit to a large number of discrete, discontinuous events. Each discrete event is the probability of m heads and n-m tails in n coin tosses. The height is the coefficient in the binomial expansion of $(p + q)^n$ where $p = q = \frac{1}{2}$.

In mathematics, we can say that a finite number of discrete points approaches a continuum as we let the number approach infinity. This is the "law of large numbers" and the "central limit theorem."

But in physics, the continuous appearance of material things is only because the discrete atoms that make it up are too small to see. The analytic perfection of the Gaussian curve cannot be realized by any finite number of events.

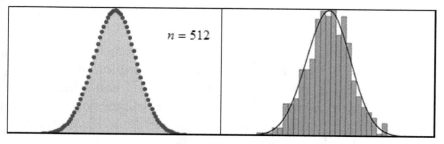

Figure 1-2. The appearance of a continuous curve and actual finite events.

Is the Nature of Reality Continuous or Discrete?

Is it possible that the physical world is made up of nothing but discrete discontinuous *particles*? Are continuous *fields* with well-defined values for matter and energy at all places and times simply theoretical constructs, averages over large numbers of particles?

Space and time themselves have well-defined values everywhere, but are these just the abstract information of the *ideal* coordinate system that allows us to keep track of the positions and motions of particles? Space and time are physical, but they are not *material*.

We use material things, rulers and clocks, to measure space and time. We use the abstract mathematics of real numbers and assume there are an *infinite number* of real points on any line segment and an infinite number of moments in any time interval. But are these continuous functions of space and time nothing but *immaterial* ideas with no material substance?

The two great physical theories at the end of the nineteenth century, Isaac Newton's classical mechanics and James Clerk Maxwell's electrodynamics, are *continuous field theories*.

Solutions of their field equations determine precisely the exact forces on any material particle, providing complete information about their past and future motions and positions. Field theories are generally regarded as *deterministic* and *certain*.

Although the dynamical laws are "free inventions of the human mind," as Einstein always said,[1] and although they ultimately depend on experimental evidence, which is always *statistical*, the field theories have been considered superior to merely statistical laws. Dynamical laws are thought to be *absolute*, based on *principles*.

We will find that the continuous, deterministic, and analytical laws of classical dynamics and electromagnetism, expressible as differential equations, are idealizations that "go beyond experience."

These continuous laws are to the discontinuous and discrete particles of matter and electricity (whose motions they describe perfectly) as the analytical normal distribution above is to the finite numbers of heads and tails. A *continuum* is approached in the limit of large numbers of particles, when the random fluctuations of individual events can be averaged over.

Experiments that support physical laws are always finite in number. Experimental evidence is always *statistical*. It always contains *errors* distributed randomly around the most probable result. And the distribution of those errors is often *normal*.

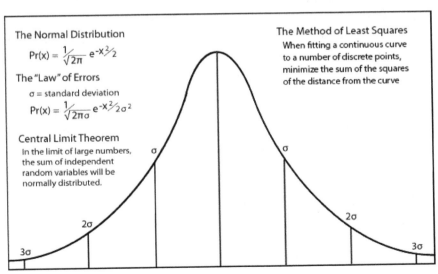

Figure 1-3. Random errors are normally distributed around the mean value.

---

1   EInstein, 1934, p.234

The Absolute Principles of Physics

There are of course *absolute* principles in physics, such as the conservation laws for mass/energy, momentum, angular momentum, and electron spin. The constant velocity of light is another.

The great mathematician EMMY NOETHER proposed a theorem that conservation principles are the consequence of deep *symmetry* principles of nature. She said for any property of a physical system that is symmetric, there is a corresponding conservation law.

Noether's theorem allows physicists to gain insights into any general theory in physics, by analyzing the various transformations that would make the form of the laws involved *invariant*.

For example, if a physical system is symmetric under rotations, its angular momentum is conserved. If it is symmetric in space, its momentum is conserved. If it is symmetric in time, its energy is conserved. Now locally there is time symmetry, but cosmically the expansion of the universe gives us an arrow of time connected to the increase of entropy and the second law of thermodynamics.

The conservation of energy was the *first law* of thermodynamics.

The famous *second law* says entropy rises to a maximum at thermal equilibrium. It was thought by most scientists to be an absolute law, but we shall see in chapter 3 that Maxwell and LUDWIG BOLTZMANN considered it a statistical law. Boltzmann thought it possible that a system that had reached equilibrium might spontaneously back away, if only temporarily, from the maximum. Assuming that the universe had an infinite time to reach equilibrium, he thought it might be that the non-equilibrium state we find ourselves in might be a giant *fluctuation*. Given his assumption of infinite time, even such an extremely improbable situation is at least *possible*.

In his early work on statistical mechanics, Einstein showed that small *fluctuations* in the motions of gas particles are constantly leading to departures from equilibrium. Somewhat like the departures from the smooth analytic bell curve for any finite number of events, the entropy does not rise smoothly to a maximum and then stay there indefinitely. The second law is not *continuous* and *absolute*.

The second law of thermodynamics is unique among the laws of physics because of its *irreversible* behavior. Heat flows from hot into cold places until they come to the same equilibrium temperature. The one-direction nature of *macroscopic* thermodynamics (with its gross "phenomenological" variables temperature, energy, entropy) is in fundamental conflict with the assumption that *microscopic* collisions between molecules, whether fast-moving or slow, are governed by dynamical, deterministic laws that are time-reversible. But is this correct?

The microscopic second law suggests the "arrow of time" does not apply to the time-reversible dynamical laws. At the atomic and molecular level, there appears to be no arrow of time, but we will see that Einstein's work shows particle collisions are not reversible

The first statistical "laws" grew out of examples in which there are very large numbers of entities. Large numbers make it impractical to know much about the individuals, but we can say a lot about *averages* and the probable distribution of values around the averages.

Probability, Entropy, and Information

Many scientists and philosophers of science say that the concept of entropy is confusing and difficult to understand, let alone explain. Nevertheless, with the help of our diagrams demonstrating probability as the *number of ways* things have happened or been arranged, divided by the total number of ways they might have happened or been arranged, we can offer a brief and visual picture of entropy and its important connection to information.

We begin with LUDWIG BOLTZMANN's definition of the entropy S in terms of the number of ways W that gas particles can be distributed among the cells of "phase space," the product of ordinary coordinate space and a momentum space.

$S = k \log W$

Let's greatly simplify our space by imagining just two cubicle bins separated by a movable piston. Classical thermodynamics was developed studying steam engines with such pistons.

Now let's imagine that a thousand molecules are dropped *randomly* into the two bins. In this very artificial case, imagine that they all land up on the left side of the piston. Assuming the probabilities of falling into the left or right bin are equal, this is again the binomial expansion with $(p + q)^{1000}$ with $p = q = \frac{1}{2}$. All molecules on the left would have probability $(1/2)^{1000}$. This is of course absurdly improbable if each events were random, but steam engines do this all the time, and calculating the improbability gives us a measure of the machine's available energy.

Figure 1-4. An ideal piston with gas on the left and a perfect vacuum on the right.

To see how this very improbable situation corresponds to very low entropy, how low entropy corresponds to maximum information,

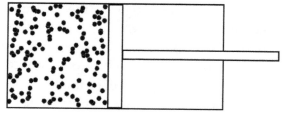

and how low entropy means energy available to do work, let's consider the number of yes/no questions needed to figure out the chessboard square where a single pawn is located.

  1) Is it in the top half? No.
     Of the remaining half,
  2) is it in the left half? No.
     Of the remaining half,
  3) Is it in the right half? No.
     Of the remaining half,
  4) Is it in the top half? Yes.
     Of the remaining half,
  5) Is it in the left half? Yes.
     Of the remaining half,
  6) Is it in the top half? Yes.

In CLAUDE SHANNON's 1948 theory of the communication of information, the answer to a yes/no question communicates one bit (a binary digit can be 1 or 0) of information. So, as we see, it takes

6 bits of information to communicate the particular location of the pawn on one of the 64 possible squares on the chessboard.

Shannon and his mentor, the great mathematical physicist JOHN VON NEUMANN, noticed that the information I is the logarithm of the number of possible ways W to position the pawn. Two raised to the 6th power is 64 and the base 2 logarithm of 64 is 6. Thus

$I = log_2 W$ and $6 = log_2 64$

The parallel with Boltzmann's entropy formula is obvious. His formula needs a constant with the physical dimensions of energy divided by temperature (ergs/degree). But Shannon's information has no physical content and does not need Boltzmann's constant k. Information is just a dimensionless number.

For Shannon, entropy is the *number of messages* that can be sent through a communications channel in the presence of noise. For Boltzmann, entropy was proportional to the *number of ways* individual gas particles can be distributed between cells in phase space, assuming that all cells are equally probable.

So let's see the similarity in the case of our piston. How many ways can all the 1000 gas particles be found *randomly* on the left side of the piston, compared to all the other ways, for example only 999 on the left, 1 on the right, 998 on the left, 2 on the right, etc.

Out of 2^{1000} ways of distributing them between two bins, there is only *one way* all the particles can be on the left. [2] The logarithm of 1 is zero ($2^0 = 1$). This is the minimum possible entropy and the maximum of available energy to do work pushing on the piston.

Boltzmann calculated the likelihood of random collisions resulting in the *unmixing* of gases, so that noticeably fewer are in the left half of a 1/10 liter container, as of the order of $10^{10^{10}}$ years. [3] Our universe is only of the order of 10^{10} years old.

It seems most unlikely that such chance can lead to the many interesting information structures in the universe. But chance will play a major role in Einstein's description of what he called "objective reality," as we shall see.

---

2   1000! (factorial) is 1000 x 999 x 998 ... x 2 x 1. (really big)

3   Boltzmann, 2011, *p.444*

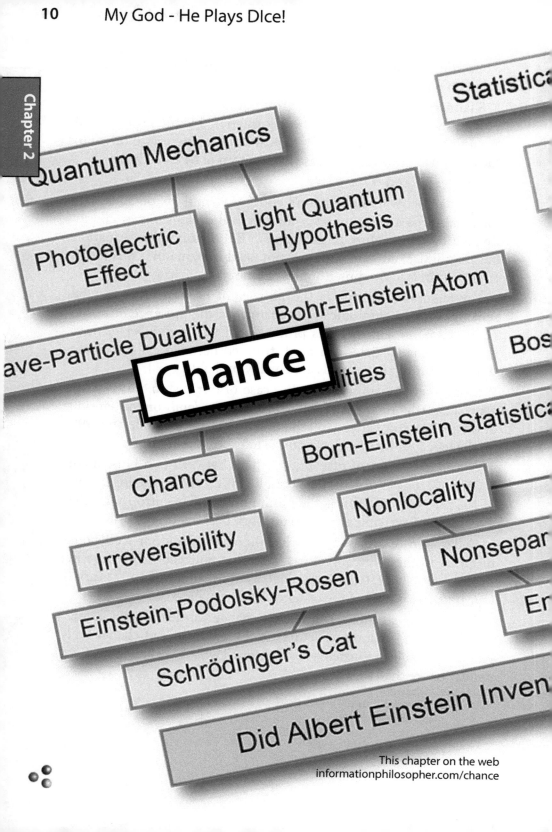

This chapter on the web
informationphilosopher.com/chance

Chance

We hope to develop your ability to *visualize* actual *chance* events and distinguish them clearly from the *continuous* mathematical equations that predict very large numbers of them so perfectly. This will be critical if you are to visualize the quantum wave function and see it the way Einstein saw it.

A continuous "bell curve" is an *ideal* analytic function with values for each of the infinite number of points on the horizontal axis. In the real *material* world of particles, a *discrete* histogram approaches that ideal curve in the limit of large numbers of events. A finite number of particles never gets there.

The "binomial coefficients" in figure 1.1 were arranged by BLAISE PASCAL in what is known as Pascal's triangle. Each number is the sum of the two numbers above, giving us the *number of ways* from the top to reach each point in the lower rows.

```
                              1
                           1     1
                        1     2     1
                     1     3     3     1
                  1     4     6     4     1
               1     5    10    10     5     1
            1     6    15    20    15     6     1
         1     7    21    35    35    21     7     1
      1     8    28    56    70    56    28     8     1
   1     9    36    84   126   126    84    36     9     1
1    10    45   120   210   252   210   120    45    10    1
1   11    55   165   330   462   462   330   165    55   11    1
1   12   66   220  495  792  924  792  495  220   66   12   1
1   13  78  186  715 1287 1716 1716 1287 715  186  78   13   1
1   14  91  364 1001 2002 3003 3432 3003 2002 1001 364  91  14  1
1  15  105 455 1365 3003 5005 6435 6435 5005 3003 1365 455 105 15  1
```

Figure 2-1. Pascal's triangle. Plotting the numbers in the bottom row would show how sharp and peaked the normal distribution is for 16 coin flips.

To illustrate *physically* how random events approach the normal distribution in the limit of large numbers, the sociologist and statistician Francis Galton designed a *probability machine*, with balls bouncing randomly left or right in an array of pins.

Chapter 2

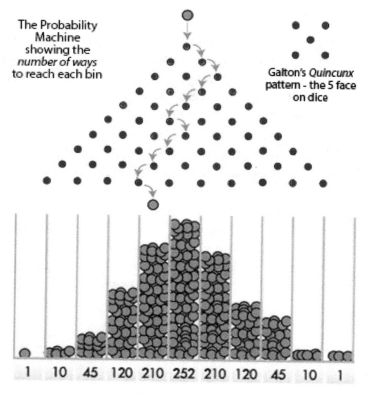

The Probability Machine showing the *number of ways* to reach each bin

Galton's *Quincunx* pattern - the 5 face on dice

| 1 | 10 | 45 | 120 | 210 | 252 | 210 | 120 | 45 | 10 | 1 |

Figure 2-2. Galton's "Quincunx." The number of ways to a bin in the bottom row is the binomial coefficient.

The *probability* of reaching a bin is the *number of ways* to the bin divided by the total number of ways, $2^{10} = 1024$.

Generalizing now to the cases of shuffling decks of cards, or throwing pairs of dice, the most probable outcomes are those that can be accomplished in the largest *number of ways*.

Now we test your physical understanding of probability. Do you consider each bounce of a ball above as random? As really random? Or is it *determined* by the laws of nature, by the laws of classical mechanics?

Is the use of probability just because we cannot know the exact details of the initial conditions, as the proponents of deterministic chaos maintain? Is the randomness only human ignorance, thus subjective and epistemological? Or is it objective and ontological?

You may be surprised to learn that many physicists, and perhaps most philosophers of science, think physics is deterministic, despite the evidence for quantum indeterminism, following centuries of tradition which were deep beliefs of ALBERT EINSTEIN.

To deny ontological chance is to commit to just one possible future and to the belief that if we could reverse the velocities and directions of all material particles from their current positions, Newton's laws say that all the particles would retrace their paths back in time to the beginning of the universe.

The History of Chance

For most of the history of philosophy and physics, ontological chance has been strictly denied. LEUCIPPUS (440 B.C.E.) stated the first dogma of determinism, an absolute necessity.

> "Nothing occurs by chance (*maton*), but there is a reason (*logos*) and necessity (*ananke*) for everything."[1]

Chance is regarded as inconsistent with reasons and causes.

The first thinker to suggest a physical explanation for chance in the universe was EPICURUS. Epicurus was influenced strongly by ARISTOTLE, who regarded chance as a possible fifth cause. Epicurus said there must be cases in which the normally straight paths of atoms in the universe occasionally bend a little and the atoms "swerve" to prevent the universe and ourselves from being completely determined by the mechanical laws of DEMOCRITUS.

For Epicurus, the chance in his atomic swerve was simply a means to deny the fatalistic future implied by determinism. As the Epicurean Roman LUCRETIUS explained the idea,

> "...if all motion is always one long chain, and new motion arises out of the old in order invariable, and if the first-beginnings do not make by swerving a beginning of motion such as to break the decrees of fate, that cause may not follow cause from infinity, whence comes this freedom in living creatures all over the earth."[2]

Epicurus did not say the swerve was directly involved in decisions so as to make them random. His critics, ancient and

---

1   Fragment 569 - from Fr. 2 *Actius* I, 25, 4

2   *De Rerum Natura*, Book 2, lines 251-256

modern, have claimed mistakenly that Epicurus did assume "one swerve - one decision." Some recent philosophers call this the "traditional interpretation" of Epicurean free will.

On the contrary, following ARISTOTLE, Epicurus thought human agents have an autonomous ability to transcend the necessity and chance of some events. This special ability makes us morally *responsible* for our actions.

Epicurus, clearly following Aristotle, finds a *tertium quid*, beyond the other two options, necessity (Democritus' and Leucippus' determinism) and chance (Epicurus' swerve).

The *tertium quid* is agent autonomy. Epicurus wrote:

> "...some things happen of necessity (ἀνάγκη), others by chance (τύχη), others through our own agency (παρ' ἡμᾶς)...necessity destroys responsibility and chance is uncertain; whereas our own actions are autonomous, and it is to them that praise and blame naturally attach."[3]

Despite abundant evidence, many philosophers deny that real chance exists. If a single event is determined by chance, then indeterminism would be true, they say, undermining the very possibility of reasoning to certain knowledge. Some go to the extreme of saying that chance makes the state of the world totally independent of any earlier states, which is nonsense, but it shows how anxious they are about chance.

The Stoic CHRYSIPPUS (200 B.C.E.) said a single uncaused cause could destroy the universe (*cosmos*), a concern shared by some modern philosophers, for whom reason itself would fail. He wrote:

> "Everything that happens is followed by something else which depends on it by causal necessity. Likewise, everything that happens is preceded by something with which it is causally connected. For nothing exists or has come into being in the cosmos without a cause. The universe will be disrupted and disintegrate into pieces and cease to be a unity functioning as a single system, if any uncaused movement is introduced into it."[4]

The core idea of chance and indeterminism is closely related to the idea of *causality*. Indeterminism for some is simply an event without a cause, an uncaused cause or *causa sui* that starts

3   *Letter to Menoeceus*, §133
4   Plutarch, *Stoic. Rep.*, 34, 1050A

a new causal chain. If we admit some uncaused causes, we can have an adequate causality without the physical necessity of strict determinism - which implies complete predictability of events and only one possible future.

An example of an event that is not strictly caused is one that depends on chance, like the flip of a coin. If the outcome is only probable, not certain, then the event can be said to have been caused by the coin flip, but the head or tails result itself was not predictable. So this "soft" causality, which recognizes prior uncaused events as causes, is undetermined and to some extent the result of chance.

Even mathematical theorists of games of chance found ways to argue that the chance they described was somehow necessary and chance outcomes were actually determined. The greatest of these, PIERRE-SIMON LAPLACE, preferred to call his theory the "calculus of probabilities." With its connotation of approbation, probability was a more respectable term than chance, with its associations of gambling and lawlessness. For Laplace, the random outcomes were not predictable only because we lack the detailed information to predict. As did the ancient Stoics, Laplace explained the appearance of chance as the result of human ignorance. He said,

> "The word 'chance,' then expresses only our ignorance of the causes of the phenomena that we observe to occur and to succeed one another in no apparent order." [5]

As we have seen, decades before Laplace, ABRAHAM DE MOIVRE discovered the normal distribution (the bell curve) of outcomes for ideal random processes, like the flip of a coin or throw of dice. But despite this de Moivre did not believe in chance. It implies events that God can not know. De Moivre labeled it *atheistic*.

> Chance, in atheistical writings or discourse, is a sound utterly insignificant: It imports no determination to any mode of existence; nor indeed to existence itself, more than to non existence; it can neither be defined nor understood...it is a mere word. [6]

We have seen that random processes produce a regular distribution pattern for many trials (the law of large numbers). Inexplicably, the discovery of these regularities in various social phenomena led Laplace and others to conclude that the phenomena are *determined*, not random. They simply denied chance in the world.

---

5   *Memoires de l'Academie des Sciences* 1783, p. 424.

6   *The Doctrine of Chances*, 1756, p.253.

A major achievement of the Ages of Reason and Enlightenment was to banish absolute chance as unintelligible and atheistic. Newton's Laws provided a powerful example of deterministic laws governing the motions of everything. Surely Leucippus' and Democritus' original insights had been confirmed.

As early as 1784, IMMANUEL KANT had argued that the regularities in social events from year to year showed that they must be determined.

> "Thus marriages, the consequent births and the deaths, since the free will seems to have such a great influence on them, do not seem to be subject to any law according to which one could calculate their number beforehand. Yet the annual (statistical) tables about them in the major countries show that they occur according to stable natural laws."[7]

In the early 1800's, the social statisticians ADOLPHE QUÉTELET and HENRY THOMAS BUCKLE argued that these regularities in social physics proved that individual acts like marriage and suicide are determined by natural law. Quételet and Buckle thought they had established an absolute deterministic law behind all statistical laws. Buckle went so far as to claim it established the lack of free will.

The argument for determinism of Quételet and Buckle is quite illogical. It appears to go something like this:

- As we saw above, random, unpredictable individual events (like the throw of dice in games of chance or balls in a probability machine) have a normal distribution that becomes more and more certain with more events (the law of large numbers).
- Human events are normally distributed.
- Therefore, human events are determined.

They might more reasonably have concluded that individual human events are unpredictable and random. Were they in fact determined, the events might show a non-random pattern, perhaps a signature of the Determiner?

In the next chapter, we shall see that Quételet and Buckle had a major influence on the development of statistical physics.

In the nineteenth century in America, CHARLES SANDERS PEIRCE coined the term "tychism" for his idea that absolute chance is the first step in three steps to "synechism" or continuity.

---

7   *Idea for a Universal History*, introduction

Peirce was influenced by Buckle and Quételet, by the French philosophers CHARLES RENOUVIER and ALFRED FOUILLEE, who also argued for some absolute chance, but most importantly Peirce was influenced by Kant and GEORG W. F. HEGEL, who saw things arranged in the triads that Peirce so loved.

Renouvier and Fouillee introduced chance or indeterminism simply to contrast it with determinism, and to discover some way, usually a dialectical argument like that of Hegel, to reconcile the opposites. Renouvier argues for human freedom, but nowhere explains exactly how chance might contribute to that freedom, other than negating determinism.

Peirce does not explain much with his tychism, and with his triadic view that adds continuity, then evolutionary love, which is supreme, he may have had doubts about the importance of chance. Peirce did not propose chance as directly or indirectly providing free will. He never mentions the ancient criticisms that we cannot accept responsibility for chance decisions. He does not really care for chance as the origin of species, preferring a more deterministic and continuous lawful development, under the guidance of evolutionary love. Peirce called Darwinism "greedy." But he does say clearly that the observational evidence simply does not establish determinism.

It remained for WILLIAM JAMES, Peirce's close friend, to assert that chance can provide random unpredictable alternatives from which the will can choose or determine one alternative. James was the first thinker to enunciate clearly a two-stage decision process, with chance in a present time generating random alternatives, leading to a choice which selects one alternative and transforms an equivocal ambiguous future into an unalterable determined past. There are free and undetermined alternatives followed by adequately determined choices made by the will.

Chance allows alternative futures. The deep question is how the one *actual* present is realized from *potential* alternative futures.

CLAUDE SHANNON, creator of the mathematical theory of the communication of information, said the information in a message depends on the number of possibilities. If there is only one possibility, there can be no new information. If information in the universe is a conserved constant quantity, like matter and energy, there is only one possible future.

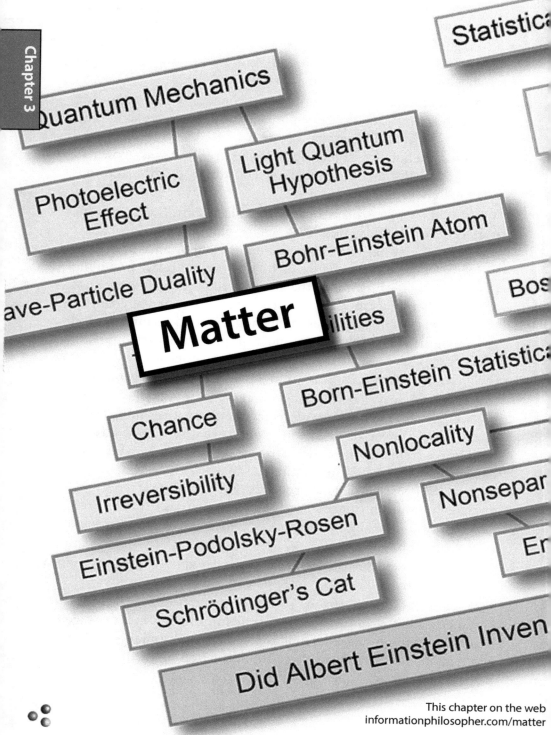

Quantum Mechanics

Statistica

Photoelectric Effect

Light Quantum Hypothesis

ave-Particle Duality

Bohr-Einstein Atom

Bos

Matter

ilities

Born-Einstein Statistica

Chance

Nonlocality

Irreversibility

Einstein-Podolsky-Rosen

Nonsepar

Er

Schrödinger's Cat

Did Albert Einstein Inven

This chapter on the web
informationphilosopher.com/matter

Matter

JAMES CLERK MAXWELL and LUDWIG BOLTZMANN were atomists who accepted the idea that the apparently continuous pressure of a gas on the walls of its container is caused by a number of atomic collisions so vast that the individual discrete bumps against the walls are simply not detectable.

Maxwell's great contribution to the kinetic theory of gases was to find the velocity (or energy) distribution of the gas particles. From simple considerations of symmetry and the assumption that motions in the y and z directions are not dependent on motions in the x direction, Maxwell in 1860 showed that velocities are distributed according to the same normal distribution as the "law of errors" found in games of chance. Boltzmann in 1866 derived Maxwell's velocity distribution dynamically, putting it on a firmer ground than Maxwell.

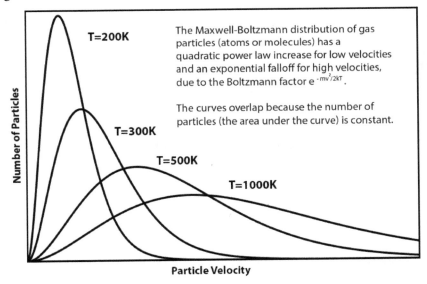

T=200K

The Maxwell-Boltzmann distribution of gas particles (atoms or molecules) has a quadratic power law increase for low velocities and an exponential falloff for high velocities, due to the Boltzmann factor $e^{-mv^2/2kT}$.

The curves overlap because the number of particles (the area under the curve) is constant.

T=300K

T=500K

T=1000K

Number of Particles

Particle Velocity

Maxwell derived his velocity distribution law using math that he found in a review of ADOLPH QUÉTELET's work on social statistics, but he did not accept the conclusion of Quételet and

THOMAS HENRY BUCKLE that the normal distribution seen in large numbers of random events implies that they are *determined*.[1]

Maxwell's criticism of his English colleague Buckle was clear.

> We thus meet with a new kind of regularity — the regularity of averages — a regularity which when we are dealing with millions of millions of individuals is so unvarying that we are almost in danger of confounding it with absolute uniformity.

> Laplace in his theory of Probability has given many examples of this kind of statistical regularity and has shown how this regularity is consistent with the utmost irregularity among the individual instances which are enumerated in making up the results. In the hands of Mr Buckle facts of the same kind were brought forward as instances of the unalterable character of natural laws. But the stability of the averages of large numbers of variable events must be carefully distinguished from that absolute uniformity of sequence according to which we suppose that every individual event is *determined* by its antecedents.[2]

Six years after his derivation of the velocity distribution from classical dynamics, Boltzmann found a mathematical expression he called H that appears to decrease as particle collisions occur. He identified it as the negative of the thermodynamic entropy that always increases according to the second law of thermodynamics.

In 1874, Boltzmann's mentor JOSEF LOSCHMIDT criticized his younger colleague's attempt to derive from classical dynamics the increasing entropy required by the second law of thermodynamics. Loschmidt's criticism was based on the simple idea that the laws of classical dynamics are time reversible. Consequently, if we just turn the time around, the time evolution of the system should lead to decreasing entropy.

Of course we cannot turn time around, but a classical dynamical system will evolve in reverse if all the particles could have their velocities exactly reversed. Apart from the practical impossibility of doing this, Loschmidt had showed that systems could exist for which the entropy should decrease instead of increasing. This is called Loschmidt's reversibility objection or "Loschmidt's paradox."

It is also known as the problem of *microscopic reversibility*. How can the macroscopic entropy be irreversibly increasing when microscopic collisions are time reversible?

1   See chapter 2 for such arguments beginning with Immanuel Kant.
2   Draft Lecture on *Molecules*, 1874 (our italics)

Maxwell too was critical of Boltzmann's 1872 dynamical result based on Newton's deterministic laws of motion. The kinetic theory of gases must be purely *statistical*, said Maxwell.

In 1877, Boltzmann followed Maxwell's advice. He counted the number of ways W that N particles can be distributed among the available cells of "phase-space," a product of ordinary coordinate space and "momentum space."

Boltzmann showed that some distributions of particles are highly improbable, like all the balls in our probability machine landing in one of the side bins. In nature, he said, the tendency of transformations is always to go from less probable to more probable states. [3]

There are simply many more ways to distribute particles randomly among cells than to distribute them unevenly. Boltzmann counted each unique distribution or arrangement of particles as a "microstate" of the system. Arguing from a principle of indifference, he assumed that all microstates are equally probable, since we have no reasons for any differences.

Boltzmann then gathered together microstates that produce similar macroscopic descriptions into "macrostates." For example, having all the particles in a single cell in a corner of a container would be a macrostate with a single microstate, and thus minimum entropy. Boltzmann's idea is that macrostates with few microstates will evolve statistically to macrostates with large numbers of microstates. For example, taking the top off a bottle of perfume will allow the molecules to expand into the room and never return.

Figure 3-3. Entropy increases when the number of possible microstates W increases. The likelihood of all the molecules returning to the bottle is vanishingly small. [4]

---

3   Boltzmann, 2011, p.74
4   Layzer, 1975, p.57

In the mid 1890's, some British scientists suggested that there must be some low-level mechanism maintaining what Boltzmann had called "molecular chaos" or "molecular disorder." Since classical microscopic dynamical laws of physics are time reversible, collisions between material particles can not explain the *macroscopic* irreversibility seen in classical thermodynamics and in the statistical mechanical explanations developed by Boltzmann.

Boltzmann himself did not take the need for microscopic irreversibility very seriously, because even his classical dynamical analysis showed that collisions quickly randomize a large number of gas particles and his calculations indicated it would be astronomical times before any departure from randomness would return.

For Boltzmann, microscopic irreversibility is needed only to defeat the Loschmidt paradox. See chapter 12.

Boltzmann's Philosophy

In his 1895 *Lectures on Gas Theory*, read by ALBERT EINSTEIN as a student, Boltzmann raised questions about the continuum and its representation by partial differential equations, which were to be questions Einstein struggled with all his life. Boltzmann wrote,

> Whence comes the ancient view, that the body does not fill space continuously in the mathematical sense, but rather it consists of discrete molecules, unobservable because of their small size. For this view there are philosophical reasons. An actual continuum must consist of an infinite number of parts; but an infinite number is undefinable. Furthermore, in assuming a continuum one must take the partial differential equations for the properties themselves as initially given. However, it is desirable to distinguish the partial differential equations, which can be subjected to empirical tests, from their mechanical foundations (as Hertz emphasized in particular for the theory of electricity). Thus the mechanical foundations of the partial differential equations, when based on the coming and going of smaller particles, with restricted average values, gain greatly in plausibility; and up to now no other mechanical explanation of natural phenomena except atomism has been successful...

> Once one concedes that the appearance of a continuum is more clearly understood by assuming the presence of a large number of adjacent discrete particles, assumed to obey the laws of mechanics,

then he is led to the further assumption that heat is a permanent motion of molecules. Then these must be held in their relative positions by forces, whose origin one can imagine if he wishes. But all forces that act on the visible body but not equally on all the molecules must produce motion of the molecules relative to each other, and because of the indestructibility of kinetic energy these motions cannot stop but must continue indefinitely...

We do not know the nature of the force that holds the molecules of a solid body in their relative positions, whether it is action at a distance or is transmitted through a medium, and we do not know how it is affected by thermal motion. Since it resists compression as much as it resists dilatation, we can obviously get a rather rough picture by assuming that in a solid body each molecule has a rest position...

If each molecule vibrates around a fixed rest position, the body will have a fixed form; it is in the solid state of aggregation...

However, when the thermal motion becomes more rapid, one gets to the point where a molecule can squeeze between its two neighbors... It will no longer then be pulled back to its old rest position... When this happens to many molecules, they will crawl among each other like earthworms, and the body is molten.

In any case, one will allow that when the motions of the molecules increase beyond a definite limit, individual molecules on the surface of the body can be torn off and must fly out freely into space; the body evaporates.

A sufficiently large enclosed space, in which only such freely moving molecules are found, provides a picture of a gas. If no external forces act on the molecules, these move most of the time like bullets shot from guns in straight lines with constant velocity. Only when a molecule passes very near to another one, or to the wall of the vessel, does it deviate from its rectilinear path. The pressure of the gas is interpreted as the action of these molecules against the wall of the container. [5]

---

5   Boltzmann, 2011 §1, p.27

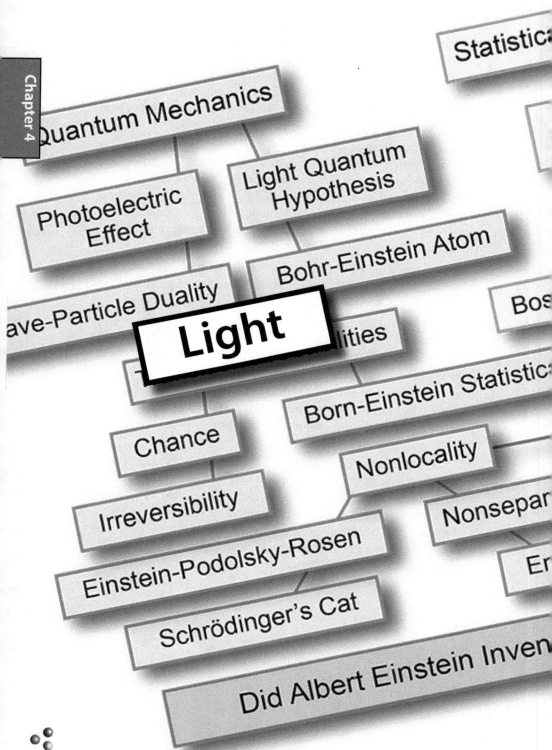

Chapter 4

Light

MAX PLANCK was just twelve years younger than LUDWIG BOLTZMANN. He wrote his 1879 doctoral thesis two years after Boltzmann's statistical defense of his H-Theorem and entropy increase. In his thesis, Planck vowed to show that the second law of thermodynamics (with its *irreversible* increase of entropy) is an absolute law, fully as deterministic as the first law, the conservation of energy. An absolute law cannot be statistical, he said.

Planck was called to Berlin in 1889 to take GUSTAV KIRCHHOFF's chair in theoretical physics. Over the next five years he edited Kirchhoff's lengthy *Lectures on Heat Theory* and came to appreciate the universal (and perhaps absolute?) function K_λ that Kirchhoff had found for the distribution of so-called "blackbody" radiation energy as a function of wavelength λ in conditions of thermal equilibrium. Blackbody radiation is independent of the specific kind of material, a universal fact that impressed Planck deeply.

Kirchhoff showed that the amount of radiation absorbed by a material body at a given wavelength must exactly equal the amount emitted at that wavelength, or else the body would heat up or cool down, providing an energy difference that could run a perpetual motion machine. If the absorbed energy $\alpha_\lambda K_\lambda$ and the emitted energy $\varepsilon_\lambda K_\lambda$ are equal, then the emissity and absorbtivity coefficients must be equal,

$\varepsilon_\lambda = \alpha_\lambda$, which is Kirchhoff's law.

Planck set out to determine the universal function K_λ. And he further hypothesized that the *irreversibility* of the second law might be the result of an *interaction* between matter and radiation. We shall see in chapter 12 that Planck's intuition about irreversibility was correct.

In his lectures, Kirchhoff noted that in a perfectly reflecting cavity, there is no way for monochromatic rays of one frequency to change to another frequency. But he said that a single speck of material would be enough to produce blackbody radiation. His student Planck said that a single carbon particle would be enough to change perfectly arbitrary radiation into black radiation.[1]

---

1   Planck, 1991, p.44

Chapter 4

Planck asked whether radiation absorbed by an electrical oscillator coming in as a plane wave from one direction could be emitted by the oscillator as a spherical wave in all directions, producing an irreversible change, since incoming spherical waves are never seen in nature.

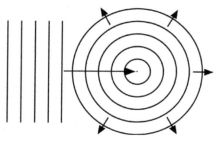

Note that Planck was attempting to locate the source of *macroscopic* irreversibility in the *microscopic* interaction of light with single particles of matter. But his suggestion was not received well. Planck described the strong reaction by Boltzmann.

> [My] original silent hope that the radiation emitted by the oscillator would differ, in some characteristic way, from the absorbed radiation, turned out to have been mere wishful thinking. Moreover, my suggestion that the oscillator was capable of exerting a unilateral, in other words irreversible, effect on the energy of the surrounding field, drew a vigorous protest from Boltzmann, who, with his wider experience in this domain, demonstrated that according to the laws of classical dynamics, each of the processes I considered could also take place in the opposite direction; and indeed in such a manner, that a spherical wave emitted by an oscillator could reverse its direction of motion, contract progressively until it reached the oscillator and be reabsorbed by the latter, so that the oscillator could then again emit the previously absorbed energy in the same direction from which the energy had been received. [2]

This "vigorous protest" from Boltzmann was a pivotal moment in the history of microscopic irreversibility. It led to the eventual understanding of the interaction of matter and light.

It began in 1895 when Planck's brilliant student ERNST ZERMELO (who later developed the basis for axiomatic set theory) challenged Boltzmann's idea of irreversible entropy increase with still another objection, now known as Zermelo's recurrence paradox. Using the

---

2   *Planck, 1949,*, pp.36-37

recurrence theorem of HENRI POINCARÉ, Zermelo said an isolated mechanical system must ultimately return to a configuration arbitrarily close to the one from which it began.

"Hence," Zermelo wrote, "in such a system *irreversible processes are impossible* since no single-valued continuous function of the state variables, such as entropy, can continuously increase; if there is a finite increase, then there must be a corresponding decrease when the initial state recurs."[3]

Beginning in 1897, Planck wrote a series of seven articles all titled "On Irreversible Radiation Processes." In the first three articles, he did not yet think a statistical or probabilistic approach could be the answer. Planck wrote to a friend that reconciling the second law with mechanics is "the most important with which theoretical physics is currently concerned."

> On the main point I side with Zermelo, in that I think it altogether hopeless to derive the speed of irreversible processes...in a really rigorous way from contemporary gas theory. Since Boltzmann himself admits that even the direction in which viscosity and heat conduction act can be derived only from considerations of probability, how can it happen that under all conditions the magnitude of these effects has an entirely determinate value. Probability calculus can serve, if nothing is known in advance, to determine the most probable state. But it cannot serve, if an improbable [initial] state is given, to compute the following state. That is determined not by probability but by mechanics. To maintain that change in nature always proceeds from lower to higher probability would be totally without foundation.[4]

But after Boltzmann's criticism, Planck's fourth article defined irreversible radiation with a maximum of entropy or disorder as "natural radiation," very much analogous to Boltzmann's molecular disorder.

> It will be shown that all radiation processes which possess the characteristic of natural radiation are necessarily irreversible.[5]

---

3   *Annalen der Physik*, 57 (1896). cited in Kuhn, *1978*, p26.
4   Kuhn, op. cit., p.27
5   *On Irreversible Radiation Processes*, IV, 1898, Kuhn, op. cit., p.78

Planck thus apparently began in 1898 to study carefully Boltzmann's approach to entropy and irreversibility, but he did not explicitly employ Boltzmann's identification of entropy with probability and his counting of microstates until late 1900, when Planck stumbled upon his formula for Kirchhoff's universal radiation law and then hastily sought a physical justification for it.

Planck's Discovery of the Blackbody Radiation Law

In 1896, a year before Planck tried to connect Kirchhoff's universal function with the *irreversibility* of his "natural radiation," Willy Wien had formulated an expression for the radiation law that agreed reasonably well with the experimental data at that time. The intensity I of energy at each frequency v, Wien wrote as

$$I_v(v, T) = a'v^3 e^{-av/T}$$

Wien's radiation "distribution" law agreed with his "displacement" law that the wavelength λ of maximum intensity λ_{max} is inversely proportional to the temperature T or that $\lambda_{max} T = $ constant.

Wien said that his law was inspired by the shape of the Maxwell-Boltzmann velocity (or energy) distribution law, which as we saw in chapter 3 has a negative exponential factor for increasing energy.

Wien also proposed that the distribution over different frequencies might be the result of fast-moving gas particles emitting radiation with Doppler shifts toward higher and lower frequencies.

In May 1899, Planck *derived* the entropy for Wien's energy distribution in his fifth article on irreversible radiation. He used the fact that classical thermodynamic entropy S is defined by a change in entropy equaling the change in energy U divided by the absolute temperature. $\partial S = \partial U/T$. He solved Wien's distribution law for $1/T$ by first taking its logarithm,

$$\log I_v(v, T) = \log(a'v^3) - av/T,$$

then solving for $1/T$,

$$1/T = \partial S/\partial U = -(1/av) \log(U/ea'v).$$

He then took the second derivative of entropy with respect to energy to find

$$\partial^2 S/\partial U^2 = -(1/av)(1/U).$$

When the second derivative of a function is negative, it must have a maximum. Confident that he had thus shown Wien's law to be consistent with the entropy increase to a maximum as required by the second law, Planck called for further experimental tests. But these tests proved to be a shock for him. Measurements for long wavelengths (small v) disagreed with Wien's law and showed a dependence on temperature.

On October 7, 1900, one of the experimenters, Heinrich Rubens, who was a close friend, came to dinner at Planck's home and showed him a comparison of their latest data with five proposed curves, one of which was Lord Rayleigh's proposal of June 1900 that long-wavelength radiation should be proportional to the temperature T. Rubens' graphs showed that the termperature dependence at long wavelength agreed with the recently published theory of Lord Rayleigh.

Planck described his attempt to find an interpolation formula that would include two terms, "so that the first term becomes decisive for small values of the energy and the second term for large values."[6]

His task was to find an equation that approaches Wien's law at high frequencies and Rayleigh's law at low frequencies (long wavelengths). Initially, he may have simply rewritten Wien's law, putting the exponential in the denominator and added a -1 term to the exponential term,

$$I_v (v, T) = a'v^3 / (e^{av/T} - 1) \qquad (1)$$

When av/T is large, we can ignore the -1 and this reduces to Wien's law at high frequencies.

For small av/T, we can expand the exponential as a series,

$$e^{av/T} = 1 + av/T + 1/2 (av/T)^2 + ...$$

Ignoring the squared and higher order terms, the 1 and -1 cancel and we have

$I_v (v, T) = (a'/a) v^2 T$, which is the Rayleigh expression.

By the evening of October 7, Planck had the new equation with -1 in the denominator, which he called a lucky guess at an interpolation

---

6   Planck, 1949. p.40.

formula (*eine glückliche Interpolationformel*). He sent a messenger with his new formula to Rubens, who replied the very next morning that Planck's equation was an excellent fit to his experimental data.

Planck submitted his new radiation formula for examination to the Berlin Physical Society at its meeting on October 19, 1900. Rubens and Kurlbaum presented their confirming experimental data and the new Planck radiation law has been accepted ever since.

Theoretical physicists describe the radiation law as a function of frequency ν. Experimenters plot against the wavelength λ.

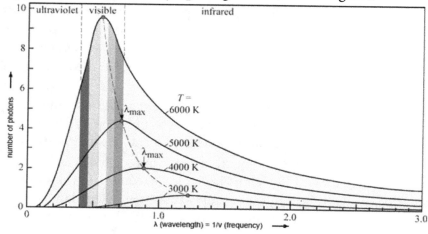

Figure 4-4. Planck's radiation law plotted against wavelength λ, showing Wien's displacement law, $\lambda_{max} T$ = constant.

Planck immediately started searching for the physical meaning of his new law, which at that point he said had been just a lucky guess.

> This quest automatically led me to study the interrelation of entropy and probability—in other words, to pursue the line of thought inaugurated by Boltzmann. Since the entropy S is an additive magnitude but the probability W is a multiplicative one, I simply postulated that $S = k \cdot \log W$, where k is a universal constant; and I investigated whether the formula for W, which is obtained when S is replaced by its value corresponding to the above radiation law, could be interpreted as a measure of probability.[7]

---

7   Planck, 1949., p.41

Planck probably pulled his hair out until he found that he could add a second term to $\partial^2 S/\partial U^2$ proportional to $1/U^2$. and then *derive* his new formula by integration. Planck's former student K. A. G. Mendelssohn wrote in 1969 (before the major analyses of Planck's thinking were published) that integrating Planck's radiation law yielded this "semi-empirical formula" for the entropy.

$S = (a'/a)\{(1 + U/a'v)log (1 + U/a'v) - (U/a'v)log (U/a'v)\}$

Mendelssohn says Planck used Boltzmann's method, that the entropy is simply the probability, calculated as the number of ways particles can be distributed or arranged.

> by considering a number N of equal oscillators with average energy U and by assuming the total energy to be made up of a number P of equal energy elements ε so that $NU=P\varepsilon$. Forming the complexion which gives the number of ways in which the P energy elements can be distributed over N, and which is the required probability, he calculated the entropy of the oscillator system as:

$NS = k\ log\ \{(P + N)!/P!\ N!\}$

> which can be written in the form

$S = k\ \{(1 + P/N)log (1 + P/N) - (P/N)log (P/N)\}$

> This theoretical expression is identical with the semi-empirical interpolation formula if a'/a is set equal to k and a' becomes the new universal constant h.[8]

Substituting these values for a and a' in equation 1, and multiplying by the classical density of states with frequency v $(8\pi v^2/c^3)$ we have Planck's radiation law, the hoped for universal function for blackbody radiation first described by Kirchhoff forty years earlier.

$\rho_v\ (v,\ T) = (8\pi h v^3/c^2)\ (1\ /\ (e^{\ hv/kT} - 1).$ (2)

This was the introduction of Planck's quantum of action h and also "Boltzmann's constant" k. Boltzmann himself never used this constant, but a combination of the number N of particles in a standard volume of matter and the universal gas constnt R.

---

8   "Max Planck," in *A Physics Anthology*, ed. Norman Clarke p.71

The Significance of Planck's Quantum of Action

Planck's quantum of action h restricts the energy in oscillators to integer multiples of hv, where v is the radiation frequency.

Planck could not really justify his statistical assumptions following Boltzmann. They were in conflict with his own deep beliefs that the laws of thermodynamics are absolute laws of nature like the dynamic laws of Newton and the electromagnetic laws of Maxwell.

Planck stopped looking for a *continuous*, deterministic, dynamical, and *absolute* explanation for the second law of thermodynamics and embraced a *discrete*, statistical view that was to lead to the quantization of the physical world, the birth of the quantum theory.

It is important to realize that Planck never *derived* his laws from first principles. In his 1920 Nobel lecture, he said "the whole deduction of the radiation law was in the main illusory and represented nothing more than an empty non-significant play on formulae." [9] In 1925, he called his work "a fortunate guess at an interpolation formula" and "the quantum of action a fictitious quantity... nothing more than mathematical juggling." [10]

Despite the many modern textbooks and articles claiming that he did, Planck did not suggest that the emission and absorption of radiation itself actually came in quantized (discrete) bundles of energy. We shall see in chapter 6 that that was the work of ALBERT EINSTEIN five years later in his photoelectric effect paper (for which he won the Nobel Prize). For Einstein, the particle equivalent of light, a "light quantum" (now called a "photon") contains hv units of energy.

Einstein hypothesized that light quanta do not radiate as a spherical wave but travel in a single direction as a localized bundle of energy that can be absorbed only in its entirety by an electron. Einstein assumed the light quanta actually have momentum. Since the momentum of a material particle is the energy divided by velocity, the momentum p of a photon is $p = hv/c$, where c is the velocity of light. To make the dual aspect of light as both waves and particles (photons) more plausible, Einstein interpreted the *continuous* light wave intensity as the probable density of *discrete* photons.

---

9   *The Genesis and Present State of Development of the Quantum Theory*, Planck's Nobel Prize Lecture, June 2, 1920

10   Planck, 1993. pp.106, 109.

Despite the "light-quantum hypothesis," Planck refused for many years to believe that light radiation itself existed as quanta. Planck's quantization assumption was for an ensemble of "oscillators" or "resonators" with energy values limited to *hv, 2 hv, 3 hv*, etc.

In 1906 Einstein showed that the Planck radiation law could be derived by assuming light too is quantized. He argued that Planck had essentially made the light quantum hypothesis in his work without realizing it.

Note that in NIELS BOHR's theory of the atom thirteen years later, where Bohr postulated stationary states of the electron and transitions between those states with the emission or absorption of energy equal to *hv*, but in *continuous* waves, because, just like Planck, Bohr denied the existence of light quanta (photons)!

It is unfair to Einstein that today so many books and articles give credit to Planck for the light quantum hypothesis and to Bohr for the idea that quantum jumps between his stationary states are accompanied by the absorption and emission of photons!

Comparison of Matter and Light Distribution Laws

Planck was pleased to find that his blackbody radiation law was the first known connection between the mechanical laws of matter and the laws of electromagnetic energy. He knew this was a great step in physical understanding, "the greatest discovery in physics since Newton," he reportedly told his seven-year-old son in 1900.

It took many years to see the deep connection between matter and light, namely that they both have wave and particle properties. But if we look carefully at the distribution laws for matter and radiative energy, we can begin to see some similarities

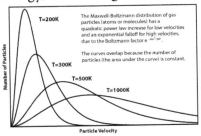 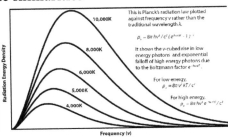

Figure 4-5. Distribution laws for radiation and matter

Here we plot both matter and light with energy (frequency) increasing to the right to emphasize the similarities and differences.

Both curves increase from zero with a power law (v^2, v^3). Both curves decrease exponentially at higher energies with the Boltzmann factor $e^{-E/kT}$.

Both maxima move to higher energies, matter to higher velocities, just as peak radiation moves to higher frequencies. But matter distribution curves overlap, where light curves do not.

The reason for the different looks is that when temperature increases, the number of gas particles does not change, so the Maxwell-Boltzmann distribution flattens out, preserving the area under the curve.

By contrast, when the temperature of radiation increases, the added energy creates more photons, and the Planck curve gets higher for all frequencies.

We shall argue that the spectroscopic analysis of light has been the most fundamental tool elucidating the atomic structure of matter. The similarity between the velocity distribution of matter and the energy distribution of light led to an expression for the *continuous* spectrum. We will see that the *discrete* spectrum provided even deeper insight into the quantum structure of matter.

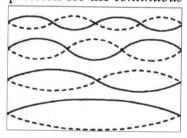

With 20-20 hindsight we will show that one can almost "read off" the atomic structure of matter just by taking a careful look at atomic spectra.

The Ultraviolet Catastrophe

Years after Planck had found a formula that included Wien's exponential decay of energy for higher frequencies of light, it was pointed out by Einstein and others that if Planck had done his calculations according to classical physics, he should have put equal amounts of energy in all the higher frequency intervals, leading to an infinite amount of energy in shorter-wavelength ultraviolet light.

The idea of equipartition of energy assumes that every "degree of freedom" or mode of vibration should get an equal amount of energy. Equipartition was a fundamental tenet of thermodynamic equilibrium in the nineteenth century. Classical electromagnetic theory claimed each vibration wavelength that could be a standing wave in a container, up to those with infinite frequency and zero wavelength, should be counted. This seems to suggest an infinite amount of energy in the radiation field.

The classical density of states with frequency v is $8\pi v^2/c^3$, and Planck used that to derive his radiation law. Today we know that a Boltzmann factor $e^{-En/kT}$ must be applied to each vibration mode to prevent an infinite amount of energy in the high frequencies.

Einstein criticized Planck for combining classical and quantum ideas, but we shall see that Einstein himself kept this classical density of states in his own derivations of the Planck law until a complete quantum derivation was made by SATYENDRA NATH BOSE and Einstein in 1924, when they discovered the indistinguishability of quantum particles and the origin of the term $8\pi v^2/c^3$ with no reference to classical electromagnetic radiation.

Planck's Accurate Determination of Natural Constants

Planck's blackbody radiation distribution law contains a surprising number of fundamental constants of nature. Some are related to light (the velocity c and the frequency λ), some to matter (Boltzmann's constant $k = R/N$), and his own constant h, important for both. Because the experimental data were quite accurate, Planck realized that he could calculate values for these constants and some others. His calculated values would prove to be more accurate than those available from direct measurements at that time.

It is ironic that a physicist who had denied the existence of discrete particles only a few years earlier would now use the constants in his new law to determine the most accurate values for N, the number of particles in a gram-molecular weight of any gas, and for e, the charge on the electron that was discovered just one year earlier.

Where Planck wrote k, Boltzmann had used R/N, where R is the universal gas constant and N is Avogadro's number. Although it is inscribed on Boltzmann's tomb as part of his famous formula relating entropy to probability, $S = k \log W$, Boltzmann himself had never described the constant k as such. It was Planck who gave "Boltzmann's constant" a symbol and a name.

Planck found $k = 1.346\times10^{-16}$ ergs/degree. He had no idea how accurate it was, Today's value is 1.3806×10^{-16}. Planck's value, which was of course dependent on the accuracy of the experimental data, was within an incredible 2.5%. Now together with the well-known gas constant R, Planck could estimate the number of molecules N in one mole (a gram molecular weight) of a gas.

Planck found $N = 6.175\times10^{23}$. The modern value is 6.022×10^{23}.

Planck's estimate of the fundamental unit of electrical charge e. was 4.69×10^{-10} esu. The modern value is 4.80×10^{-10} esu. Other estimates in Planck's day were 1.29 and 1.65×10^{-10} esu showing how his numbers were so much more accurate than any others made at that time. These results gave Planck great confidence that his "fictitious," wholly theoretical, radiation formula must be correct.

Planck's discovery of "natural constants" led to the effort to define the constants with no reference to human values for mass, length, time, electric charge, etc.

We describe these new "dimensions" as Planck units. For example, we can set the velocity of light c to 1. Now a particle moving at one-tenth light velocity is moving at 0.1 Planck unit.

Familar and famous equations now look different and may hide some important physical relations. Einstein's $E = mc^2$ becomes $E = m$.

Similar to light velocity $c = 1$, other important constants are defined as 1.

Planck units are defined by

$\hbar = G = k_e = k_B = c = 1,$

$\hbar$ is the reduced Planck constant $h/2\pi$, G is the gravitational constant, k_e the Coulomb constant, and k_B the Boltzmann constant.

In cosmology and particle physics, Planck's natural constants describe the so-called "Planck scale." They can be combined to describe a "Planck time" of 5.4×10^{-44} seconds, a "Planck length" of 1.6×10^{-35} meters, and a "Planck energy" of 1.22×10^{19} GeV.

They are thought to best describe the earliest moment of the Big Bang, the first 10^{-43} seconds of the universe.

No Progress on Microscopic Irreversibility

Although Planck was the first to find equations that involve both matter and light, and though he thought for many years that their interaction could explain microscopic irreversibility, this nineteenth-century problem is thought by many physicists to be still with us today.

Planck's intuition was good that irreversibility depends on both light and matter, but true irreversibility must depart from physical determinism, and that had to wait for ALBERT EINSTEIN's discovery of ontological chance in 1916.

In the intervening years, Planck rightly thought his greatest achievement was not just the discovery of equations involving both discrete material particles and continuous wavelike radiation, both matter and energy. He also found and named the natural constants in these equations, both his h and Boltzmann's k.

Planck knew from his mentor Kirchhoff that monochromatic radiation can not thermally equilibrate to all the frequencies in his new distribution law without at least a tiny bit of matter.

We have seen in chapter 3 that matter by itself can approximate thermal equilibrium with Boltzmann's classical statistics, but deterministic physics leaves it open to the reversibility and recurrence objections of Loschmidt and Zermelo.

So we shall see in chapter 12 that the collision of Einstein's light quanta with particles that have internal quantum structures adds the necessary element of indeterminacy for microscopic irreversibility.

Planck initially hoped for a second law of thermodynamics that was as absolute as the first law. What we now find is only a statistical law, but his insight that it would depend on both matter and energy was confirmed, and their roles are oddly symmetric.

Even a tiny bit of matter will equilibrate radiation. Even a tiny bit of radiation can equilibrate matter. And both are the result of quantum mechanics.

Statistic

Quantum Mechanics

Light Quantum
Hypothesis

Photoelectric
Effect

Bohr-Einstein Atom

ave-Particle Duality

Bos

Statistical Mechanics

Tr

Born-Einstein Statistic

Chance

Nonlocality

Irreversibility

Nonsepa

Einstein-Podolsky-Rosen

E

Schrödinger's Cat

Did Albert Einstein Inven

Statistical Mechanics

Statistical mechanics and thermodynamics are nineteenth-century *classical* physics, but they contain the seeds of the ideas that ALBERT EINSTEIN would use to create *quantum* theory in the twentieth, especially the work of his *annus mirabilis* of 1905.

Einstein wrote three papers on statistical mechanics between 1902 and 1904 He put earlier ideas on a firmer basis. Einstein claimed that although JAMES CLERK MAXWELL's and LUDWIG BOLTZMANN's theories had come close, they had not provided a foundation for a general theory of heat based on their kinetic theory of gases, which depend on the existence of microscopic atoms and molecules. In his 1902 paper, Einstein did so, deriving the equipartition theory of the distribution of energy among the degrees of freedom of a system that is in equilibrium with a large heat reservoir that maintains the system temperature.

But, Einstein said in his second paper (1903), a general theory of heat should be able to explain both thermal equilibrium and the second law of thermodynamics *independent of* the kinetic theory. The laws of *macroscopic* phenomenological thermodynamics do not depend on the existence of microscopic atoms and molecules. His second paper derived the second law based solely on the probability of distributions of states, Boltzmann's entropy, $S = k \log W$, which Einstein redefined, as the fraction of time the system spends in each state. This work, he said, bases thermodynamics on general *principles* like the impossibility of building a perpetual motion machine.

In his third paper (1904), Einstein again derived the second law and the entropy, using the same statistical method used by Boltzmann in his theory of the ideal gas and by Planck in his derivation of the radiation law. Einstein investigated the significance of what Planck had called Boltzmann's constant k. With the dimensions of ergs/degeree, as a multiplier of the absolute temperature T, $\frac{1}{2}kT$ gives us a measure of the average energy in each degree of freedom. But Einstein showed that k is also a measure of the thermal stability of the system, how much it departs from equilibrium in the form of energy *fluctuations*.

What Did Statistics Mean for Einstein?

In 1904, Einstein was only 25 years old, but in two years he had independently derived or rederived the work of the previous three decades in the kinetic theory of gases and statistical mechanics.

As we saw in chapter 2 on chance, most scientists did not believe that the appearance of randomly distributed events is any proof that there is ontological chance in the universe. For them, regularities in the "normal" distribution implied underlying unknown laws determining events. And Einstein was no exception.

The use of "statistical" methods is justified by the apparent impossibility of knowing the detailed paths of an incredibly large number of particles. One might think that increasing the number of particles would make their study increasingly complex, but the opposite is true. The regularities that appear when averaging over their large numbers gives us mean values for the important quantities of classical thermodynamics like energy and entropy.

In principle, the motions of individual particles obey the laws of classical mechanics. They are *deterministic*, and *time reversible*. In 1904, Einstein certainly subscribed to this view, until 1909 at least.

So when Boltzmann's H-theorem had shown in 1872 that the entropy in an isolated system can only increase, it was that the increase in entropy is only *statistically irreversible*.

Before Boltzmann, we saw in chapter 3 it was Maxwell who first derived a mathematical expression for the distribution of gas particles among different velocities. He assumed the particles were distributed at random and used probabilities from the theory of errors to derive the shape of the distribution. There is some evidence that Maxwell was a skeptic about determinism and may thus have accepted that randomness as ontological chance.

But Boltzmann clearly accepted that his macroscopic irreversibility did not prove the existence of *microscopic irreversibility*. He had considered the possibility of some "molecular chaos." But even without something microscopically random, Boltzmann's statistical irreversibility does explain the increase in entropy, despite his critics JOSEF LOSCHMIDT and ERNST ZERMELO.

What Then Are the Fluctuations?

In the last of his papers on statistical mechanics, Einstein derived expressions for expected *fluctuations* away from thermal equilibrium. Fluctuations would be examples of entropy decreasing slightly, proving that the second law is not an *absolute* law, but only a statistical one, as both Maxwell and Boltzmann had accepted.

Boltzmann had calculated the size of fluctuations and declared them to be *unobservable* in normal gases. One year after his 1904 paper, Einstein would demonstrate that molecular fluctuations are *indirectly* observable and can explain the Brownian motion. Einstein's prediction and its experimental confirmation by Jean Perrin a few years later would prove the existence of atoms.

Einstein also expressed the possibility in his 1904 paper that a general theory of physical systems would apply equally to matter and radiation. He thought fluctuations would be even more important for radiation, especially for radiation with wavelengths comparable to the size of their container. He showed that the largest fluctuations in energy would be for particles of average energy.

Einstein argued that the general principle of *equipartition of energy* among all the degrees of freedom of a system should be extended to radiation. But he was concerned that radiation, as a continuous theory, might have infinite degrees of freedom. A system of N gas particles has a finite number of degrees of freedom, which determines the finite number of states W and the system's entropy.

Einstein's speculation that the kinetic-molecular theory of statistical mechanics should also apply to radiation shows us an Einstein on the verge of discovering the particulate or "quantum" nature of radiation, which most physicists would not accept for another one or two decades at least.

We saw in chapter 4 that the term "quantum" was introduced into physics in 1900 by MAX PLANCK, who hypothesized that the total energy of the mechanical oscillators generating the radiation field must be limited to integer multiples of a quantity hv, where v is the radiation frequency and h is a new constant with the dimensions of action (energy x time *or* momentum x distance). Planck did not think the radiation itself is quantized. But his quantizing the energy states of the matter did allow him to avoid infinities and use Boltzmann's definition of entropy as disorder and probability.

Einstein saw that Planck had used Boltzmann's probabilistic and statistical methods to arrive at an equation describing the distribution of frequencies in blackbody radiation. [1]

But Einstein also saw that Planck did not think that the radiation field itself could be described as particles. Nevertheless, Planck clearly had found the right equation. His radiation law fit the experimental data perfectly. But Einstein thought Planck had luckily stumbled on his equation for the wrong physical reasons. Indeed, a proper derivation would not be given for two more decades, when Einstein himself finally explained it in 1925 as the result of *quantum statistics* that have no place in classical statistical mechanics. [2]

Had Gibbs Done Everything Before Einstein?

Some historians and philosophers of science think that JOSIAH WILLARD GIBBS completed all the important work in statistical mechanics before Einstein. Gibbs had worked on statistical physics for many decades. Einstein had not read Gibbs, and when he finally did, he said his own work added little to Gibbs. But he was wrong.

Gibbs earned the first American Ph.D. in Engineering from Yale in 1863. He went to France where he studied with the great Joseph Liouville, who formulated the theorem that the phase-space volume of a system evolving under a conservative Hamiltonian function is a constant along the system's trajectory. This led to the conclusion that entropy is a conserved quantity, like mass, energy, momentum, etc.

In his short text *Principles in Statistical Mechanics,* published the year before his death in 1903, Gibbs coined the English term phase space and the name for the new field - statistical mechanics. This book brought him his most fame. But it was not his first work. Gibbs had published many articles on thermodynamics and was well known in Europe, though not by Einstein. Einstein independently rederived much of Gibbs's past work.

Einstein, by comparison, was an unknown developing his first ideas about a molecular basis for thermodynamics. His readings were probably limited to Boltzmann's *Lectures on Gas Theory.*

1   See chapter 4.

2   See chapter 22.

Gibbs transformed the earlier work in "kinetic gas theory" by Boltzmann, making it more mathematically rigorous. Gibbs made kinetic gas theory obsolete, but he lacked the deep physical insight of either Boltzmann or Einstein.

Perhaps inspired by the examples of other conservation laws in physics discovered during his lifetime, Gibbs disagreed with Boltzmann's view that information is "lost" when the entropy increases. For Gibbs, every particle is in principle distinguishable and identifiable. For Boltzmann, two gases on either side of a partition with particles distinguishable from one another, but otherwise identical, will increase their entropy when the partition is removed and they are allowed to mix.

For Gibbs, this suggested a paradox, what if the gases on both side were identical? On Boltzmann's view, the entropy would not go up, because there would be no "mixing." Entropy seems to depend on what we know about the particles? For Gibbs, complete information about every particle, their identities, their classical paths, would give us a constant entropy, essentially zero.

For Gibbs, information is conserved when macroscopic order disappears because it simply changes into microscopic (thus invisible) order as the path information of all the gas particles is preserved. As Boltzmann's mentor JOSEF LOSCHMIDT had argued in the early 1870's, if the velocities of all the particles could be reversed at an instant, the future evolution of the gas would move in the direction of decreasing entropy. All the original order would reappear.

Nevertheless, Gibbs's idea of the conservation of information is still widely held today by mathematical physicists. And most texts on statistical mechanics still claim that microscopic collisions between particles are reversible. Some explicitly claim that quantum mechanics changes nothing, because they limit themselves to the unitary (conservative and deterministic) evolution of the Schrödinger equation and ignore the collapse of the wave function.

So if Gibbs does not calculate the permutations of molecules in "microstates" and their combinations into the "complexions" of Boltzmann's "macrostates," what exactly is his statistical thinking?

It is the statistics of a large number of identical thermodynamic systems that he calls "ensembles." Boltzmann had also considered such large numbers of identical systems, averaging over them and assuming the averages give the same results as time averages over a single system. Such systems are called *ergodic*.

Maxwell thought that Boltzmann's ergodic hypothesis requires that the time evolution of a system pass through every point consistent with the energy. If the system is continuous, there are an infinite number of such points.

Boltzmann relaxed the ergodic requirement, dividing what Gibbs later called "phase space" into finite cells that Boltzmann described as "coarse graining." Quantum mechanics would later find reasons for particles being confined to phase-space volumes equal to the cube of Planck's quantum of action h^3. This is not because space is quantized but because material particles cannot get closer together than Heisenberg's uncertainty principle allows. $\Delta p\, \Delta x = h$.

Both Boltzmann and Gibbs considered two kinds of ensembles. Boltzmann called his ensembles *monodes*. Boltzmann's *ergode* is known since Gibbs as the microcanonical ensemble, in which energy is constant. In Gibbs's canonical ensemble energy may change. Boltzmann called it a *holode*. Einstein's focus was on the canonical ensemble. For him, the canonical is one where energy may be exchanged with a very large connected heat reservoir, which helped Einstein to define the absolute temperature T.

Where Gibbs ignored the microscopic behavior of molecules, Einstein followed Boltzmann in considering the motions and behavior of molecules, atoms, even electrons, and then photons.

Gibbs' statistical mechanics provided a formal basis for all the classical results of thermodynamics. But he discovered nothing new in atomic and molecular physics.

By contrast, Einstein's statistical mechanics gave him insight into things previously thought to be unobservable - the motions of molecules that explain the Brownian motion,[3] the behavior of electrons in metals as electrical and thermal conductors, the existence of energy levels in solids that explains anomalies in their specific heat,[4] and even let him discover the particle nature of light.[5]

---

3   Chapter 7.
4   Chapter 8.
5   Chapter 6.

Einstein's study of *fluctuations* let him see both the particle nature and the wave nature of light as separate terms in his analysis of entropy. In the final section of his 1904 paper, Einstein applied his calculations to radiation.

He thought that energy fluctuations would be extreme if the radiation is confined to a volume of space with dimensions of the same order of magnitude as the wavelength of the radiation.

While Einstein may or may not be correct about the maximum of fluctuations, he did derive the wavelength of the maximum of radiation λ_{max}, showing it is inversely proportional to the absolute temperature T. Einstein estimated theoretically that

$\lambda_{max} = 0.42/T$

Wien had discovered this relationship ten years earlier empirically as his displacement law. Wien had found

$\lambda_{max} = 0.293/T.$

Einstein wrote

> One can see that both the kind of dependence on the temperature and the order of magnitude of λ_m can be correctly determined from the general molecular theory of heat, and considering the broad generality of our assumptions, I believe that this agreement must not be attributed to chance.[6]

Einstein's work on statistical mechanics thus goes well beyond that of Boltzmann and Gibbs. The work of Gibbs did not depend on the existence of material particles and that of Boltzmann had nothing to do with radiation.

The tools Einstein developed in his three papers on statistical mechanics, especially his ability to calculate microscopic fluctuations, gave him profound insights into both matter and light.

All this work may be largely forgotten today, especially in many modern texts on quantum physics that prefer the conservative Gibbs formalism to that of Einstein. But Einstein's next three papers, all published in just one year often called his *annus mirabilis*, were all based on his young ability to see far beyond his older colleagues.

In particular, Einstein had a knack for seeing what goes on at the microscopic level that he called an "objective reality."

6   On the General Molecular Theory of Heat, §5 Application to Radiation, *Annalen der Physik*, 14 (1904) pp.354-362.

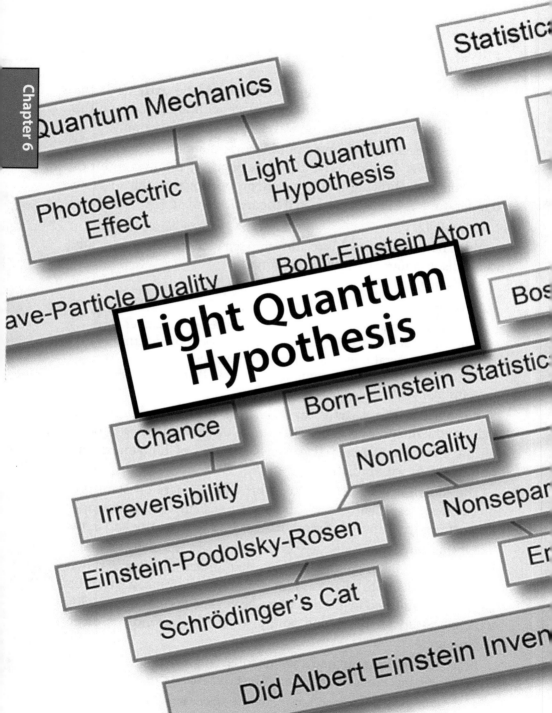

Light Quantum Hypothesis

In his "miracle year" of 1905, Einstein wrote four extraordinary papers, one of which won him the 1921 Nobel prize in physics. Surprisingly, the prize was not for his third paper, on the theory of relativity. Special relativity was accepted widely, but it remained controversial for some conservative physicists on the Nobel committee. Nor was it for the second paper, in which Einstein showed how to prove the existence of material particles. Nor even the fourth, in which the famous equation $E = mc^2$ first appeared.

Einstein's Nobel Prize was for the first paper of 1905. In it he hypothesized the existence of light particles. The prize was not for this hypothesis he called "very revolutionary." The prize was for his explanation for the photoelectric effect (as quanta of light!).

The idea that light consists of discrete "quanta," which today we call *photons*, was indeed so revolutionary that it was not accepted by most physicists for *nearly two decades*, and then reluctantly, because it leaves us with the mysterious dual aspect of light as sometimes a particle, and sometimes a wave.

A close reading of Einstein's work will give us the tools to resolve this quantum mystery and several others. But we begin with trying to see today what Einstein already saw clearly in 1905.

We must keep in mind that the model of a physical theory for Einstein was a "field theory." A field is a *continuous* function of four-dimensional space-time variables such as Newton's gravitational field and Maxwell's electrodynamics.

For Einstein, the theories and principles of physics are *fictions* and "free creations of the human mind." Although they must be tested by experiment, one cannot *derive* the basic laws from experience, he said. And this is particularly true of field theories, like his dream of a "unified field theory." They are thought to have continuous values at every point in otherwise empty space and time. Listen to Einstein's concern in his first sentence of 1905...

> There exists a profound formal distinction between the theoretical concepts which physicists have formed regarding gases and other ponderable bodies and the Maxwellian theory of electromagnetic processes in so-called empty space. [1]

---

1   Einstein, 1905a.. p.86

According to the Maxwellian theory, energy is to be considered a continuous spatial function in the case of all purely electro-magnetic phenomena including light, while the energy of a ponderable object should, according to the present conceptions of physicists, be represented as a sum carried over the atoms and electrons...[2]

Here Einstein first raises the deep question that we hope to show he struggled with his entire life. *Is nature continuous or discrete?*

Is it possible that the physical world is made up of nothing but discrete discontinuous *particles*? Are continuous *fields* with well-defined values for matter and energy at all places and times simply *fictional* constructs, *averages* over large numbers of particles?

The energy of a ponderable body cannot be subdivided into arbitrarily many or arbitrarily small parts, while the energy of a beam of light from a point source (according to the Maxwellian theory of light or, more generally, according to any wave theory) is continuously spread over an ever increasing volume.

It should be kept in mind, however, that the optical observations refer to time averages rather than instantaneous values. In spite of the complete experimental confirmation of the theory as applied to diffraction, reflection, refraction, dispersion, etc., it is still conceivable that the theory of light which operates with continuous spatial functions may lead to contradictions with experience when it is applied to the phenomena of emission and transformation of light.[3]

One should keep in mind, Einstein says, that our observations apply to averages (over a finite number of particles) and that a continuum theory leads to contradictions with emission and absorption processes. In particular, the continuum has an infinite number of "degrees of freedom," while matter and energy quanta are finite. We saw in chapter 3 that LUDWIG BOLTZMANN had made this point,

"An actual continuum must consist of an infinite number of parts; but an infinite number is undefinable... Thus the mechanical foundations of the partial differential equations, when based on the coming and going of smaller particles, with restricted average values, gain greatly in plausibility."[4]

---

2   *ibid.,. p.86

3   *ibid.,. p.86-87

4   Boltzmann, 2011, *p.27*

The Photoelectric Effect

Continuing his investigations into a single theory that would describe both matter and radiation, Einstein proposed his "very revolutionary" hypothesis to explain a new experiment that showed a direct connection between radiation and electrons.

Before Einstein, it was thought that the oscillations of electrons in a metal are responsible for the emission of electromagnetic waves, but Einstein argued that it is the *absorption* of light that is causing the ejection of electrons from various metal surfaces.

It is called the *photoelectric effect*.

HEINRICH HERTZ had shown in 1889 that high-voltage spark gaps emit electromagnetic waves that are light waves obeying Maxwell's equations. He also noticed that ultraviolet light shining on his spark gaps helped them to spark. In 1902, the Hungarian physicist PHILIPP LENARD confirmed that light waves of sufficiently high frequency v shining on a metal surface cause it to eject electrons.

To Lenard's surprise, below a certain frequency, no electrons are ejected no matter how strong he made the intensity of the light. Assuming that the energy in the light wave was simply being converted into the energy of moving electrons, this made no sense.

Furthermore, when Lenard increased the frequency of the incident light (above a critical frequency v_c) the ejected electrons appeared to move faster for higher light frequencies.

These strange behaviors gave Einstein very strong reasons for imagining that light must be concentrated in a physically localized bundle of energy. He wrote:

The usual conception, that the energy of light is continuously distributed over the space through which it propagates, encounters very serious difficulties when one attempts to explain the photoelectric phenomena, as has been pointed out in Herr Lenard's pioneering paper.

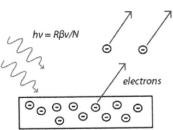

$hv = R\beta v/N$

electrons

According to the concept that the incident light consists of energy quanta of magnitude $R\beta v/N$ [hv], however, one can conceive of the ejection of electrons by light in the following way. Energy quanta penetrate into the surface layer of the body, and their energy is transformed, at least in part, into kinetic energy of electrons. The simplest way to imagine this is that a light quantum delivers its entire energy to a single electron; we shall assume that this is what happens...

An electron to which kinetic energy has been imparted in the interior of the body will have lost some of this energy by the time it reaches the surface. Furthermore, we shall assume that in leaving the body each electron must perform an amount of work P characteristic of the substance...

If each energy quantum of the incident light, independently of everything else, delivers its energy to electrons, then the velocity distribution of the ejected electrons will be independent of the intensity of the incident light; on the other hand the number of electrons leaving the body will, if other conditions are kept constant, be proportional to the intensity of the incident light... [5]

Einstein shows here that the whole energy of an incident light quantum is absorbed by a single electron.

Some of the energy absorbed by the electron becomes P, the work needed to escape from the metal. The rest is the kinetic energy $E = \frac{1}{2} mv^2$ of the electron. Einstein's "photoelectric equation" thus is

$$E = h\nu - P.$$

Einstein's equation predicted a linear relationship between the frequency of Einstein's light quantum $h\nu$, and the energy E of the ejected electron. It was more than ten years later that R. A. MILLIKAN confirmed Einstein's photoelectric equation. Millikan nevertheless denied that his experiment proved Einstein's radical but clairvoyant ideas about light quanta! He said in 1916

Einstein's photoelectric equation... cannot in my judgment be looked upon at present as resting upon any sort of a satisfactory theoretical foundation. [6]

---

5   Einstein, 1905a. p.99.
6   A Direct Photoelectric Determination of Planck's "h". *Physical Review*, 7(3), 355.

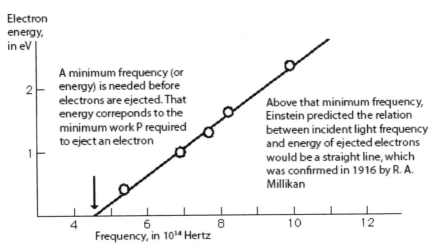

Figure 6-6. The Photoelectric Effect.

The Entropies of Radiation and Matter

Einstein clearly recognized the well-established difference between matter and energy, but he hoped to find some kind of symmetry between them in a general theory that describes them both.

Within the 1905 year, he writes the most famous equation in physics that connects the two, $E = mc^2$. But Einstein discovers a symmetry by calculating the entropy of matter and radiation, using the methods he developed in his three papers on statistical mechanics.[7]

Einstein begins by asking for the probability W that a particular movable point (an abstract property of a molecule) would be randomly found in a small volume v in a large container with volume v_0. He then asks "how great is the probability that at a randomly chosen instant of time all n independently movable points in a given volume v_0 will be contained (by chance) in volume v?"

The probability of independent events is the product of the individual probabilities, so $W = [v/v_0]^n$. Einstein then uses "Boltzmann's Principle, that the entropy $S = k \log W$.

$$S - S_0 = k \log [v/v_0]^n = k\, n \log [v/v_0]$$

Einstein derived a similar expression for the entropy of radiation with energy E and frequency v as

---

7   See chapter 5.

$$S - S_0 = k \ (E/h\nu) \log [\nu/\nu_0]$$

If we compare the two expressions, it appears that $E/h\nu$ is the *number* of independent light particles. Einstein concluded

> Monochromatic radiation of low density (within the range of validity of Wien's radiation formula) behaves thermodynamically as if it consisted of mutually independent energy quanta of magnitude $h\nu$ [Einstein wrote $R\beta\nu/N$]. [8]

Einstein showed that thermodynamically, radiation behaves like gas particles. It seems reasonable, he said,

> "to investigate whether the laws of generation and conversion of light are also so constituted as if light consisted of such energy quanta. Light can not be spread out continuously in all directions if individual energy quanta can be absorbed as a unit that ejects a photoelectron in the photoelectric effect."

Nonlocality

How can energy spread out continuously over a large volume yet later be absorbed in its entirety at one place, without contradicting his principle of relativity? Einstein clearly describes here what is today known as *nonlocality*, but he does not describe it explicitly until 1927, and then only in comments at the fifth Solvay conference. He does not publish his concerns until the EPR paper in 1935!

If the energy travels as a spherical light wave radiated into space in all directions, how can it instantaneously collect itself together to be absorbed into a single electron. Einstein already in 1905 sees something *nonlocal* about the photon. What is it that Einstein sees?

It is events at two points in a spacelike separation occurring "simultaneously," a concept that his new special theory of relativity says is impossible in any absolute sense.

He also sees that there is both a wave aspect and a particle aspect to electromagnetic radiation. He strongly contrasts the finite number of variables that describe *discrete* matter with the assumption of *continuous* radiation.

> While we consider the state of a body to be completely determined by the positions and velocities of a very large, yet finite, number of

---

8   Einstein, 1905a., p.97.

atoms and electrons, we make use of continuous spatial functions to describe the electromagnetic state of a given volume, and a finite number of parameters cannot be regarded as sufficient for the complete determination of such a state.

The wave theory of light, which operates with continuous spatial functions, has worked well in the representation of purely optical phenomena and will probably never be replaced by another theory.

It seems to me that the observations associated with blackbody radiation, fluorescence, the production of cathode rays by ultraviolet light, and other related phenomena connected with the emission or transformation of light are more readily understood if one assumes that the energy of light is discontinuously distributed in space.

In accordance with the assumption to be considered here, the energy of a light ray spreading out from a point source is not continuously distributed over an increasing space but consists of a finite number of energy quanta which are localized at points in space, which move without dividing, and which can only be produced and absorbed as complete units.

We therefore arrive at the conclusion: the greater the energy density and the wavelength of a radiation, the more useful do the theoretical principles we have employed turn out to be; for small wavelengths and small radiation densities, however, these principles fail us completely. [9]

As late as the Spring of 1926, perhaps following NIELS BOHR, WERNER HEISENBERG could not believe in the reality of light quanta.

Whether or not I should believe in light quanta, I cannot say at this stage. Radiation quite obviously involves the discontinuous elements to which you refer as light quanta. On the other hand, there is a continuous element, which appears, for instance, in interference phenomena, and which is much more simply described by the wave theory of light. But you are of course quite right to ask whether quantum mechanics has anything new to say on these terribly difficult problems. I believe that we may at least hope that it will one day. [10]

---

9   Einstein, 1905a.,
10   Heisenberg, 1971, p. 67

Statistic

Quantum Mechanics

Light Quantum
Hypothesis

Photoelectric
Effect

Bohr-Einstein Atom

ave-Particle Duality

Bos

Brownian Motion and Relativity

Born-Einstein Statistic

Chance

Nonlocality

Irreversibility

Nonsepar

Einstein-Podolsky-Rosen

Er

Schrödinger's Cat

Did Albert Einstein Inven

Brownian Motion and Relativity

In this chapter we describe two of Einstein's greatest works that have little or nothing to do with his amazing and deeply puzzling theories about quantum mechanics. The first, Brownian motion, provided the first quantitative proof of the existence of atoms and molecules. The second, special relativity in his miracle year of 1905 and general relativity eleven years later, combined the ideas of space and time into a unified space-time with a non-Euclidean curvature that goes beyond Newton's theory of gravitation.

Einstein's relativity theory explained the precession of the orbit of Mercury and predicted the bending of light as it passes the sun, confirmed by ARTHUR STANLEY EDDINGTON in 1919. He also predicted that galaxies can act as gravitational lenses, focusing light from objects far beyond, as was confirmed in 1979. He also predicted gravitational waves, only detected in 2016, one century after Einstein wrote down the equations that explain them..

What are we to make of this man who could see things that others could not? Our thesis is that if we look very closely at the things he said, especially his doubts expressed privately to friends, today's mysteries of quantum mechanics may be lessened.

As great as Einstein's theories of Brownian motion and relativity are, they were accepted quickly because measurements were soon made that confirmed their predictions. Moreover, contemporaries of Einstein were working on these problems. Marion Smoluchowski worked out the equation for the rate of diffusion of large particles in a liquid the year before Einstein. He did not publish, hoping to do the experimental measurements himself.

In the development of special relativity, Hendrik Lorentz had assumed the constancy of the velocity of light and developed the transformation theory that predicted the apparent contraction of space and/or time when measured by moving clocks. Henri Poincaré used the Lorentz transformation and had described a "principle of relativity" in which the laws of physics should be the same in all frames unaccelerated relative to the ether (which Poincaré continued to believe in for years). Hermann Minkowski combined space and time into a four-dimensional "space-time."

With regard to general relativity, the mathematician David Hilbert took a great interest in Einstein's ideas about a general relativity. He invited Einstein to give six lectures in Göttingen several months before Einstein completed his work. Einstein stayed at Hilbert's home and they began an extensive exchange of ideas which led Hilbert close to a theory unifying gravitation and electromagnetism.

Einstein was very concerned that Hilbert might beat him to the correct equations, which Hilbert knew Einstein had been working on since 1913. In the end, Hilbert stated clearly that Einstein had been the original author of general relativity.

A excellent survey of these priority debates is on Wikipedia. [1]

Einstein's 1905 explanation for the motions of tiny visible particles in a gas or liquid, that they are caused by the motions of *invisible* particles - atoms or molecules - was hardly new, having been suggested exactly as such by LUCRETIUS in his *De Rerum Natura* at the dawn of the theory of atoms.

> It clearly follows that no rest is given to the atoms in their course through the depths of space... This process, as I might point out, is illustrated by an image of it that is continually taking place before our very eyes. Observe what happens when sunbeams are admitted into a building and shed light on its shadowy places. You will see a multitude of tiny particles mingling in a multitude of ways in the empty space within the light of the beam...From this you may picture what it is for the atoms to be perpetually tossed about in the illimitable void...their dancing is an actual indication of underlying movements of matter that are hidden from our sight. [2]

The importance of Einstein's work is that he *calculated* and *published* the motions of molecules in ordinary gases, predictions confirmed by experiment just a few years later by Jean Perrin.

Now chemists and many physicists had *believed* in atoms for over a century in 1905 and they had excellent reasons. But we must understand Einstein's work as leading to experimental evidence for the existence of atoms, that is to say material particles. But it was the first of Einstein's insights into the *discrete* nature of reality that conflicted with his deeply held *beliefs* about reality as *continuous*.

---

1   en.wikipedia.org/wiki/Relativity_priority_dispute
2   *On the Nature of Things*, Book II, lines 115-141

Chapter 7

The goal of this book is to show that many things Einstein clearly saw provide a better picture of reality than those of most of today's physicists and philosophers of science, many of whom pursue physical theories that Einstein *believed*, not what he *saw*.

We will study what Einstein thought went on in "objective reality."

For Einstein, the model of a physical theory was a "field theory." A field is a *continuous* function of four-dimensional space-time variables such as Newton's gravitational field and Maxwell's electro-dynamics. Einstein said "The most difficult point for such a field-theory at present is how to include the atomic structure of matter and energy." [3] It is the question of the nature of reality we raised in the introduction - is the nature of reality *continuous* or *discrete*. Does nature consist primarily of *particles* or *fields*?

Einstein could never see how to integrate the discrete particles of matter and of light into his ideas for a "unified field theory." He hoped all his life to show that the light particles he discovered and all material particles are singularities in his unified field.

Einstein said many time that the theories of physics are *fictions* and "free creations of the human mind." Although theories must be tested by experiment, one cannot *derive* or *construct* the basic laws from experience. They must depend on *principles.*

In his 1905 article "On the Movement of Small Particles Suspended in a Stationary Liquid Demanded by the Molecular Kinetic Theory of Heat," Einstein wrote

> In this paper it will be shown that according to the molecular-kinetic theory of heat, bodies of microscopically-visible size suspended in a liquid will perform movements of such magnitude that they can be easily observed in a microscope, on account of the molecular motions of heat. It is possible that the movements to be discussed here are identical with the so-called "Brownian molecular motion"; however, the information available to me regarding the latter is so lacking in precision, that I can form no judgment in the matter.[4]

Because Einstein published, leaving experiments to others, the credit is his rather than Smoluchowski's. But more important than credit, Einstein saw these particles, and the light quanta of the last chapter, though he could never integrate them into his field theory.

---

3   "On the Method of Theoretical Physics," p.168
4   CPAE, vol. 2, p.123

Statistic

Quantum Mechanics

Light Quantum
Hypothesis

Photoelectric
Effect

Bohr-Einstein Atom

ave-Particle Duality

Bos

Specific Heat

Born-Einstein Statistic

Chance

Nonlocality

Irreversibility

Nonsepar

Einstein-Podolsky-Rosen

E

Schrödinger's Cat

Did Albert Einstein Inven

Specific Heat

A few months after the three famous papers of his miracle year, Einstein published in September 1905 a three-page paper showing that energy and matter are interconvertible according to the famous equation $E = mc^2$. This result greatly strengthened his belief in the light quantum hypothesis of March. He now saw that radioactive decay involves the liberation of a vast amount of radiation which is a consequence of the conversion of mass into energy. This was forty years before the first atomic bomb.

In 1906 and early 1907, Einstein published two more papers on the Planck radiation law and the deeper physical connections that must exist between matter and radiation. The first was on the emission and absorption of radiation by matter, the second on the specific heat of different materials.

In the first paper, Einstein was puzzled how Planck had arrived at his law for the distribution of energy in blackbody radiation, especially the exponential factor in the denominator and the added -1. He concluded (ironically?) that Planck had effectively, without understanding it, "introduced into physics a new hypothetical element: the hypothesis of light quanta." He wrote

> the energy of an elementary resonator can only assume values that are integral multiples of $(R/N)\beta v$: by emission and absorption, the energy of a resonator changes by jumps of integral multiples of $(R/N)\beta v$. (In modern notation, hv.)

Einstein thus introduced "quantum jumps" inside atoms six years before NIELS BOHR's atomic model with Bohr's proposal for "stationary states" or energy levels. Forty-five years later, ERWIN SCHRÖDINGER denied quantum jumps in two articles,[1] JOHN BELL questioned them again in 1986,[2] and decoherence theorists deny the "collapse of the wave function" to this day.

Einstein's paper of 1907 was an extraordinary investigation into the *specific heat* of solid materials. In this paper, Einstein again

---

[1] "Are There Quantum Jumps?," *British Journal for the Philosophy of Science* 3.10 (1952):

[2] "Are There Quantum Jumps?" in *Schrödinger, Centenary of a Polymath* ed. C. Kilmister, Cambridge University Press (1987)

took the implications of Planck's quantum theory more seriously than had Planck himself. Matter must have internal quantum states.

Internal quantum states at energies higher than the ground state will not be populated unless there is enough energy available to cause a jump from the ground state to one or more of the "excited" states. The populations of higher states are proportional to the "Boltzmann factor" $e^{-E/kT}$.

There are many kinds of states in atoms, molecules, and in the so-called "solid state," atoms arranged in lattice structures like crystals and metals. The quantum states correspond to classical "degrees of freedom." A molecule can rotate in two orthogonal directions. It can vibrate in one dimension, the distance between the atoms. Atoms and molecules have excited electronic states. In general, rotational states have the lowest energy separations, vibrational states next, and electronic states the highest energies above the ground state. And bulk matter vibrates like a violin string or a sound wave (phonons).

Specific heat is the amount of energy that must be added to raise the temperature of material one degree. It is closely related to the entropy, which has the same dimensions - ergs/degree. It depends on the quantum internal structure of the material, as first understood by Einstein, who is sometimes recognized as the first solid-state physicist.

As the temperature increases, the number of degrees of freedom, and thus the number of states (whose logarithm is the entropy), may all increase suddenly, in so-called phase changes (the number of available cells in phase space changes).

Conversely, as temperature falls, some degrees of freedom are said to be "frozen out," unavailable to absorb energy. The specific heat needed to move one degree is reduced. And the entropy of the system approaches zero as the temperature goes to absolute zero.

Some diatomic molecular gases were known to have anomalously low specific heats. It had been one of the strong arguments against the kinetic-molecular theory of heat. In a monatomic gas, each atom has three degrees of freedom, corresponding to the three independent dimensions of translational motions, x, y, and z.

A diatomic molecule should have six degrees of freedom, three for the motion of the center of mass, two for rotations, and one for vibrations along the intramolecular axis.

While some diatomic materials appear to have the full specific heat expected if they can move, rotate, and vibrate, it was Einstein who explained why many molecules can not vibrate at ordinary temperatures. The vibrational states are quantized and need a certain minimum of energy before they can be excited.

Einstein's research into specific heats suggested that internal molecular quantum states could account for emission and absorption lines and the continuous bands seen in spectroscopy.

Einstein speculated that the vibrational states for some molecules were too far above the ground state to be populated, thus not absorbing their share of energy when heat is added). Most diatomic molecules were known to have a specific heat c of 5.94, but Einstein said that according to Planck's theory of radiation, their specific heat would vary with temperature. He found

$$c = 5.94 \, \beta v \, / e^{\beta v /T} - 1.$$

Einstein plotted a graph to show his increase in specific heat with temperature, along with a few experimental measurements. [3]

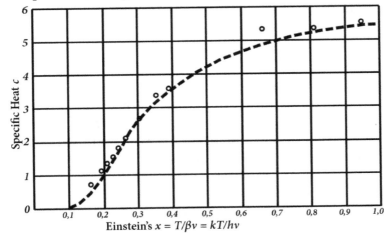

Figure 8-7. Einstein's prediction for specific heats.

In 1913, Niels Bohr would identify the internal quantum states of excited electrons as responsible for the spectral lines in atomic hydrogen. This was a direct extension of Einstein's discoveries.

3   CPAE, vol,2, Doc.38, p.220.

WALTHER NERNST was one of the first physicists to embrace the quantum theory of Einstein. He did not support the light quantum hypothesis. No one but Einstein himself took it seriously for decades, but Nernst accepted Einstein's idea of quantized energy levels in matter as the explanation of the anomalous specific heats.

We saw in chapter 4 that Planck assumed the energy of radiating oscillators was limited to multiples of hv, but this was just a lucky guess at a mathematical formula matching the experimental data.

Planck himself did not believe in the *reality* of this hypothesis about quantized energy levels, but Einstein in 1906 showed that the Planck radiation law required such energy levels, and that they explained the specific heat approaching zero for low temperatures.

In 1905 Nernst proposed a radical theory for the specific heats and entropy of liquids and solids at what he called absolute zero. He began a program of detailed measurements of specific heat at extremely low temperatures.

A few years later Nernst announced a postulate that later became known as the "third law" of thermodynamics - the entropy of a perfect crystal at absolute zero (zero degrees Kelvin) is exactly equal to zero. He wrote

> one gains the clear impression that the specific heats become zero or at least take on very small values at very low temperatures. This is in qualitative agreement with the theory developed by Herr Einstein. [4]

Nernst was thus one of the few supporters of Einstein's contributions to quantum theory to appear in the long years from 1905 to 1925. To be sure, it must have been terribly frustrating for Einstein to see his critically important light quantum hypothesis ignored for so long. But the idea that atoms and molecules contained energy levels was about to be taken very seriously (by BOHR in 1913), and Einstein was the first proponent of discrete energy levels.

Nernst organized the first international meeting of scientists that took Einstein's quantum theory seriously. It was financed by the Belgian industrialist Ernst Solvay. The topic of the first Solvay conference, in 1911, was specific heats. Nernst gave Einstein the privilege of being the last speaker. His paper was called "The Current Status of the Specific Heat Problem."

---

4   Pais, 1982, p.398.

Einstein included a very lengthy recapitulation of all his earlier arguments for the light quantum hypothesis. His paper is twenty-three pages long[5] and is followed by an eleven-page discussion by Poincaré, Lorentz, Wien, Planck, and of course, Einstein and Nernst.

Although Nernst was the earliest supporter of quantum theory, as applied to matter, he was very frank at the first Solvay conference that it still needed a lot of experimental research.

> At this time, the quantum theory is essentially a computational rule, one may well say a rule with most curious, indeed grotesque, properties. However,...it has borne such rich fruits in the hands of Planck and Einstein that there is now a scientific obligation to take a stand in its regard and to subject it to experimental test.[6]

Unfortunately, Einstein did no more work on quantum theory for the next five years as he focused all his energy on publishing his general theory of relativity.

As Abraham Pais said, one hopes that Einstein got some small satisfaction from the fact that his work on the specific heats of solids was a step in the right direction. He deserves the title of first solid state physicist. But as he wrote to a friend in 1912, Einstein was at least as puzzled as he was pleased with his ideas about specific heat,

> In recent days, I formulated a theory on this subject. *Theory* is too presumptuous a word — it is only a groping without correct foundations. The more success the quantum theory has, the sillier it looks. How nonphysicists would scoff if they were able to follow the course of its development.[7]

Albert Messiah's classic text makes Einstein's contribution clear.

> Historically, the first argument showing the necessity of "quantizing" material systems was presented by Einstein in the theory of the specific heat of solids (1907).[8]

Nernst and others extended Einstein's ideas on specific heat to liquids, but made no progress with gases at temperature absolute zero. That problem had to wait for nearly two decades and Einstein's discovery of quantum statistics. See chapter 15.

---

5   CPAE, vol 3, Doc.26.
6   Pais, 1982, p.399.
7   Pais, *ibid*.
8   Messiah, 1961, p.21

Statistic

Quantum Mechanics

Light Quantum
Hypothesis

Photoelectric
Effect

Bohr-Einstein Atom

ave-Particle Duality

Wave-Particle Duality

Bos

Born-Einstein Statistic

Chance

Nonlocality

Irreversibility

Nonsepar

Einstein-Podolsky-Rosen

E

Schrödinger's Cat

Did Albert Einstein Inven

Wave-Particle Duality

Einstein greatly expanded his light-quantum hypothesis in his presentation at the Salzburg conference in September, 1909. He argued that the interaction of radiation and matter involves elementary processes that have no inverse, a deep insight into the *irreversibility* of natural processes. While incoming spherical waves of radiation are mathematically possible, they are not practically achievable. Nature appears to be asymmetric in time. Einstein speculates that the continuous electromagnetic field might be made up of large numbers of light quanta - singular points in a field that superimpose collectively to create the wavelike behavior.

Although Einstein could not yet formulate a mathematical theory that does justice to both the *continuous* oscillatory and *discrete* quantum structures - the wave and particle pictures, he argued that they are compatible. This was more than fifteen years before WERNER HEISENBERG's particle matrix mechanics and ERWIN SCHRÖDINGER's wave mechanics in the 1920's. Because gases behave statistically, Einstein thought that the connection between waves and particles may involve probabilistic behavior.

> Once it had been recognized that light exhibits the phenomena of interference and diffraction, it seemed hardly doubtful any longer that light is to be conceived as a wave motion. Since light can also propagate through vacuum, one had to imagine that vacuum, too, contains some special kind of matter that mediates the propagation of light waves. [the ether] However, today we must regard the ether hypothesis as an obsolete standpoint. It is even undeniable that there is an extensive group of facts concerning radiation that shows that light possesses certain fundamental properties that can be understood far more readily from the standpoint of Newton's emission theory of light than from the standpoint of the wave theory. [1]

Einstein's 1905 relativity theory requires that the inertial mass of an object decreases by L/c^2 when that object emits radiation

---

[1]   CPAE, vol.2. p. 379

Chapter 9

of energy L. The inertial mass of an object is diminished by the emission of light. Einstein now says in 1909,

> The energy given up was part of the mass of the object. One can further conclude that every absorption or release of energy brings with it an increase or decrease in the mass of the object under consideration. Energy and mass seem to be just as equivalent as heat and mechanical energy.

Indeed, in 1905, Einstein had shown that $E = mc^2$. He had found a *symmetry* between light and matter. They are both particles. But in 1909 Eintsein finds the wave nature of light emerging from his equations and suggests a "fusion" of wave and particle theories

> It is therefore my opinion that the next stage in the development of theoretical physics will bring us a theory of light that can be understood as a kind of fusion of the wave and emission theories of light. To give reasons for this opinion and to show that a profound change in our views on the nature and constitution of light is imperative is the purpose of the following remarks.[2]

On the other hand, Einstein identified an important *asymmetry*.

> In the kinetic theory of molecules, for every process in which only a few elementary particles participate (e.g., molecular collisions), the inverse process also exists. But that is not the case for the elementary processes of radiation. In the foregoing it has been assumed that the energy of at least some of the quanta of the incident light is delivered completely to individual electrons.

> According to our prevailing theory, an oscillating ion generates a spherical wave that propagates outwards. The inverse process does not exist as an elementary process. A converging spherical wave is mathematically possible, to be sure; but to approach its realization requires a vast number of emitting entities. The elementary process of emission is not invertible. In this, I believe, our oscillation theory does not hit the mark. Newton's emission theory of light seems to contain more truth with respect to this point than the oscillation theory since, first of all, the energy given to a light particle is not scattered over infinite space, but remains available for an elementary process of absorption.[3]

Recall from chapter 4 that Planck had argued the interaction of light with matter might explain the *irreversibility* of the increase in

2   *ibid.*, p.379

3   *ibid.*, p.387

entropy of the second law of thermodynamics. Planck thought a plane wave might be converted to a spherical wave going outward from the oscillator. But Boltzmann had talked him out of the idea, because time reversal would produce the incoming wave that Einstein here says is impossible. We shall see that Einstein's insight can explain the origin of *microscopic irreversibility*. See chapter 12.

From Matter to Light to Matter

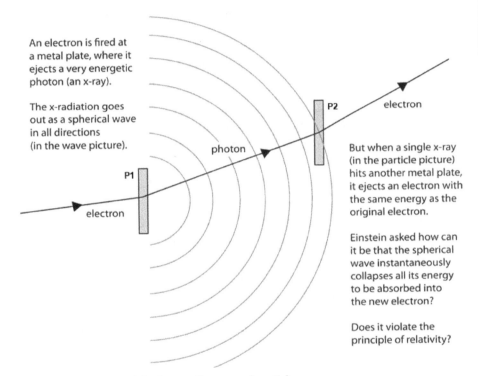

An electron is fired at a metal plate, where it ejects a very energetic photon (an x-ray).

The x-radiation goes out as a spherical wave in all directions (in the wave picture).

P1

electron

photon

P2

electron

But when a single x-ray (in the particle picture) hits another metal plate, it ejects an electron with the same energy as the original electron.

Einstein asked how can it be that the spherical wave instantaneously collapses all its energy to be absorbed into the new electron?

Does it violate the principle of relativity?

Figure 9-8. Einstein's picture of waves and particles.

Einstein imagined an experiment in which the energy of an electron (a cathode ray) is converted to a light quantum and back.

Consider the laws governing the production of secondary cathode radiation by X-rays. If primary cathode rays impinge on a metal plate P1, they produce X-rays. If these X-rays impinge on a second metal plate P2, cathode rays are again produced whose speed is of the same order as that of the primary cathode rays.

As far as we know today, the speed of the secondary cathode rays depends neither on the distance between P1 and P2, nor on the intensity of the primary cathode rays, but rather entirely on the speed of the primary cathode rays. Let's assume that this is strictly true. What would happen if we reduced the intensity of the primary cathode rays or the size of P1 on which they fall, so that the impact of an electron of the primary cathode rays can be considered an isolated process?

If the above is really true then, because of the independence of the secondary cathode rays' speed on the primary cathode rays' intensity, we must assume that an electron impinging on P1 will either cause no electrons to be produced at P2, or else a secondary emission of an electron whose speed is of the same order as that of the initial electron impinging on P1. In other words, the elementary process of radiation seems to occur in such a way that it does not scatter the energy of the primary electron in a spherical wave propagating in every direction, as the oscillation theory demands.[4]

Extending his 1905 hypothesis, Einstein shows energy can not spread out like a wave continuously over a large volume, because it is absorbed in its entirety to produce an ejected electron at P2, with essentially the same energy as the original electron absorbed at P1.

Rather, at least a large part of this energy seems to be available at some place on P2, or somewhere else. The elementary process of the emission of radiation appears to be directional. Moreover, one has the impression that the production of X-rays at P1 and the production of secondary cathode rays at P2 are essentially inverse processes...Therefore, the constitution of radiation seems to be different from what our oscillation theory predicts.

The theory of thermal radiation has given important clues about this, mostly by the theory on which Planck based his radiation formula...Planck's theory leads to the following conjecture. If it is really true that a radiative resonator can only assume energy values that are multiples of $h\nu$, the obvious assumption is that the emission and absorption of light occurs only in these energy quantities.[5]

4   ibid., p.387
5   ibid., p.390

This important conjecture by Einstein, that light is emitted and absorbed in units of hv, is often misattributed to MAX PLANCK, who never fully accepted Einstein's "very revolutionary" hypothesis..

Einstein found theoretical evidence for his "fusion of wave and emission theories of light" in his study of statistical fluctuations of the gas pressure (collisions with gas particles) and the radiation pressure (collisons with light quanta) on a metal plate suspended in a cavity.

Using results from his years deriving the laws of statistical mechanics, and assuming the plate, the cavity walls, the gas and the light particles are all in equilibrium at temperature T, he derived an expression for the fluctuations in the radiation pressure in the frequency interval dv as containing two terms.

$$<\varepsilon^2> = (Vdv) \{hv\rho + (c^3/8\pi v^2)\, \rho^2\}.$$

> The wave theory provides an explanation only for the second term... That the expression for this fluctuation must have the form of the second term of our formula can be seen by a simple dimensional analysis.

> But how to explain the first term of the formula?... If radiation consisted of very small-sized complexes of energy hv,... a conception that represents the very roughest visualization of the hypothesis of light quanta—then the momenta acting on our plate due to fluctuations of the radiation pressure would be of the kind represented by the first term alone. [6]

In a second independent analysis using Boltzmann's principle to calculate the mean squared energy fluctuation in terms of the density of radiation ρ with frequency v, and substituting Planck's radiation law for $\rho(v)$, Einstein once again derived the two-term expression for fluctuations in the radiation pressure. [7]

Einstein can again see the first (particle) term with light quanta hv and the second (wave) term with the classical expression for the number of modes $8\pi v^2/c^3$ in the radiation field with frequency v. The first term describes light with high frequencies (Wien's Law), the second light with long wavelengths (Rayleigh-Jeans Law).

---

6   *ibid,,* p.393

7   See Klein, 1964, p.11

Statistic

Quantum Mechanics

Light Quantum
Hypothesis

Photoelectric
Effect

Bohr-Einstein Atom

ave-Particle Duality

Bos

Bohr-Einstein
Atom

Born-Einstein Statistic

Chance

Nonlocality

Irreversibility

Nonsepar

Einstein-Podolsky-Rosen

Er

Schrödinger's Cat

Did Albert Einstein Inven

Bohr-Einstein Atom

NIELS BOHR is widely, and correctly, believed to be the third most important contributor to quantum mechanics, after MAX PLANCK and ALBERT EINSTEIN. Bohr is said to have introduced quantum numbers, quantization of properties, and "quantum jumps" between his postulated energy states in the atom.

But we have seen that Einstein made predictions of such "jumps" between energy levels in solid state matter several years earlier. The "quantum condition" for Bohr was quantization of the angular momentum, following a suggestion of J. W. Nicholson. Angular momentum has the same dimensions as Planck's "quantum of action" h. And we shall see that the integer numbers of quantum mechanics could be seen decades earlier in the empirical formulas for spectral-line frequencies.

Today the "Bohr atom" is described in many textbooks as making quantum jumps between energy levels, with the emission and absorption of *photons*. But this is a serious anachronism, because Bohr denied the existence of Einstein's localized light quanta for well over a decade after his 1913 model of the atom.

For Bohr, as for Planck, radiation was always a *continuous* wave, without which it was thought that one can not possibly explain the interference and diffraction phenomena of light. Planck himself did not accept Einstein's 1905 hypothesis of light quanta, although in 1913 Bohr suggested that "Planck's theory" did so.

> Now the essential point in Planck's theory of radiation is that the energy radiation from an atomic system does not take place in the continuous way assumed in the ordinary electro-dynamics, but that it, on the contrary, takes place in distinctly separated emissions, the amount of energy radiated out from an atomic vibrator of frequency v in a single emission being equal to $\tau h v$, where τ is an entire number, and h is a universal constant. [1]

This mistake is a source of much confusion about Einstein. Bohr did mention Einstein, but not his light quanta. His remarks indicate that Bohr knows about Einstein's work on specific heats, which showed in 1907 that there are energy levels in matter.

---

[1]　Bohr, 1913, p.4

The general importance of Planck's theory for the discussion of the behaviour of atomic systems was originally pointed out by Einstein. The considerations of Einstein have been developed and applied on a number of different phenomena, especially by Stark, Nernst, and Sommerfeld.[2]

This theory is not the work of Planck, who denied Einstein's light quantum hypothesis, but of Einstein, in the 1905 paper cited by Bohr as "considerations". Planck had only quantized the energy of his radiating oscillators. And as we saw in chapter 4, Planck's quantum of action was just a "fortunate guess" at a mathematical formula that fit experimental spectroscopic data for the *continuous spectrum* of electromagnetic radiation in thermal equilibrium.

Bohr had been invited by ERNEST RUTHERFORD to study in England, where Rutherford had shown that the nucleus of an atom is confined to a small central mass of positive charge, suggesting that the electrons might orbit about this center as planets orbit the sun. Rutherford's model conflicted with the fact that accelerated electrons should radiate a *continuous* stream of radiation of increasing frequency, as the electron spirals into the nucleus.

Bohr made two radical hypotheses about orbits, one of which Einstein would *derive* from quantum principles in 1916.

1) Orbits are limited to what Bohr called "stationary states," discrete energy levels in which the electrons do not radiate energy.

2) Electrons can emit or absorb radiation with energy hv only when they "jump" between energy levels where $E_m - E_n = hv$.

It is most odd that Bohr maintained for the next ten years that the energy radiated in a quantum jump is *continuous* radiation, not Einstein's *discrete* and localized quanta. Bohr would only accept Einstein's photons after the failure of the Bohr-Kramers-Slater proposal of 1925, which claimed energy is only statistically conserved in the emission and absorption of continuous radiation. Einstein insisted energy is conserved for individual quantum interactions, and experiments showed he was correct.

Apart from these mistakes in his physics, Bohr's atomic model was a work of genius at the same level as Planck's radiation law. They both are deservedly famous as introducing quantum theory to the world. Strangely, they both began as *fitting* their theory

2   *ibid.*, p.5

directly to spectroscopic data, Planck to the *continuous* spectrum of light, Bohr to the *discrete* spectroscopic lines of matter.

Einstein regarded their work as *constructive* theories, based primarily on experimental observations. His idea of the best theories are those based on *principles*, like the constant velocity of light, conservation laws, or Boltzmann's Principle, that entropy is *probability*, a function of the number of available *possibilities*.

Planck had spent the last three decades of the nineteenth century in search of a fundamental *irreversibility* that might establish the second law of thermodynamics as an *absolute* and not a statistical law. He wanted an absolute radiation law independent of matter. Where Planck took years, Bohr spent only several months refining the Rutherford atomic model of lightweight electrons orbiting a heavy central nucleus.

Yet for both Planck and Bohr, it was a matter of only a few weeks between the time they first saw the spectroscopic data and the final development of their expressions that fit the data perfectly. Although the experimental data on the continuous spectrum was accurate to only a few percent, Planck nevertheless was able to calculate the natural physical constants far more accurately than had been done before him. And it was his accurate estimates of the natural constants that made physicists accept his radical ideas.

By contrast, the data on spectroscopic lines was accurate to a few parts in ten thousand, so Bohr could calculate spectral line frequencies in hydrogen to four decimal places, starting with the values of m_e, the mass of the electron, e, the electron charge and especially h, Planck's new quantum of action, all of which greatly impressed Bohr's colleagues.[3]

But it was *not* Bohr who discovered the highly accurate fit of a simple theoretical expression to the experimental data. That was the work of the Swiss mathematical physicist JOHANN BALMER, who in the 1880's carefully studied the wavelength measurements by the Swedish inventor of spectroscopy, ANDERS ÅNGSTRÖM.

Ångström had in 1862 discovered three hydrogen lines in the solar spectrum and in 1871 found a fourth, all to several significant figures of accuracy. He named the tiny Ångström unit (10^{-8}cm) after himself as a unit of length. And he measured hydrogen wavelengths to one thousandth of an Ångström!

---

3   Sommerfeld, 1923, p.217.

4101.74 Å 4340.47Å   4861.33 Å   6562.852 Å

Hδ   Hγ   Hβ   Hα

With just these four hydrogen-line wavelengths, and by extraordinary trial and error, Balmer in 1885 found a simple formula that represents all four spectral lines to a high degree of accuracy.

λ (in Å) = 3645.6 n^2 / ($n^2 - 2^2$), where n = 3, 4, 5, and 6.

Note that it would be four decades before these arbitrary integers of Balmer's formula would acquire a physical meaning, becoming the quantum numbers in Bohr's energy levels with $E_m - E_n = h\nu$.

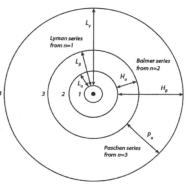

Balmer's colleagues entered n = 7 and 8 into the formula, then looked in the spectrum for lines at those wavelengths and found them! Note that when n = ∞, Balmer's formula predicts the wavelength of the series limit. Shorter wavelengths ionize hydrogen.

In 1886, the Swede JOHANNES RYDBERG generalized Balmer's formula as a reciprocal of the wavelength,

$1/\lambda = R_H (1/m^2 - 1/n^2)$.

This reduces to the Balmer formula for m = 2, but it describes all possible electronic transitions in hydrogen. R_H is the Rydberg constant that Bohr calculated theoretically. Bohr's result amazed physicists as well beyond the accuracy normally achieved in the lab.

Now the reciprocal of wavelength (multiplied by the velocity of light) is a frequency, and Bohr surely saw that multiplying by Planck's constant *h* would make it an energy. The right hand side of the Balmer formula looks like the difference between two energies that are functions of integer numbers. This was the first appearance of *quantum numbers.* They point directly to the discrete nature of reality that Einstein saw in Planck's work nearly a decade earlier.

Bohr would also have seen in the Balmer formula the obvious fact that radiation is the consequence of something involving not one state, but the *difference between two states.* Just looking at Rydberg's version of the Balmer formula, Bohr could "read off" both of his hypotheses or what he called his "quantum postulates."

Bohr's writings nowhere say how one can visualize the energy levels as being implicit in what spectroscopists call the "terms" in their diagrams. Bohr seems to create them out of thin air. He says:

1) There are "stationary" states with integer quantum numbers n that do not radiate energy.

2) Quantum "jumps" between the states, with $E_m - E_n = h\nu$ yielding the precise energies of the discrete spectral lines.[4]

As with Planck, Bohr's discovery of a perfect fit with an experimental spectroscopic formula now needed a more physically satisfying interpretation. What can explain the integer numbers and implicit discreteness of Balmer's formula? Bohr set out to find a *derivation.* Otherwise it would appear to be another case of a "lucky guess" like that Planck had called his "fortunate interpolation."

What needs to be derived from fundamental principles is the *origin* of the *discreteness,* the so-called "quantum condition." As we saw in chapter 3, chemists had thought since the early nineteenth century that the chemical elements come in discrete units, though the "atoms" remained controversial for many physicists.

Ludwig Boltzmann's statistical mechanics (chapter 5) showed that atoms can explain the second law of thermodynamics. And Einstein extended his statistical mechanics to explain Brownian motions, proving that the atoms are real. It was therefore Einstein who established the fact that matter comes in discrete particles, just a year before Boltzmann's death. And it was also Einstein who hypothesized that energy comes in discrete particles the same year.

Now we must give some credit to James Clerk Maxwell, the author of electromagnetic theory and its continuous waves, for

---

4   As we saw in chapter 8, Einstein had pointed out that Planck's theories implied "jumps" between energy levels as early as 1907 in his work on specific heats.

seeing the *stability* of the atoms that underlies Bohr's notion of "stationary." Maxwell's famous equations require that an electron going around in a circular orbit should be generating electromagnetic waves at the orbital frequency. The energy radiating away from the atom should cause the electron to lose energy and spiral into the nucleus. Maxwell knew that did not happen. He marvelled that the microscopic atoms do not wear out, like macroscopic matter. They seem to be indestructible.

And the spectral lines of the hydrogen atom are *discrete* frequencies, not the continuously varying values of Maxwell's theory.

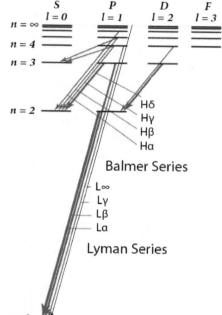

To "quantize" energy levels, Bohr used the original suggestion of J. W. NICHOLSON that the angular momentum of the electron in its orbit is an integer multiple of Planck's constant divided by 2π. Quantization of angular momentum is key to the future development of quantum theory. We shall see that this is the heart of the discreteness seen by LOUIS DE BROGLIE, WERNER HEISENBERG, and ERWIN SCHRÖDINGER, though all three for different reasons!

Atomic Spectra and Atomic Structure

Bohr's atomic model explains how spectroscopy can be transformed from a vast catalogue of thousands of measurements of spectral line wavelengths into a visual image of the stationary states that are the starting and ending points for quantum jumps.

The "term diagrams" of spectroscopists that reduce a huge number of spectral lines to the differences between a much smaller number of "terms," show plainly that the "terms" correspond to

Bohr's energy levels and his stationary states, from which we can "read off" the Bohr model.

Although it does not yield precise calculations for atoms with more than a single electron, Bohr's model gave us a theory of atomic structure that predicts electronic transitions between higher orbits with principal quantum number n out to infinity. Later an angular momentum number l between 0 and $n-1$, a magnetic quantum number m between l and $-l$, and ultimately an electronic spin, $s = \pm\frac{1}{2}$ added greatly to understanding the digital and discrete nature of quantum reality.

Bohr's picture led to a complete theory of the periodic table. He explained isotopes as atoms with the same atomic number (number of protons), but different atomic weights (numbers of neutrons). He convinced Rutherford that radioactivity comes from changes in the nucleus and not electrons, that α-particles reduce the atomic number by 2 and the emission of β-particles (electrons) increases it by 1.

Chance in Atomic Processes

When Rutherford received the draft version of Bohr's theory, he asked Bohr the deep question about *causality* that would be answered just a few years later by Einstein,[5]

> There appears to me one grave difficulty in your hypothesis,
> which I have no doubt you fully realize, how does an electron
> decide which frequency it is going to vibrate at when it passes
> from one stationary state to the other? It seems to me that the
> electron knows beforehand where it is going to stop?[6]

We don't have Bohr's reply, but it might have been the answer he would give years later when asked what is going on in the microscopic world of quantum reality, "We don't know" or "Don't ask!" Or perhaps he would offer his positivist and analytic language philosophy answer - "That's a meaningless question."

But we are getting ahead of the story. We must ask why the young Bohr did not connect his work more clearly in 1913 to that of Einstein, and why he gave so much credit to Planck that clearly

5   See the next chapter.
6   Bohr, *Collected Works*, vol.2, p. 583.

belongs to Einstein. This was the beginning of decades of sidelining Einstein's contributions to quantum mechanics.

Bohr especially ignores Einstein's hopes to see what is going on at the microscopic quantum level, something Einstein called "objective reality," while Bohr maintained "There is no quantum world."

An Independent Criticism of Bohr on Einstein

As I was finishing editing this book and returning the ten volumes of Bohr's *Collected Works* to Widener library, a tiny slip of paper fell out. On it were notes by some unknown person who appears to have detected an effort by the editors of the *Collected Works* to minimize Bohr's references to Einstein's extraordinary original work on the light quantum hypothesis and on specific heat, at least in the English translations.

This unknown critic noticed that a very significant paragraph in Bohr's original Danish had not been translated in the English version, effectively hiding it from all but native Danish speakers.

It does not mention Einstein by name but does reference specific heat and radiation at high frequencies, where the particle nature of light became clear to Einstein

We quote this short note in its entirety, including the critic's rough translation.

Bohr on "non-mechanical forces"...

Den omtalte Antagelse er ikke paa Forhaand selvfølgelig, idet man maa antage, at der i Naturen ogsaa findes Kræfter af ganske anden Art end de almindelige mekaniske Kræfter; medens man nemlig paa den ene Side har opnaaet overordentlig store Resultater i den kinetiske Lufttheori ved at antage, at Kræfterne mellem de enkelte Molekyler er af almindelig mekanisk Art, er der paa den anden Side mange af Legemernes Egenskaber, det ikke er muligt at forklare, dersom man antager, at de Kræfter, der virker indenfor de enkelte Molekyler (der efter den almindelig antagne Opfattelse bestaar af Systemer, i hvilke indgaar et stort Antal »bundne« Elektroner), er af en saadan Art. Foruden for- skellige almindelig kendte Eksempler herpaa, f. Eks. Beregningen af Legemernes Varmefylde og Beregningen af Varmestraaling- sloven for korte Svingningstider, skal vi i det følgende ogsaa se et yderligere Eksempel herpaa, nemlig ved Omtalen af Legemernes magnetiske Forhold.

In this important paragraph Bohr cites Einstein's work on specific heat and high frequency radiation. Specific heat (Einstein 1907) is regarded as first establishing the quantum nature of matter. At high frequencies, the particle nature of light becomes apparent (Einstein 1905) For no apparent reason this paragraph is eliminated in the English translation of Bohr's thesis (presumably by Leon Rosenfeld, the collected works editor, or J. Rud Nielsen, the editor of volume 1.)

Bohr clearly knows that Einstein has established quantum properties that he will exploit in his landmark atomic models with only vague references to Planck's merely heuristic quantum and less often, the real quantum of Einstein.

Here is a very rough translation...

The aforementioned assumption is not obvious, of course, assuming that in nature there are also forces of a very different nature than mechanical forces; While, on the one hand, one has achieved very great results in the kinetic theory of gases by assuming that the forces between the individual molecules are of a common mechanical nature, there are on the other hand many properties of bodies it is not possible to explain by assuming that the forces that work within the individual molecules (which, according to the generally accepted perception, consist of systems in which a large number of "bound" electrons belong), are of such a kind. In addition to various common known examples herein, e.g., the calculation of the specific heat capacity and the calculation of thermal radiation for high frequencies, we will also see a further example, namely the mention of the magnetic properties of the bodies.

We will see in later chapters that Leon Rosenfeld was a fierce defender of the Copenhagen Interpretation of quantum mechanics, especially its most extreme idea that particles lack any properties when they are not being observed in a physical experiment.

Despite the fact that Einstein was first to prove that matter is discrete particles (atoms) and that light consists of discrete quanta (now *photons*), Bohr and his colleagues worked hard to establish Copenhagen as the originators of the atomic theory.

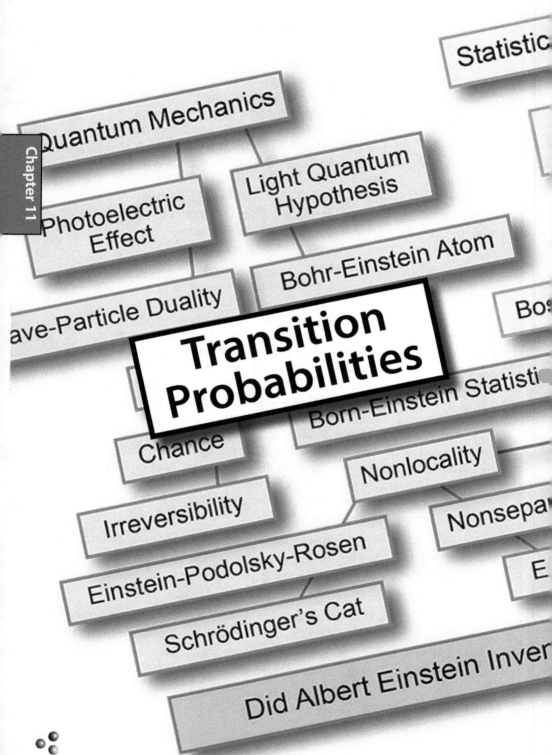

Statistic

Quantum Mechanics

Light Quantum
Hypothesis

Photoelectric
Effect

Bohr-Einstein Atom

ave-Particle Duality

Bos

**Transition
Probabilities**

Chance

Born-Einstein Statisti

Nonlocality

Irreversibility

Nonsepa

Einstein-Podolsky-Rosen

E

Schrödinger's Cat

Did Albert Einstein Inver

Transition Probabilities

When he finished the years needed to complete his general theory of relativity, Einstein turned back to quantum theory and to Bohr's two postulates about 1) electrons in stationary (non-radiating) states and 2) radiating energy $E_m - E_n = hv$ when "jumping" (Einstein's word from 1907) between two energy levels.

Bohr's two postulates provided amazingly accurate explanations of the spectroscopic lines in the hydrogen spectrum. They became the basis for a theory of atomic structure that is still taught today as the introduction to quantum chemistry.

But Bohr, and Planck, *used* expressions that cleverly fit known spectroscopic data. In 1916, Einstein showed how to *derive* Bohr's second postulate from more fundamental physical *principles*, along with Einstein's latest, and thus far simplest, *derivation* of the Planck radiation law that demonstrated its discrete nature.

Where Bohr and Planck manipulated mathematical expressions to make them fit experimental data, Einstein derived the *transition probabilities* for absorption and emission of light quanta when an electron jumps between Bohr's energy levels. Starting with "Boltzmann's Principle" that defines entropy S as probability, calculated as the number of possible states W, and using fundamental conservation laws for energy and momenta, Einstein showed his deep physical understanding of interactions between electrons and radiation that went back over ten years., but had not been accepted by his colleagues, not even Planck or Bohr.

Planck had speculated for many years that the *irreversibility* of the entropy increase somehow depends on the interaction of radiation and matter. Now Einstein's expressions for the absorption and emission of light quanta showed how they maintain thermodynamical equilibrium between radiation and matter as well as how some interactions are indeed *irreversible*.

In addition, Einstein predicted the existence of the unidirectional "stimulated emission" of radiation, the basis for today's lasers.

Chapter 11

Most amazingly, Einstein showed that quantum theory implies the existence of ontological *chance* in the universe.

At this time, Einstein felt very much alone in believing the reality (his emphasis) of light quanta:

> I do not doubt anymore the *reality* of radiation quanta,
> although I still stand quite alone in this conviction. [1]

In two papers, "Emission and Absorption of Radiation in Quantum Theory," and "On the Quantum Theory of Radiation," he again derived the Planck law. For Planck it had been a "lucky guess" at the formula needed to fit spectroscopic measurements.

Einstein derived "transition probabilities" for quantum jumps, describing them as A and B coefficients for the processes of absorption, spontaneous emission, and (his newly predicted) stimulated emission of radiation.

In these papers, Einstein *derived* what had been only a postulate for Planck' ($E = h\nu$). He also derived Bohr's second postulate $E_m - E_n = h\nu$. Einstein did this by exploiting the obvious relationship between the Maxwell-Boltzmann distribution of gas particle velocities and the distribution of radiation in Planck's law. [2]

> The formal similarity between the curve of the chromatic distribution of thermal radiation and the Maxwellian distribution law of velocities is so striking that it could not have been hidden for long. As a matter of fact, W. Wien was already led by this similarity to a farther-reaching determination of his radiation formula in his theoretically important paper, where he derives his displacement law... Recently I was able to find a derivation of Planck's radiation formula which I based upon the fundamental postulate of quantum theory, and which is also related to the original considerations of Wien such that the relation between Maxwell's curve and the chromatic distribution curve comes to the fore. This derivation deserves attention not only because of its simplicity, but especially because it seems to clarify somewhat the still unclear processes of emission and absorption of radiation by matter. I made a few hypotheses about the emission and absorption of radiation by molecules,

---

1   Letter to Besso, in Pais, 1982, p.411
2   See Figure 4-3. "Distribution laws for radiation and matter" on page 33

which suggested themselves from a quantum-theoretic point of view, and thus was able to show that molecules under quantum theoretically distributed states at temperature equilibrium are in dynamical equilibrium with Planck's radiation. By this procedure, Planck's formula followed in an amazingly simple and general manner. It resulted from the condition that the distribution of molecules over their states of the inner energy, which quantum theory demands, must be the sole result of absorption and emission of radiation. If the hypotheses which I introduced about the interaction between radiation and matter are correct, they must provide more than merely the correct statistical distribution of the inner energy of the molecules. Because, during absorption and emission of radiation there occurs also a transfer of momentum upon the molecules. This transfer effects a certain distribution of velocities of the molecules, by way of the mere interaction between radiation and the molecules. This distribution must be identical to the one which results from the mutual collision of the molecules, i.e., it must be identical with the Maxwell distribution...

When a molecule absorbs or emits the energy e in the form of radiation during the transition between quantum theoretically possible states, then this elementary process can be viewed either as a completely or partially directed one in space, or also as a symmetrical (nondirected) one. *It turns out that we arrive at a theory that is free of contradictions, only if we interpret those elementary processes as completely directed processes.* [3]

If light quanta are particles with energy $E = hv$ traveling at the velocity of light c, then they should have a momentum $p = E/c = hv/c$. When light is absorbed by material particles, this momentum will clearly be transferred to the particle. But when light is emitted by an atom or molecule, a problem appears.

If a beam of radiation effects the targeted molecule to either accept or reject the quantity of energy hv in the form of radiation by an elementary process (induced radiation process), then there is always a transfer of momentum hv/c to the molecule, specifically in the direction of propagation of the beam when energy is absorbed by the molecule, in the opposite direction if the molecule releases the energy. If the

---

3   CPAE, vol.6, Doc. 38, "On the Quantum Theory of Radiation," p.220-221.

molecule is exposed to the action of several directed beams of radiation, then always only one of them takes part in an induced elementary process; only this beam alone determines the direction of the momentum that is transferred to this molecule. If the molecule suffers a loss of energy in the amount of hv without external stimulation, i.e., by emitting the energy in the form of radiation (spontaneous emission), then this process too is a directional one. There is no emission of radiation in the form of spherical waves. The molecule suffers a recoil in the amount of hv/c during this elementary process of emission of radiation; the direction of the recoil is, at the present state of theory, determined by "chance." The properties of the elementary processes that are demanded by [Planck's] equation let the establishment of a quantumlike theory of radiation appear as almost unavoidable. The weakness of the theory is, on the one hand, that it does not bring us closer to a link-up with the undulation theory; on the other hand, it also leaves time of occurrence and direction of the elementary processes a matter of "chance." Nevertheless, I fully trust in the reliability of the road taken. [4]

Conservation of momentum requires that the momentum of the emitted particle will cause an atom to recoil with momentum hv/c in the opposite direction. However, the standard theory of spontaneous emission of radiation is that it produces a spherical wave going out in all directions. A spherically symmetric wave has no preferred direction. In which direction does the atom recoil?, Einstein asked:

An outgoing light particle must impart momentum hv/c to the atom or molecule, but the direction of the momentum can not be predicted! Neither can the theory predict the time when the light quantum will be emitted. Einstein called this "weakness in the theory" by its German name - *Zufall* (chance), and he put it in scare quotes. It is only a weakness for Einstein, of course, because his God does not play dice.

Such a random time was not unknown to physics. When ERNEST RUTHERFORD derived the law for radioactive decay of unstable

---

4   CPAE, vol.6, Doc.38, "On the Quantum Theory of Radiation," p.232.

atomic nuclei in 1900, he could only give the probability of decay time. Einstein saw the connection with radiation emission:

> It speaks in favor of the theory that the statistical law assumed for [spontaneous] emission is nothing but the Rutherford law of radioactive decay.[5]

Einstein clearly saw that the element of chance that he discovered threatens *causality*. It introduces *indeterminism* into physics.

The indeterminism involved in quantizing matter and energy was known, if largely ignored, for another decade until WERNER HEISENBERG's quantum theory introduced his famous *uncertainty* (or *indeterminacy*) principle in 1927, which he said was *acausal*.

Where Einstein's indeterminism is qualitative, Heisenberg's principle is quantitative, stating that the exact position and momentum of an atomic particle can only be known within certain (*sic*) limits. The product of the position error and the momentum error is greater than or equal to Planck's constant $h/2\pi$.

$\Delta p \Delta x \geq h/2\pi$.

See chapter 21.

Irreversibility

We shall see in the next chapter that the interaction of the light quantum with matter, especially the transfer of momentum $h\nu/c$ in a random direction, introduces precisely the element of "molecular chaos" that LUDWIG BOLTZMANN speculated might exist at the level of gas particles.

Planck had always thought that the mechanism of irreversibility would be found in the interaction of radiation and matter. Planck's intuition was correct, but in the end he did not like at all the reasons why his *microscopic* quantum would be the thing that produces the *macroscopic* irreversibility of the second law of thermodynamics.

And Planck's hopes for the second law becoming an *absolute* principle were dashed when Einstein showed that the quantum world is a *statistical* and *indeterministic* world, where ontological chance plays an irreducible foundational role.

---

5   CPAE vol.6,Doc.34, p.216

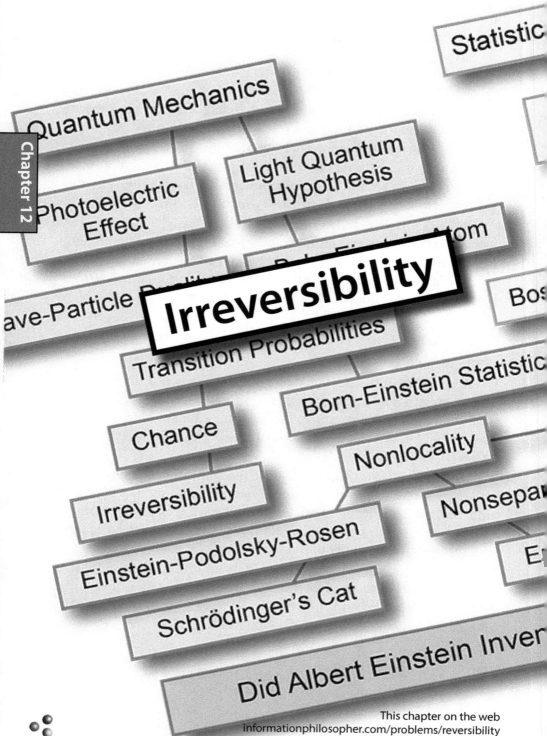

This chapter on the web
informationphilosopher.com/problems/reversibility

Microscopic Irreversibility

In the 1870's, Ludwig Boltzmann developed his transport equation and his dynamical H-theorem to show exactly how gases with large numbers of particles have *macroscopic* irreversibility.

We see this fact every day when things mix but never unmix. Imagine putting 50 white and 50 black balls into a box and shaking them, now pour out 50 each into two smaller boxes and consider the possibility that one contains all black, the other all white.

In 1876, Josef Loschmidt criticized his younger colleague Boltzmann's attempt to derive from classical dynamics the increasing entropy required by the second law of thermodynamics. Loschmidt's criticism was based on the simple idea that the laws of classical dynamics are time reversible. Consequently, if we just turn the time around, the time evolution of the system should lead to decreasing entropy.

But we cannot turn time around. This is the intimate connection between time and the increasing entropy of the second law of thermodynamics that Arthur Stanley Eddington later called the Arrow of Time.[1]

We saw in chapter 4 that Max Planck hoped for many years to show that the second law of thermodynamics and its irreversible increase in entropy are universal and absolute laws. Planck hoped some irreversibility might emerge from a study of the interaction of matter and radiation. We now know his intuition was correct about that interaction, but wrong about the *absolute* nature of the second law. Irreversibility is a *statistical* phenomenon.

Microscopic time *reversibility* remains one of the foundational assumptions of classical mechanics. This is because the classical differential equations (Newton's laws) that describe the motion are time reversible. So are Maxwell's laws of electromagnetism.

Our first problem in the preface, known since the nineteenth century, is how can we reconcile macroscopic irreversibility with microscopic reversibility? The short answer is quantum mechanics. The laws of classical mechanics are adequate only for statistical averages over a large number of quantum particles.

Chapter 12

---

1   See Doyle, 2016a, chapter 23.

A careful quantum analysis shows that microscopic reversibility fails in the case of two particles in collision - provided the quantum mechanical interaction with radiation is taken into account. Planck was looking in the right place.

As we saw in the last chapter, Einstein found that when a light quantum is emitted (or absorbed) there is a transfer of momentum hv/c to the particle. Since the direction of emission is random, the gas particle suffers a random and irreversible change in direction, because the outgoing radiation is irreversible. Einstein's discovery of ontological chance, despite the fact that he did not like it, is the basis for understanding microscopic irreversibility.

Some scientists still believe that microscopic time reversibility is true because the deterministic linear Schrödinger equation itself is time reversible. But the Schrödinger equation only describes the deterministic time evolution of the *probabilities* of various quantum events. It does not determine individual events. As Einstein knew, quantum mechanics is *statistical*. MAX BORN put this distinction concisely

> The motion of the particle follows the laws of probability, but the probability itself propagates in accord with causal laws. [2]

When a quantum event occurs, if there is a record of the event (if new information enters the universe), the previous probabilities of multiple *possible* events collapse to the occurrence of just one *actual* event. This is the collapse of the wave function that JOHN VON NEUMANN called process 1.[3]

An irreversible event that leaves a record (stable new information) may become a measurement, if and when the new information is observed. Measurements are fundamentally and irreducibly irreversible, as many quantum physicists believed.

When particles collide, even structureless particles should not be treated as individual particles with single-particle wave functions, but as a single system with a two- or multiple-particle wave function, because particles are now entangled.[4]

Treating two atoms in collision as a temporary molecule means we must use molecular, rather than atomic, wave functions. The

---

2   "Quantum mechanics of collision processes," *Zeit. Phys.*, 38, 804 (1927)
3   See chapter 23.
4   See chapter 27.

quantum description of the molecule now transforms the six independent degrees of freedom for two atoms into three for the molecule's center of mass and three more that describe vibrational and rotational quantum states.

The possibility of quantum transitions between closely spaced vibrational and rotational energy levels in the "quasi-molecule' introduces indeterminacy in the future paths of the separate atoms. The classical path information needed to ensure the deterministic dynamical behavior has been partially erased. The memory of the past needed to predict the future has been lost.

Quantum transitions, especially the random emission of radiation. erases information about the particle's past motions.

Even assuming the practical impossibility of a perfect classical time reversal, in which we simply turn the two particles around, quantum physics requires two measurements to locate the two particles, followed by two state preparations to send them in the opposite direction.

Heisenberg indeterminacy puts calculable limits on the accuracy with which perfect reversed paths could be achieved.

Let us assume this impossible task can be completed, and it sends the two particles into the reverse collision paths. But on the return path, there is still only a finite probability that a "sum over histories" calculation will produce the same (or reversed) quantum transitions between vibrational and rotational states that occurred in the first collision. Reversibility is not impossible, but extremely improbable,

Thus a quantum description of a two-particle collision establishes the microscopic irreversibility that Boltzmann sometimes described as his assumption of "molecular disorder." In his second (1877) statistical derivation of the H-theorem, Boltzmann used a statistical approach and the molecular disorder assumption to get away from the time-reversibility assumptions of classical dynamics.

The Origin of Microscopic Irreversibility

The path information required for microscopic reversibility of particle paths is destroyed or erased by local interactions with radiation and other particles in the environment. This is the origin of microscopic irreversibility.

Photon emission and absorption during molecular collisions is shown to destroy nonlocal molecular correlations, justifying Boltzmann's assumption of "molecular chaos" (*molekular ungeordnete*) as well as Maxwell's earlier assumption that molecular velocities are not correlated. These molecular correlations were retained in WILLARD GIBBS' formulation of entropy. But the microscopic information implicit in classical particle paths (which would be needed to implement Loschmidt's deterministic motion reversal) is actually *erased*. Boltzmann's physical insight was correct that his increased entropy is irreversible, not just macroscopically but microscopically.

It has been argued that photon interactions can be ignored because radiation is *isotropic* and thus there is no *net* momentum transfer to the particles. The radiation distribution, like the distribution of particles, is indeed *statistically* isotropic, but, as Einstein showed in 1916, each discrete quantum of angular momentum exchanged during individual photon collisions alters the classical paths sufficiently to destroy molecular velocity correlations.

Reversibility is closely related to the maintenance of path information forward in time that is required to assert that physics is *deterministic*. Indeterministic interactions between matter and radiation erase that information. The elementary process of the emission of radiation is not time reversible, as first noted by Einstein in 1909. He argued that the elementary process of light radiation does not have reversibility ("*Umkehrbarkeit*"). The reverse process ("*umgekehrte Prozess*") does not exist as an elementary process.

Macroscopic physics is only *statistically* determined. Macroscopic processes are adequately determined when the mass m of an object is large compared to the Planck quantum of action h (when there are large numbers of quantum particles).

But the information-destroying elementary processes of emission and absorption of radiation ensure that macroscopic processes are not individually reversible.

When interactions with a thermal radiation field and rearrangement collisions are taken into account, a quantum-mechanical treatment of collisions between material particles shows that a hypothetical reversal of all the velocities following a collision would only extremely rarely follow the original path backwards.

A rearrangement collision is one in which the internal energy of one or both of the colliding particles changes because of a quantum jump between its internal energy levels. These internal energy levels and jumps between them were first seen by Einstein in his 1907 work on specific heats (chapter 8).

Although the deterministic Schrödinger equation of motion for an isolated two-particle material system is time reversible (for conservative systems), the quantum mechanics of radiation interactions during collisions does not preserve particle path information, as does classical dynamics. Particle interactions with photons in the thermal radiation field and rearrangement collisions that change the internal states of the colliding particles are shown to be microscopically irreversible for all practical purposes. These quantum processes are involved in the irreversible "measurements" that von Neumann showed increase the entropy.

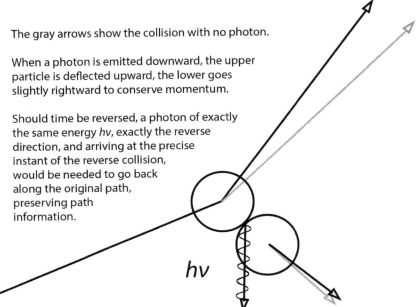

The gray arrows show the collision with no photon.

When a photon is emitted downward, the upper particle is deflected upward, the lower goes slightly rightward to conserve momentum.

Should time be reversed, a photon of exactly the same energy $h\nu$, exactly the reverse direction, and arriving at the precise instant of the reverse collision, would be needed to go back along the original path, preserving path information.

$h\nu$

Chapter 12

Consider a collision between two atoms that results in the emission of a photon.

At some time t after the collision, let's assume we can reverse the separating atoms, sending them back toward the reverse collision. If there had been no photon emission, the most likely path is an exact traversal of the original path back before the collision.

But since a photon was emitted, traversing the original path requires us to calculate the probability that at precisely the moment of a reversed collision a photon of exactly the same frequency is absorbed by the quasi-molecule, corresponding to a quantum jump back to the original rotational-vibrational state, with the photon absorption direction exactly opposite to the original emission, allowing the colliding atoms to reverse their original paths. While this is not impossible, it is extraordinarily improbable.

The uncertainty principle would prevent an experimenter from preparing the two material particles with the precise positions and reverse momenta needed to follow the exact return paths to the collision point. Moreover, the Schrödinger equation of motion for the two particles would only provide a probability that the particles would again collide.

As to the photon, let us assume with Einstein that a light quantum is "directed" and so could be somehow aimed perfectly at the collision point. Even so, there is only a probability, not a certainty, that the photon would be absorbed.

We conclude that collisions of particles that involve radiation are not microscopically reversible.

Detailed Balancing

It is mistakenly believed that the detailed balancing of forward and reverse chemical reactions in thermal equilibrium, including the Onsager reciprocal relations, for example, depend somehow on the principle of microscopic reversibility.

Einstein's work is sometimes cited as proof of detailed balancing and microscopic reversibility. The *Wikipedia* article is an example. [5] In fact, Einstein started with Boltzmann's assumption of detailed balancing, along with the assumption that the probability of states with energy E is reduced by the exponential "Boltzmann factor," $f(E) \sim e^{-E/kT}$, to derive the transition probabilities for emission and

---

5   https://en.wikipedia.org/wiki/detailed_balance

absorption of radiation. Einstein then *derived* Planck's radiation law and Bohr's "quantum postulate" that $E_m - E_n = hv$. But Einstein denied symmetry in the elementary processes of emission and absorption.

As early as 1909, he noted that the elementary process is not "invertible." There are outgoing spherical waves of radiation, but incoming spherical waves are never seen.

> "In the kinetic theory of molecules, for every process in which only a few elementary particles participate (e.g., molecular collisions), the inverse process also exists. But that is not the case for the elementary processes of radiation. According to our prevailing theory, an oscillating ion generates a spherical wave that propagates outwards. The inverse process does not exist as an elementary process. A converging spherical wave is mathematically possible, to be sure; but to approach its realization requires a vast number of emitting entities. The elementary process of emission is not invertible." [6]

The elementary process of the emission and absorption of radiation is asymmetric, because the process is "directed." The apparent isotropy of the emission of radiation, when averaged over a large number of light quanta, is only what Einstein called "pseudo-isotropy" (*Pseudoisotropie*), a consequence of time averages over large numbers of events. Einstein often substituted time averages for space averages, or averages over the possible states of a system in statistical mechanics.

Detailed balancing is thus a consequence of averaging over extremely large numbers of particles in equilibrium. This is the same limit that produces the so-called "quantum-to-classical" transition. And it is the same condition that gives us the "adequate" statistical determinism in the macroscopic, everyday world.

Neither detailed balancing nor the adequate determinism that we see in classical Newtonian experiments does anything to deny that, at the microscopic quantum level, events are completely statistical, involving ontological chance. The interaction of radiation with matter has "a 'chance'-dependent value and a 'chance'-dependent sign" (emission or absorption), said Einstein in 1917.[7]

Reversibility is remotely possible, but extraordinarily improbable.

6   "On the Development of Our Views Concerning the Nature and Constitution of Radiation," 1909, CPAE, vol.2, p.387

7   "On the Quantum Theory of Radiation," *CPAE*, vol.6, p.213

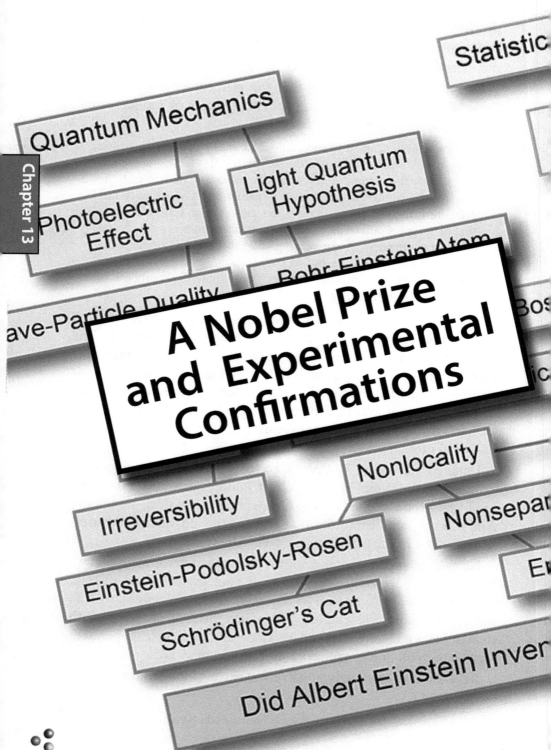

Statistic

Quantum Mechanics

Light Quantum
Hypothesis

Photoelectric
Effect

Bohr-Einstein Atom

ave-Particle Duality

Bos

A Nobel Prize
and Experimental
Confirmations

ic

Nonlocality

Irreversibility

Nonsepar

Einstein-Podolsky-Rosen

E

Schrödinger's Cat

Did Albert Einstein Inver

A Nobel Prize and Two Experimental Confirmations

In 1910 Robert A. Millikan established himself as one of the world's leading experimentalists with his "oil-drop" experiment that measured the elementary charge on the electron. The charge-to-mass ratio had been predicted by J.J.Thomson, the discoverer of the electron, so Millikan's work now provided both the charge and the mass independently.

Like most physicists, theoreticians and experimentalists, Millikan doubted Einstein's light quantum hypothesis, and he set out to build the cleanest possible surface in a vacuum that could test Einstein's prediction that the relation between light frequency and the energy of an ejected electron is linear. The graph should be a straight line (see p.51).

While admitting that Einstein's photoelectric equation "represents very accurately the behavior," Millikan wrote that it "cannot in my judgement be looked upon as resting upon any sort of satisfactory theoretical foundation." When Einstein learned of the experimental confirmation of his prediction, along with the denial of his theory, the first World War had begun and all his energies were devoted to his general theory of relativity.

At this time, Einstein felt very much alone in believing the *reality* (his emphasis) of light quanta:

> I do not doubt anymore the *reality* of radiation quanta, although I still stand quite alone in this conviction. [1]

It would be many more years before most of the physics community would accept Einstein's radical hypothesis, this despite two more dramatic confirmations of Einstein's predictions.

The first experimental confirmation was not for Einstein's work in quantum mechanics but for his 1916 theory of general . ARTHUR STANLEY EDDINGTON's eclipse expedition of 1919 made Einstein world-famous overnight. Eddington measured the angle of deflection of light from a distant star as it passed close to the surface of the darkened sun, its path curved by the sun's gravity.

---

[1]   Letter to M. Besso, quoted by Pais, 1982 p.411

Einstein's 1905 theory of special relativity had of course made him well-known among physicists and he had been frequently nominated for a Nobel Prize. But some members of the Nobel committee found Einstein's relativity theories too controversial and in 1920 they awarded him the prize for his predictions of the photoelectric effect that had been confirmed by Millikan.

Like Millikan and many others, those awarding the prize did not in any way recognize Einstein's theoretical reasoning behind his 1905 prediction, that a discrete and localized quantum of light had been completely absorbed by a single electron.

The confirmation that light has such particle properties came in 1923 when ARTHUR HOLLY COMPTON confirmed Einstein's 1916 prediction that light has the same property of momentum as a material particle. Compton showed that when light and matter interact, their collision can be described as two material particles colliding, with one scattering the direction of the other, and with the conservation of energy and momentum.

Compton measured the scattering angle after the collision between light and an electron and it agreed perfectly with Einstein's prediction that the light quantum carries momentum $p = h\nu/c$.

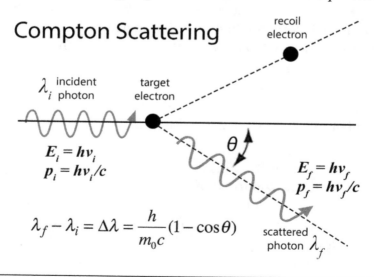

Figure 13-9. The angular measurement by Compton when a "particle" of light collides with an electron and is scattered into a new direction.

Compton scattering is "inelastic," because the energy hv_i (or hc/λ_i) of the incident photon is different from that of the scattered photon hv_f (or hc/λ_f). The lost energy is in the recoil electron.

The initial horizontal momentum is divided between the recoil electron and the scattered photon. The vertical momenta of the recoil electron and scattered photon are equal and opposite.

Compton's experiments confirmed the relation

$\lambda_f - \lambda_i = (h/m_0 c)(1 - cos\theta)$.

Depending on the angle θ, the wavelength shift $\lambda_f - \lambda_i$ varies from 0 to twice $h/m_0 c$, which is called the Compton wavelength.

This "Compton Effect" provided real support for the wave-particle duality of radiation and matter, which as we have seen Einstein had proposed as early as 1909.

Like Millikan, Compton himself initially denied that his experiment supported Einstein's idea of light quanta. Confirmations of Einstein's extraordinary predictions did not at first convince most of his colleagues of his revolutionary theoretical insights!

WERNER HEISENBERG used the Compton Effect in his gamma-ray microscope as an explanation for his uncertainty principle. Although Heisenberg denied the existence of particle paths,[2] we can visualize them using conservation principles for energy and momentum, as Einstein's "objective reality" always suggested.

WOLFGANG PAULI objected to Compton's analysis. A "free" electron cannot scatter a photon, he argued. A proper analysis, confirmed by Einstein and PAUL EHRENFEST, is that scattering should be a two-step process, the absorption of a photon of energy hv_i followed by the emission of a scattered photon hv_f, where the momentum of the photon hv/c balances the momentum of the recoil electron $m_0 v$.

Compton was awarded the Nobel Prize in Physics in 1927 for the "Compton Effect," the year that Heisenberg discovered quantum indeterminacy, by which time most physicists were accepting Einstein's light quanta, since 1924 being called photons.

A year after Compton's work, LOUIS DE BROGLIE would in his 1924 thesis propose that by symmetry, matter should show wave properties just like those of light, an idea that de Broglie said had been suggested to him by reading Einstein.

Chapter 13

---

2   See chapter 21

Chapter 14

Statistic

Quantum Mechanics

Light Quantum
Hypothesis

Photoelectric
Effect

Bohr-Einstein Atom

ave-Particle Duality

Bo

De Broglie
Pilot Waves

Born-Einstein Statistic

Chance

Nonlocality

Irreversibility

Nonsepa

Einstein-Podolsky-Rosen

E

Schrödinger's Cat

Did Albert Einstein Inven

De Broglie Pilot Waves

LOUIS DE BROGLIE was a critical link from the 1905 work of ALBERT EINSTEIN to ERWIN SCHRÖDINGER's 1926 wave mechanics and to MAX BORN's "statistical interpretation," both considered key parts of the Copenhagen Interpretation of quantum mechanics.

De Broglie is very important to our account of the slow acceptance of Einstein's work in quantum mechanics. He was very likely the first thinker to understand Einstein's case for wave-particle duality in 1909 (as we saw in chapter 9) and to take Einstein's light-quantum hypothesis seriously.

In his 1924 thesis, de Broglie argued that if light, which was thought to consist of waves, is actually discrete particles that Einstein called light quanta (later called photons), then matter, which is thought to consist of discrete particles, might also have a wave nature. He called his matter waves "pilot waves."

> The fundamental idea of [my thesis] was the following: The fact that, following Einstein's introduction of photons in light waves, one knew that light contains particles which are concentrations of energy incorporated into the wave, suggests that all particles, like the electron, must be transported by a wave into which it is incorporated... My essential idea was to extend to all particles the coexistence of waves and particles discovered by Einstein in 1905 in the case of light and photons. [1]

What Einstein had said was that the light wave at some position is a measure of the *probability* of finding a light particle there, that is, the intensity of the light wave is proportional to the number of photons there. It may have been implicit in his 1905 light quantum hypothesis, as de Broglie seems to think, but Einstein had explicitly described a "guiding field" (*Führungsfeld*) or "ghost field" (*Gespensterfeld*) a few years before de Broglie's thesis, in his private conversations.

---

1   en.wikipedia.org/wiki/Louis_de_Broglie, retrieved 03/17/2017.

Einstein had used these "field" terms privately to colleagues some time between 1918 and 1921. We don't have public quotes from Einstein until October 1927 at the fifth Solvay conference.

$|\psi|^2$ expresses the probability that there exists at the point considered a particular particle of the cloud, for example at a given point on the screen.[2]

There are subtle differences between de Broglie, Schrödinger, and Born as to the connection between a particle and a wave. Born's thinking is closest to Einstein with the idea that the wave gives us the probability of finding a particle of matter or radiation.

De Broglie thought the particle is "transported by a wave into which it is incorporated." Schrödinger is the most extreme in identifying the particle with the wave itself, to the point of denying the existence of separate particles. He strongly rejected the idea of discrete particles and the "quantum jumps" associated with them. He vehemently attacked the probabilistic interpretation of Einstein and Born. Schrödinger thought a wave alone could account for all the properties of quantum objects.

Schrödinger brilliantly showed his wave equation produced the same energy levels in the Bohr atom as WERNER HEISENBERG and WOLFGANG PAULI had found with matrix mechanics.

De Broglie used an expression for the wavelength of his "pilot wave" that followed from the expression that Einstein had used for the momentum of a light quantum, the same value that Compton had confirmed a year earlier. Since the wavelength of light is equal to the velocity of light divided by frequency, $\lambda = c/v$, and since Einstein found the momentum of a particle with energy hv is hv/c, de Broglie guessed the wavelength for a particle of matter with momentum p should be $\lambda = h/p$.

Note that this is still another case of the "quantum condition" being Planck's quantum of action. Although de Broglie began with *linear* momentum, he now could connect his hypothesis with Bohr's use of quantized *angular* momentum in the Bohr atom orbits. De Broglie showed that the wavelength of his pilot wave fits an integer number of times around each Bohr orbit and the integer is Bohr's principal quantum number.

---

2   Bacciagaluppi and Valentini, 2009. pp. 441.

Chapter 14

Once again, what is being quantized here by de Broglie is angular momentum, with the dime~~nsions of action~~.

Schrödinger was delighted that integer numbers appear naturally in wave mechanics, whereas they seem to be only *ad hoc* assumptions in Heisenberg's matrix mechanics.

De Broglie said in his Nobel lecture of 1929,

> the determination of the stable motions of the electrons in the atom involves whole numbers, and so far the only phenomena in which whole numbers were involved in physics were those of interference and of eigenvibrations. That suggested the idea to me that electrons themselves could not be represented as simple corpuscles either, but that a periodicity had also to be assigned to them too. [3]

De Broglie's hypothesis of matter waves and Einstein's insight into wave-particle duality were confirmed by Clinton Davisson and Lester Germer in the mid-1920's, following a suggestion by Walther Elsasser that electron scattering by the regular configuration of atoms in crystalline solids might reveal the wave nature, just as X-rays had been shown to be waves.

That the Davisson-Germer experiments provided evidence for matter waves was first realized by Born, who gave a talk at the 1926 summer meeting of the British Association for the Advancement of Science that was attended by the American Davisson. Davisson was surprised to see Born presenting Davisson's diffraction curves published many years earlier in *Science* magazine.

De Broglie was invited to give a major presentation on his thesis at the 1927 Solvay conference on Electrons and Photons, but his work was completely overshadowed by the presentation of Heisenberg and Born on the new quantum mechanics.

De Broglie's pilot-wave theory was largely ignored for a quarter century until DAVID BOHM revived it in 1952 in his deterministic, causal, and nonlocal interpretation of quantum mechanics using *hidden variables*. See chapter 30.

---

3   De Broglie, 1929, p.247

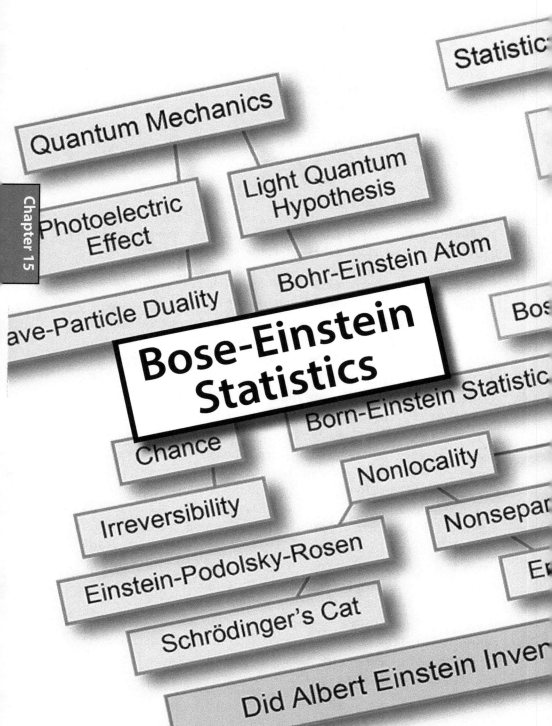

Bose-Einstein Statistics

In 1924, Einstein received an amazing very short paper sent from India by SATYENDRA NATH BOSE. Einstein must have been pleased to read the title, "Planck's Law and the Hypothesis of Light Quanta." It was more attention to Einstein's 1905 work than anyone had paid in nearly twenty years. The paper began by claiming that the "phase space" (a combination of 3-dimensional coordinate space and 3-dimensional momentum space) should be divided into small volumes of h^3, the cube of Planck's constant. By counting the number of possible distributions of light quanta over these cells, Bose claimed he could calculate the entropy and all other thermodynamic properties of the radiation.

Bose easily derived Planck's inverse exponential function $1/(e^{hv/kT}-1)$. Einstein too had derived this. Maxwell and Boltzmann derived the so-called Boltzmann factor $e^{-hv/kT}$, by analogy from the Gaussian exponential tail of probability and the theory of errors.

MAX PLANCK had simply guessed this expression from Wien's radiation distribution law $ae^{-bv/T}$ by adding the term -1 in the denominator of Wien's law in the form $a/e^{bv/T}$ to get $1/(e^{hv/kT}-1)$.

All previous derivations of the Planck law, including Einstein's of 1916-17 (which Bose called "remarkably elegant"), used classical electromagnetic theory to derive the density of radiation, the number of "modes" or "degrees of freedom" per unit volume of the radiation field,

$$\rho_v dv = (8\pi v^2 dv / c^3).$$

Bose considered the radiation to be enclosed in a volume V with total energy E. He assumed that various types of quanta are present with abundances N_i and energy hv_i ($i = 0$ to $i = \infty$).

The total energy is then

$$E = \Sigma_i N_i hv_i = V \int \rho_v dv.$$

But now Bose showed he could get ρ_v with a simple statistical mechanical argument remarkably like that Maxwell used to derive his distribution of molecular velocities. Maxwell said that the three directions of velocities for particles are independent of one another, and of course equal to the total momentum,

$$p_x^2 + p_y2 + p_z^2 = p^2,$$

Bose just used Einstein's relation for the momentum of a photon,

$$p = hv / c.$$

The momentary state of the quantum is characterized by its coordinates x, y, z and the corresponding components of the momentum p_x, p_y, p_z. These six quantities can be considered as point coordinates in a six–dimensional space, where we have the relation

$$p_x^2 + p_y^2 + p_z^2 = h^2v^2 / c^2.$$

This led Bose to calculate a frequency interval in phase space as

$$\int dx\, dy\, dz\, dp_x\, dp_y\, dp_z = 4\pi V \,(hv / c)3 \,(h\, dv / c)$$
$$= 4\pi \,(h^3\, v^2 / c^3) \, V\, dv,$$

Bose simply divided this expression by h^3, multiplied by 2 to account for two polarization degrees of freedom of light, and he had derived the number of cells belonging to dv,

$$\rho_v dv = (8\pi v^2 dv / c^3)\, E,$$

This expresion is well-known from classical electrodynamics, but Bose found this result without using classical radiation laws, a correspondence principle, or even Wien's law. His derivation was purely statistical mechanical, based only on the number of quantum cells in phase space and the number of ways N photons can be distributed among them.

When Bose calculated the number of ways of placing light quanta in these cells, i.e., the number of cells with no quanta, the number with one, two, three, etc., he put no limits on the number of quanta in a h^3 cell.

Einstein saw that unlimited numbers of particles close together implies extreme densities and low-temperature condensation of any particles with integer values of the spin. Material particles like electrons are known to limit the number of particles in a cell to two, one with spin up, one spin down. They have half-integer spin.

Particles with integer-value spins follow the new Bose-Einstein quantum statistics. This relation between spin and statistics is called the spin-statistics theorem of WOLFGANG PAULI.

When identical particles in a two-particle wave function are exchanged, the antisymmetric wave function for fermions changes sign. The symmetric boson wave function does not change sign.

PAUL DIRAC quickly developed the quantum statistics of half-integer spin particles, now called Fermi-Dirac statistics. A maximum of two particles, with opposite spins, can be found in the fundamental h^3 volume of phase space identified by Bose. This explains why there are a maximum of two electrons in the first electron shell of any atom.

Einstein's discovery led us to "Bose-Einstein condensations" as temperatures approach absolute zero, because there is no limit on the number of integer-spin particles that can be found in an h^3 volume of phase space. This work is frequently attributed to Bose instead of Einstein. Particles with integer spin are called "bosons." In a similar irony, particles with half-integer spin that obey Pauli's exclusion principle are called "fermions."

Einstein's discovery of quantum statistics is often seen as his last *positive* contribution to quantum physics. Few historians point out that Einstein was first to see the two kinds of elementary particles of today's "standard model!"

Chapter 15

Standard Model of Elementary Particles

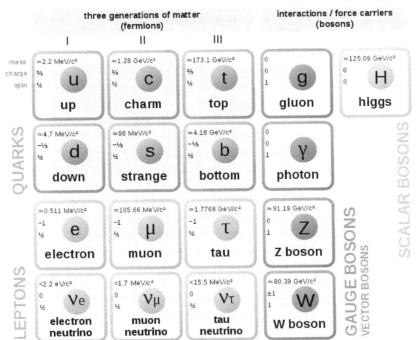

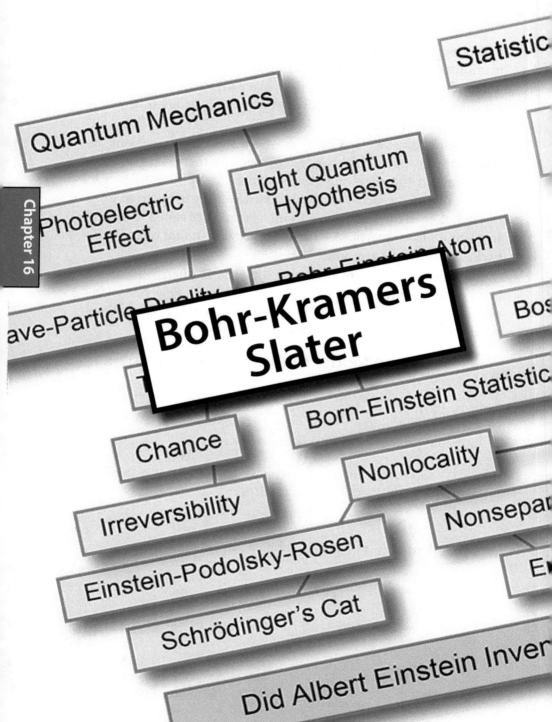

Bohr-Kramers-Slater

The 1924 paper of NIELS BOHR, HENDRIK A. KRAMERS, and JOHN CLARKE SLATER was the last major public attempt by members of the Copenhagen school to deny ALBERT EINSTEIN's light-quantum hypothesis of 1905, although we will show that Bohr's doubts continued for years, if not indefinitely.

The BKS effort was despite the fact that Einstein's most important predictions, the photoelectric effect of 1905 and that a light-quantum has momentum in 1917, had been confirmed experimentally, leading to Einstein's 1920 Nobel Prize. We must however note that the two world-famous experimenters who confirmed Einstein's predictions, ROBERT A. MILLIKAN and ARTHUR HOLLY COMPTON, both Americans, had not themselves seen the results as validating Einstein's light quanta. Nevertheless, many other physicists by that time had.

Millikan called Einstein's photoelectric idea a "bold, not to say reckless hypothesis" and said although it appears in every case to predict exactly the observed results, Einstein's "theory seems at present wholly untenable." [1]

In 1923, Compton showed that radiation (a high-energy X-ray) was being scattered by electrons, exchanging energy with them, just as if the light rays and electrons acted like colliding billiard balls. Although this was the first solid evidence for Einstein's "light-quantum hypothesis," like Millikan, Compton said his work did not support Einstein's radical hypothesis. Although by 1924 a large fraction of physicists had come to believe light had both wave and particle characteristics, there were still several holdouts. Many were found among Bohr's Copenhagen associates.

It is difficult to imagine what Einstein's feelings may have been after nearly two decades of rejection of what he called his "very revolutionary" contributions to quantum theory.

But surely the negative attitude of Bohr, who with his 1913 model for the atom was the third great thinker in quantum theory after MAX PLANCK and Einstein, was hardest for him to bear.

Chapter 16

---

1   Pais, 1982, p.357.

While the 1924 Bohr-Kramers-Slater theory may have been the most dispiriting for Einstein, it ironically grew out of an original suggestion that was based directly on Einstein's light quantum.

John Slater was a young American physicist who accepted Einstein's radical insights. He came from MIT to Copenhagen with an idea about "virtual oscillators".

But Bohr and Kramers were very explicit about their objection to Einstein's localized quantum of light. They said there is no way individual particles can explain the wave properties of light, especially its interference effects. The very idea that a light quantum has energy $h\nu$, where ν is the frequency of the light, depends on the wave theory to determine the frequency and the associated wavelength, they said.

In his 1922 Nobel Prize lecture, Planck had said,

> In spite of its heuristic value, however, the hypothesis of light-quanta, which is quite irreconcilable with so-called interference phenomena, is not able to throw light [sic] on the nature of radiation. I need only recall that these interference phenomena constitute our only means of investigating the properties of radiation and therefore of assigning any closer meaning to the frequency which in Einstein's theory fixes the magnitude of the light-quantum.

And in his popular book on the Bohr Atom in 1923, Kramers had vigorously attacked the idea of a light quantum.

> The theory of quanta may thus be compared with medicine which will cause the disease to vanish but kills the patient. When Einstein, who has made so many essential contributions in the field of the quantum theory, advocated these remarkable representations about the propagation of radiant energy, he was naturally not blind to the great difficulties just indicated. His apprehension of the mysterious light in which the phenomena of interference appear in his theory is shown in the fact that in his considerations he introduces something which he calls a 'ghost' field of radiation to help to account for the observed facts. [2]

---

2   *Kramers, 1923*, p.175

Einstein's "ghost field" or "guiding field" interpretation for the light wave, whereby the light wave gives the probability of finding a light particle, was thus well known in Copenhagen before LOUIS DE BROGLIE introduced a "pilot wave" in his 1924 thesis. Einstein may have had this view as early as 1909. See chapters 9 and 14.

What Slater brought to Copenhagen was a variation of Einstein's "ghost field." He suggested that an atom in one of Bohr's "stationary states" is continuously emitting a field that carries no energy but contains a set of frequencies corresponding to the allowed Bohr transition frequencies. Like the Einstein field, the value of the Slater field at each point gives the probability of finding a light quantum at that point. They were slightly different from Einstein's light quanta. Like our information philosophy interpretation of the quantum wave function, Slater's field was *immaterial*.

In any case, Bohr and Kramers rejected any talk of light quanta, but did embrace Slater's concept of what they called a "virtual field." Slater thought it might reconcile the continuous nature of light radiation with the discrete "quantum jumps" of the Bohr Atom. Bohr realized this could only be done if the transfer of energy did not obey the principle of conservation of energy instantaneously, but only statistically, when averaged over the emissions and absorptions of distant atoms.

In just a few weeks the BKS paper was published, written entirely by Bohr and Kramers. It met with immediate criticism from Einstein and others. Einstein objected to the violation of conservation of energy and called for experiments to test for it.

Within a year WALTHER BOTHE and HANS GEIGER, who had confirmed the Compton effect, showed that the timing of scattered radiation and an electron recoil were within a tiny fraction of a second, confirming Einstein's demand for instantaneous conservation of energy and proving the BKS theory untenable.

But Slater's notion of a virtual field of oscillators with all the frequencies of possible transitions survived as the basis of WERNER HEISENBERG's matrix mechanics, to which we now turn.

Statistic

Quantum Mechanics

Light Quantum
Hypothesis

Photoelectric
Effect

Bohr-Einstein Atom

Wave-Particle Duality

Bo

Matrix
Mechanics

T

Born-Einstein Statistic

Chance

Nonlocality

Irreversibility

Nonsepa

Einstein-Podolsky-Rosen

E

Schrödinger's Cat

Did Albert Einstein Inven

Matrix Mechanics

What the matrix mechanics of WERNER HEISENBERG, MAX BORN, and PASCUAL JORDAN did was to find another way to determine the "quantum conditions" that had been hypothesized by NIELS BOHR, who was following J.W.NICHOLSON's suggestion that the angular momentum is quantized. These conditions correctly predicted values for Bohr's "stationary states" and "quantum jumps" between energy levels.

But they were really just guesses in Bohr's "old quantum theory," validated by perfect agreement with the values of the hydrogen atom's spectral lines, especially the Balmer series of lines whose 1880's formula for term differences first revealed the existence of integer quantum numbers for the energy levels,

$$1/\lambda = R_H \, (1/m^2 - 1/n^2).$$

Heisenberg, Born, and Jordan recovered the same quantization of angular momentum that Bohr had used, but we shall see that it showed up for them as a product of non-commuting matrices.

Most important, they discovered a way to *calculate* the *energy levels* in Bohr's atomic model as well as determine ALBERT EINSTEIN's 1916 transition probabilities between levels in a hydrogen atom. They could explain the different intensities in the resulting spectral lines.

Before matrix mechanics, the energy levels were *empirically* "read off" the term diagrams of spectral lines. Matrix mechanics is a new mathematical *theory* of quantum mechanics. The accuracy of the old quantum theory came from the sharply defined spectral lines, with wavelengths measurable to six significant figures.

The new quantum theory did not try to interpret or *visualize* what is going on in transitions. Indeed, it strongly discouraged any visualizations. It even denied the existence of electron orbits, a central concept in the Rutherford-Bohr-Sommerfeld atom.

Heisenberg had worked with HENDRIK A. KRAMERS at Bohr's Institute for Physics in Copenhagen to analyze electronic orbits as Fourier series. Kramers had hoped to identify the higher harmonic frequencies in the series expansion of orbital frequencies with those of electronic transitions, but Kramer's predictions only worked for large quantum numbers where Bohr's correspondence principle applies.

Kramers' work began with estimates of what were called "dispersion laws" by RUDOLF LADENBERG. The work culminated in the Kramers-Heisenberg dispersion formula in 1925. Based on Bohr's correspondence principle, these led to accurate estimates of the intensities of spectral lines in the hydrogen atom for high quantum numbers. But the assumed orbital frequencies for low quantum numbers did not agree with observations.

Until Heisenberg in 1925, most of the work in the "old quantum theory" focused on *models* of elementary particles. For example, electrons were visualized as going around ERNEST RUTHERFORD's nucleus in orbits, like planets circling the sun. ARNOLD SOMMERFELD extended the Bohr analogy to include Keplerian elliptical orbits with differing angular momentum.

Heisenberg's great breakthrough was to declare that his theory is based entirely on "observable" quantities like the intensities and frequencies of the visible spectral lines.

The attempts by Kramers to predict observed spectral lines as higher harmonics in a Fourier analysis of the *assumed* electronic orbit frequencies ended in failure. But the methods he had developed with Heisenberg's help were adapted by Heisenberg to a Fourier analysis of the *observed* spectral line frequencies. Heisenberg assumed they originate in virtual oscillators like the simple harmonic motion of a vibrating string pinned at the ends or the more complex anharmonic oscillator.

As Kramers had done, Heisenberg identified line intensities with the square of the amplitude of vibrations, which was the classical expression for an oscillating electron. But now Heisenberg's major insight was to calculate values for the position and momentum of the particle using two states rather than one, the initial and final stationary states or energy levels, which we suggested in the chapter on the Bohr atom could simply be "read off" the empirical term diagrams.

Heisenberg's requirement for two states led to an arrangement of transitions in a two-dimensional square array. One dimension was the initial states, the other the final. The array element for i=3 and f=2 represents the transition from level 3 to level 2 with the emission of a light quantum.

When his mentor MAX BORN looked at Heisenberg's draft paper in July of 1925, he recognized the square arrays as *matrices*, a powerful mathematical tool with some unusual properties that played a decisive role in the new quantum mechanics.

Born and his assistant Pascual Jordan submitted a paper within weeks about the strange "non-commuting" of some dynamical variables in quantum mechanics. Normally the order of multiplication makes no difference, ab = ba. But the matrices for the position and momentum operators x and p exhibit what was to become the new "quantum condition," a defining characteristic of the new quantum mechanics.

As Born describes the array,

If we start from the frequencies,
$$\nu_{nm} = E_n/h - E_m/h,$$
it is a natural suggestion that we arrange them in a square array

$$
\begin{array}{cccc}
\nu_{11} = & \nu_{12} & \nu_{13} & \cdots \\
\nu_{21} & \nu_{22} = & \nu_{23} & \cdots \\
\nu_{31} & \nu_{23} & \nu_{33} = & \cdots \\
\cdots & \cdots & \cdots & \cdots
\end{array}
$$

We can proceed to define the product of two such arrays. The multiplication rule, which Heisenberg deduced solely from experimental facts, runs:
$$(a_{nm})(b_{nm}) = (\Sigma_k\, a_{nk} b_{km}).[1]$$
The central idea of matrix mechanics is that every physical magnitude has such a matrix, including the co-ordinate position and the momentum. However, the product of momentum and position is no longer commutative as in classical mechanics, where the order of multiplication does not matter.

$$p_k q_k = q_k p_k.$$

Instead, Heisenberg found that
$$p_k q_k - q_k p_k = h/2\pi i.$$

---

1   Born *Atomic Physics*, p.116

It is this purely mathematical non-commutation property that is the "quantum condition" for the new quantum mechanics, especially for Paul Dirac, see chapter 19.

But notice that Heisenberg's product of momentum and position has the dimensions of angular momentum. So we are back to Planck's original fortuitive but most insightful guess, and can now add to the answer to our opening question "what is quantized?" This Heisenberg-Born-Jordan discovery that the product of non-commuting quantities p and q leads directly to Planck's constant h, his "quantum" of action, gives us a great insight into what is going on in quantum reality.

It is always angular momentum or spin that is quantized, just as Nicholson had suggested to Bohr, including the dimensionless *isospin* of the neutrons and protons and other sub-elementary particles, which obey the same mathematics as spin and orbital angular momentum for electrons.

And it is the possible projections of the spin or angular momentum onto any preferred directions, such as an external field, that determines possible quantum states. The field is the average over all the dipole and quadrupole moments of other nearby spinning particles.

Heisenberg on Einstein's Light Quanta

Although his matrix mechanics confirmed discrete states and "quantum jumps" of electrons between the energy levels, with emission or absorption of radiation, Heisenberg did not yet accept today's standard textbook view that the radiation is also discrete and in the form of Einstein's spatially localized light quanta, which had been renamed "photons" by American chemist GILBERT LEWIS in late 1926.

Heisenberg must have known that Einstein had introduced probability and causality into physics in his 1916 work on the emission and absorption of light quanta, with his explanation of transition probabilities and prediction of stimulated emission.

But Heisenberg gives little credit to Einstein. In his letters to Einstein, he says that Einstein's work is relevant to his, but does not follow through on exactly how it is relevant. And as late as the

Spring of 1926, perhaps following Niels Bohr, he is not convinced of the reality of light quanta. "Whether or not I should believe in light quanta, I cannot say at this stage," he said. After Heisenberg's 1926 talk on matrix mechanics at the University of Berlin, Einstein invited him to take a walk and discuss some basic questions.

We only have Heisenberg's version of this conversation, but it is worth quoting at length to show how little the founders appreciated Einstein's work over the previous two decades on the fundamental concepts of quantum mechanics.:

> I apparently managed to arouse Einstein's interest, for he invited me to walk home with him so that we might discuss the new ideas at greater length. On the way, he asked about my studies and previous research. As soon as we were indoors, he opened the conversation with a question that bore on the philosophical background of my recent work. "What you have told us sounds extremely strange. You assume the existence of electrons inside the atom, and you are probably quite right to do so. But you refuse to consider their orbits, even though we can observe electron tracks in a cloud chamber. I should very much like to hear more about your reasons for making such strange assumptions."
>
> "We cannot observe electron orbits inside the atom," I must have replied, "but the radiation which an atom emits during discharges enables us to deduce the frequencies and corresponding amplitudes of its electrons. After all, even in the older physics wave numbers and amplitudes could be considered substitutes for electron orbits. Now, since a good theory must be based on directly observable magnitudes, I thought it more fitting to restrict myself to these, treating them, as it were, as representatives of the electron orbits."
>
> "But you don't seriously believe," Einstein protested, "that none but observable magnitudes must go into a physical theory?"
>
> "Isn't that precisely what you have done with relativity?" I asked in some surprise. "After all, you did stress the fact that it is impermissible to speak of absolute time, simply because absolute time cannot be observed; that only clock readings, be it in the moving reference system or the system at rest, are relevant to the determination of time."
>
> "Possibly I did use this kind of reasoning," Einstein admitted, "but it is nonsense all the same. Perhaps I could put it more diplomatically by saying that it may be heuristically useful to keep in mind what one has actually observed. But on principle, it is quite wrong to try founding a theory on observable magnitudes alone. In reality the very opposite happens. It is the theory which decides what we can observe.

You must appreciate that observation is a very complicated process. The phenomenon under observation produces certain events in our measuring apparatus. As a result, further processes take place in the apparatus, which eventually and by complicated paths produce sense impressions and help us to fix the effects in our consciousness. Along this whole path - from the phenomenon to its fixation in our consciousness — we must be able to tell how nature functions, must know the natural laws at least in practical terms, before we can claim to have observed anything at all. Only theory, that is, knowledge of natural laws, enables us to deduce the underlying phenomena from our sense impressions. When we claim that we can observe something new, we ought really to be saying that, although we are about to formulate new natural laws that do not agree with the old ones, we nevertheless assume that the existing laws — covering the whole path from the phenomenon to our consciousness—function in such a way that we can rely upon them and hence speak of 'observations'...

"We shall talk about it again in a few years' time. But perhaps I may put another question to you. Quantum theory as you have expounded it in your lecture has two distinct faces. On the one hand, as Bohr himself has rightly stressed, it explains the stability of the atom; it causes the same forms to reappear time and again. On the other hand, it explains that strange discontinuity or inconstancy of nature which we observe quite clearly when we watch flashes of light on a scintillation screen. These two aspects are obviously connected. In your quantum mechanics you will have to take both into account, for instance when you speak of the emission of light by atoms. You can calculate the discrete energy values of the stationary states. Your theory can thus account for the stability of certain forms that cannot merge continuously into one another, but must differ by finite amounts and seem capable of permanent re-formation. But what happens during the emission of light?

"As you know, I suggested that, when an atom drops suddenly from one stationary energy value to the next, it emits the energy difference as an energy packet, a so-called light quantum. In that case, we have a particularly clear example of discontinuity. Do you think that my conception is correct? Or can you describe the transition from one stationary state to another in a more precise way?"

In my reply, I must have said something like this: "Bohr has taught me that one cannot describe this process by means of the traditional concepts, i.e., as a process in time and space. With that, of course, we have said very little, no more, in fact, than that we do not know. Whether or not I should believe in light quanta, I cannot say at this stage. Radiation quite obviously involves the discontinuous elements to which you refer as light quanta. On the other hand, there is a continuous element, which appears, for instance, in interference phenomena, and which is much more simply described by the wave theory of light. But you are of course quite right to ask whether quantum mechanics has anything new to say on these terribly difficult problems. I believe that we may at least hope that it will one day.

"I could, for instance, imagine that we should obtain an interesting answer if we considered the energy fluctuations of an atom during reactions with other atoms or with the radiation field. If the energy should change discontinuously, as we expect from your theory of light quanta, then the fluctuation, or, in more precise mathematical terms, the mean square fluctuation, would be greater than if the energy changed continuously. I am inclined to believe that quantum mechanics would lead to the greater value, and so establish the discontinuity. On the other hand, the continuous element, which appears in interference experiments, must also be taken into account. Perhaps one must imagine the transitions from one stationary state to the next as so many fade-outs in a film. The change is not sudden—one picture gradually fades while the next comes into focus so that, for a time, both pictures become confused and one does not know which is which. Similarly, there may well be an intermediate state in which we cannot tell whether an atom is in the upper or the lower state."

"You are moving on very thin ice," Einstein warned me. "For you are suddenly speaking of what we know about nature and no longer about what nature really does. In science we ought to be concerned solely with what nature does. It might very well be that you and I know quite different things about nature. But who would be interested in that? Perhaps you and I alone. To everyone else it is a matter of complete indifference. In other words, if your theory is right, you will have to tell me sooner or later what the atom does when it passes from one stationary state to the next"

"Perhaps," I may have answered. "But it seems to me that you are using language a little too strictly. Still, I do admit that everything that I might now say may sound like a cheap excuse. So let's wait and see how atomic theory develops."

Einstein gave me a skeptical look. "How can you really have so much faith in your theory when so many crucial problems remain completely unsolved?"[2]

Heisenberg (with Bohr) "cannot say at this stage" (1926) whether or not they can "believe in light quanta." Nor do they understand at all Einstein's hope of understanding "objective reality," what nature really does and not just what we can say about it.

2   Heisenberg, 1971, p. 67

Statistic

Quantum Mechanics

Light Quantum
Hypothesis

Photoelectric
Effect

Bohr-Einstein Atom

ave-Particle Duality

Bos

Wave
Mechanics

Born-Einstein Statistic

Chance

Nonlocality

Irreversibility

Nonsepa

Einstein-Podolsky-Rosen

E

Schrödinger's Cat

Did Albert Einstein Inver

Wave Mechanics

ERWIN SCHRÖDINGER's creation of his quantum wave function ψ followed LOUIS DE BROGLIE's 1925 suggestion that a wave can be associated with a particle of matter - just as ALBERT EINSTEIN had associated a particle of energy with a light wave.

De Broglie predicted that the wavelength λ of a matter particle wave would be $\lambda = h/p$, since the wavelength of a photon is related to its frequency by $\lambda = c/v$. and Einstein had shown that the momentum of a light quantum should be $p = hv/c$.

In November, 1925, Schrödinger wrote to Einstein,

> A few days ago I read with the greatest interest the ingenious thesis of Louis de Broglie, which I finally got hold of; with it section 8 of your second paper on degeneracy has also become clear to me for the first time.

A colleague pointed out to Schrödinger that to explain a wave, one needs a wave equation. With his extraordinary mathematical abilities, Schrödinger found his equation within just a few weeks.

Schrödinger started with the well-known equation for the amplitude ψ of a wave with wavelength λ in three dimensions,

$$\nabla^2 \psi - (4\pi^2/\lambda)\,\psi = 0.$$

This equation gives us the density of classical electromagnetic waves $(8\pi v^2/c^3)$ used by Planck and Einstein to derive the black-body radiation law.

In 1925, Bose and Einstein had eliminated classical theory completely, replacing the expression by the number of identical light quanta in a phase-space volume of h^3. (See chapter 15.)

Schrödinger quickly converted from rectangular to spherical coordinates, R, Θ, Φ, because of the spherical symmetry of the nuclear electric charge potential $V = -e^2/r$. He could then replace the equation for $\psi(x, y, z)$ with one for $\psi(r, \theta, \varphi) = R(r)\,\Theta(\theta)\,\Phi(\varphi)$, which separates into three ordinary differential equations.

The angular functions lead to the spherical harmonics that correspond to different angular momentum states, visualized as the familiar electronic clouds in every chemistry textbook.

Chapter 18

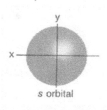

s orbital

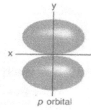

p orbital

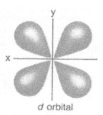

d orbital

You can clearly imagine the nodes around electron orbits as they were seen by de Broglie, but now the waves are space filling.

The radial equation solves the time-independent Schrödinger equation with the electrostatic potential of the atomic nucleus as boundary conditions. It is important to note that the resulting wave is a *standing* wave, though it was inspired by de Broglie's concept of a *traveling* "pilot wave," with a particle riding on top.

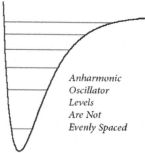

Now WERNER HEISENBERG was familiar with standing waves. He looked first for solutions to the linear harmonic oscillator and the anharmonic oscillator, whose energy levels are not evenly spaced.

Anharmonic Oscillator Levels Are Not Evenly Spaced

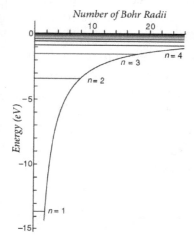

Compare the energy levels in the electrostatic potential $V = -e^2/r$ with the hydrogen atom term diagram in chapter 10.

Schrödinger's results for the bound energy levels in hydrogen matched Heisenberg's calculations exactly, but Schrödinger's math was much easier. All physicists, including Heisenberg himself, quickly replaced the awkward matrix mechanics with wave mechanics for all their calculations.

In December, 1925, Schrödinger wrote,

> I think I can specify a vibrating system that has as eigen-frequencies the hydrogen *term* frequencies - and in a relatively natural way, not through *ad hoc* assumptions.

But Schrödinger went well beyond his standing wave eigen-functions for bound states in hydrogen. He assumed that his wave mechanics could also describe *traveling* waves in free space.

Schrödinger wanted to do away with the idea of particles. He was convinced that a wave description could be a complete description of all quantum phenomena. He formulated the idea of a *wave packet*, in which a number of different frequencies would combine and interfere to produce a localized object. Where de Broglie, following Einstein, thought the wave was guiding the particle, Schrödinger wanted the wave *to be* the particle. But he soon learned that those different frequency components would cause the wave packet to rapidly disperse, not act at all like a localized particle.

Solving the Schrödinger equation for its eigenvalues works perfectly when it is a boundary value problem. Without boundary conditions, the idea of a wave as a particle has proved a failure.

All his life, Schrödinger denied the existence of particles and "quantum jumps" between energy levels, although the solution to his wave equation is a mathematical method of calculating those energy levels that is far simpler than the Heisenberg-Born-Jordan method of matrix mechanics, with its emphasis on particles.

The time-dependent Schrödinger equation is deterministic. Many physicists today think it restores determinism to physics. Although Einstein was initially enthusiastic that a wave theory might do so, he ultimately argued that the statistical character of quantum physics would be preserved in any future theory.[1]

If determinism is restored, he said, it would be at a much deeper level than quantum theory, which "unites the corpuscular and undulatory character of matter in a logically satisfactory fashion."

1   Schilpp, 1949, p.667

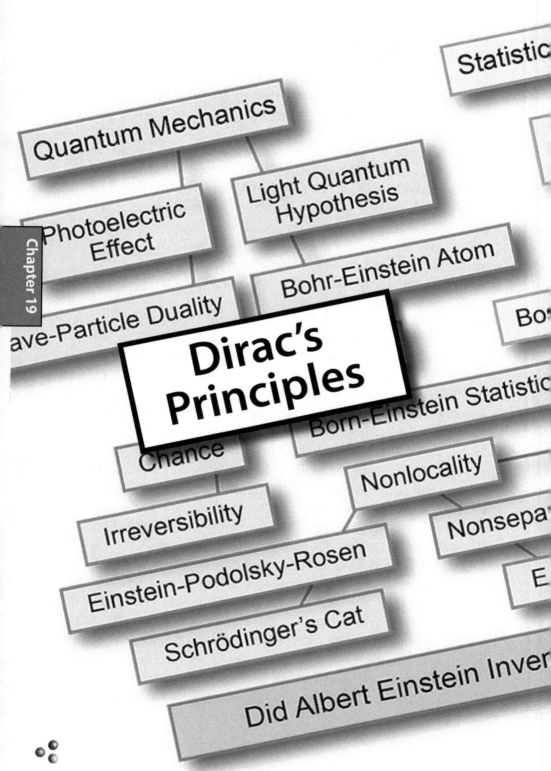

Chapter 19

Statistic

Quantum Mechanics

Light Quantum Hypothesis

Photoelectric Effect

Bohr-Einstein Atom

Bo

ave-Particle Duality

Dirac's Principles

Born-Einstein Statistic

Chance

Nonlocality

Irreversibility

Nonsepa

Einstein-Podolsky-Rosen

E

Schrödinger's Cat

Did Albert Einstein Inver

Dirac's Principles of Quantum Mechanics

In 1926 PAUL (P.A.M.) DIRAC combined the matrix mechanics of WERNER HEISENBERG and the wave mechanics of ERWIN SCHRÖDINGER into his beautifully symmetric transformation theory of quantum mechanics.

A year earlier, Dirac had been given a copy of Heisenberg's first paper on quantum mechanics. Heisenberg's work implied that some quantum-mechanical equivalents of classical entities like position and momentum do not commute with one another, as we saw in chapter 17. But Heisenberg himself did not understand that he was using a matrix . It was Heisenberg's mentor MAX BORN and Born's assistant PASCUAL JORDAN that recognized the matrices.

Independently of Born and Jordan, Dirac saw the non-commutation property of matrices implicit in Heisenberg"s work. He made it the central concept in his mathematical formulation of quantum physics. He called non-commuting quantities q-numbers (for "quantum" or "queer" numbers) and called regular numbers c-numbers (for "classical" or "commuting" numbers).

Dirac grounded his quantum mechanics on three basic ideas, the *principle of superposition*, the *axiom of measurement*, and the *projection postulate*, all of which have produced strong disagreements about the interpretations of quantum mechanics.

But there is complete agreement today that Dirac's theory is the standard tool for quantum-mechanical calculations.

In 1931, ALBERT EINSTEIN agreed,

> Dirac, to whom, in my opinion, we owe the most perfect
> exposition, logically, of this [quantum] theory, rightly points
> out that it would probably be difficult, for example, to give a
> theoretical description of a photon such as would give enough
> information to enable one to decide whether it will pass a
> polarizer placed (obliquely) in its way or not. [1]

---

1   Einstein, 1931, p.270

This is to remind us that Einstein had long accepted the controversial idea that quantum mechanics is a *statistical* theory, despite the claims of some of his colleagues, notably Born, that Einstein's criticisms of quantum mechanics were all intended to restore determinism and eliminate chance and probabilities.

Einstein's reference to photons passing through an oblique polarizer is taken straight from chapter 1 of Dirac's classic 1930 text, *The Principles of Quantum Mechanics*. Dirac uses the passage of a photon through an oblique polarizer to explain his *principle of superposition*, which he says "forms the fundamental new idea of quantum mechanics and the basis of the departure from the classical theory." [2]

Dirac's principle of superposition is very likely the most misunderstood aspect of quantum mechanics, probably because it is the departure from the deterministic classical theory. Many field-theoretic physicists believe that individual quantum systems can be in a superposition (e.g., a particle in two places at the same time, or going through both slits, a cat "both dead and alive.")

This is the source of much of the "quantum nonsense" in today's popular science literature.

Dirac's projection postulate, or collapse of the wave function, is the element of quantum mechanics most often denied by various "interpretations." The sudden discrete and discontinuous "quantum jumps" are considered so non-intuitive that interpreters have replaced them with the most outlandish alternatives.

DAVID BOHM's "pilot-wave" theory (chapter 30) introduces hidden variables moving at speeds faster than light to restore determinism to quantum physics, denying Dirac's projection probabilities.

HUGH EVERETT's "many-worlds interpretation" (chapter 31) substitutes a "splitting" of the entire universe into two equally large universes, massively violating the most fundamental conservation principles of physics, rather than allow a diagonal photon arriving at a polarizer to "collapse" into a horizontal or vertical state.

Decoherence theorists (chapter 35) simply deny quantum jumps and even the existence of particles!

JOHN BELL's inequality theorem explaining nonlocality and entanglement depends critically on a proper understanding of

---

2   Dirac, 1930, p.2

Dirac's principles. It is not clear that Bell fully accepts Dirac's work, as we shall see in chapter 32. The experimental tests of Bell's inequality depend on measuring the polarization or spin of two entangled particles.

Dirac gave a most clear description of the interaction of light particles (photons) with polarizers at various angles in the first chapter of his classic text, *The Principles of Quantum Mechanics.*

To explain his fundamental principle of superposition, Dirac considers a photon which is plane-polarized at a certain angle α and then gets resolved into two components at right angles to one another. How do photons in the original state change into photons at the right-angle states. He says

> "This question cannot be answered without the help of an entirely new concept which is quite foreign to classical ideas... The result predicted by quantum mechanics is that sometimes one would find the whole of the energy in one component and the other times one would find the whole in the other component. One would never find part of the energy in one and part in the other. Experiment can never reveal a fraction of a photon." [3]

At this point Dirac explains how many experiments have confirmed the quantum mechanical predictions for the probabilities of being found in the two components.

> If one did the experiment a large number of times, one would find in a fraction $\cos^2\alpha$ of the total number of times that the whole of the energy is in the α-component and in a fraction $\sin^2\alpha$ that the whole of the energy is in the $(\alpha + \pi/2)$-component. One may thus say that a photon has a probability $\cos^2\alpha$ of appearing in the a-component and a probability $\sin^2\alpha$ of appearing in the $(\alpha + \pi/2)$- component. These values for the probabilities lead to the correct classical distribution of energy between the two components when the number of photons in the incident beam is large. [4]

We can illustrate the passage of photons through polarizers turned at different angles, as used in tests of Bell's inequality.

---

3   ibid., pp.3-4
4   ibid., p.4

Dirac's Three Polarizers

We can use three squares of polarizing sheet material to illustrate Dirac's explanation of the quantum superposition of states and the collapse of a mixture of states to a pure state upon measurement or state preparation.

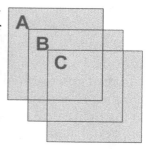

Here are the three polarizing sheets. They are a neutral gray color because they lose half of the light coming though them. The lost light is absorbed by the polarizer, converted to heat, and this accounts for the (Boltzmann) entropy gain required by our new information (Shannon entropy) about the exact polarization state of the transmittted photons.

When polarizers A and B are superimposed we see that the same amount of light comes through two polarizers, as long as the polarizing direction is the same. The first polarizer A prepares the photon in a given state of polarization. The second is then certain to find it in the same state. Let's say the direction of light polarization is vertical when the letters are upright.

If one polarizer, say B, turns 90°, its polarization direction will be horizontal and if it is on top of vertical polarizer A, no light will pass through it.

The Mystery of the Oblique Polarizer

As you would expect, any quantum mechanics experiment must contain an element of "Wow, that's impossible!" or we are not getting to the non-intuitive and unique difference between quantum mechanics and the everyday classical mechanics. So let's look at the amazing aspect of what Dirac is getting to, and then we will see how quantum mechanics explains it.

We turn the third polarizer C so its polarization is along the 45° diagonal. Dirac tells us that the wave function of light passing through this polarizer can be regarded as in a mixed state, a superposition of vertical and horizontal states.

As Einstein said, the information as to the exact state in which the photon will be found following a measurement does not exist.

We can make a measurement that detects vertically polarized photons by holding up the vertical polarizer A in front of the oblique polarizer C. Either a photon comes through A or it does not. Similarly, we can hold up the horizontal polarizer B in front of C. If we see a photon, it is horizontally polarized.

If our measuring apparatus (polarizer B) is measuring for horizontally polarized photons, the probability of detecting a photon diagonally polarized by C is 1/2. Similarly, if we were to measure for vertically polarized photons, we have the same 50% chance of detecting a photon.

Going back to polarizers A and B crossed at a 90° angle, we know that no light comes through when we cross the polarizers.

If we hold up polarizer C along the 45° diagonal and place it in front of (or behind) the 90° cross polarizers, nothing changes. No light is getting through.

But here is the amazing, impossible part. If you insert polarizer C at 45° between A and B, some light gets through. Note C is slipped between A (in the rear) and B (in front).

What is happening here quantum mechanically? If A crossed with B blocks all light, how can adding another polarization filter add light?

It is somewhat like the two-slit experiment where adding light by opening a second slit creates null points where light that was seen with one slit open now goes dark.

Here adding another polarizer allows more photons to pass.

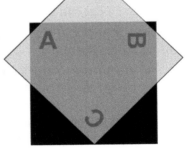

Dirac has now introduced the ideas of probability and statistics as a consequence of his principle of superposition. And he now introduces what he calls a "manner of speaking" which is today the source of much confusion interpreting quantum mechanics. He says this way of speaking will help us to "remember the results of experiments," but that "one should not try to give too much meaning to it." Einstein was looking for that deep meaning in reality.

In our polarizing experiment, Dirac suggests that we might speak *as if* a single photon is partly in each of the two states, that it is "distributed" over the two (horizontal and vertical) states.

> When we say that the photon is distributed over two or more given states the description is, of course, only qualitative, but in the mathematical theory it is made exact by the introduction of numbers to specify the distribution, which determine the *weights* with which different states occur in it. [5]

These *weights* are just the probabilities (actually the complex square roots of the probabilities). As Einstein's "objective reality" sees it, an individual photon is always in a single quantum state!

> The description which quantum mechanics allows us to give is merely a manner of speaking which is of value in helping us to deduce and to remember the results of experiments and which never leads to wrong conclusions. One should not try to give too much meaning to it...

Dirac's "manner of speaking" has given the false impression that a single particle can actually be in two states at the same time. This is seriously misleading. Dirac expresses the concern that some would be misled - don't "give too much meaning to it."

But this is something that bothered Einstein for years as he puzzled over "nonlocality." Schrödinger famously used superposition to argue that a cat can be simultaneously dead and alive! (chapter 28).

Many interpretations of quantum mechanics are based on this unfortunate mistake.

> Let us consider now what happens when we determine the energy in one of the components. The result of such a determination must be either the whole photon or nothing at all. Thus the photon must change suddenly from being partly in one beam and partly in the other to being entirely in one of the beams... It is impossible to predict in which of the two beams the photon will be found. Only the probability of either result can be calculated from the previous distribution of the photon over the two beams. [6]

5   ibid., p.5
6   ibid., p.6

One cannot picture in detail a photon being partly in each of two states; still less can one see how this can be equivalent to its being partly in each of two other different states or wholly in a single state. We must, however, get used to the new relationships between the states which are implied by this *manner of speaking* and must build up a consistent mathematical theory governing them.[7] [our italics]

Objective Reality and Dirac's "Manner of Speaking"

Dirac's "transformation theory" allows us to "represent" the initial wave function (before an interaction) in terms of a "basis set" of "eigenfunctions" appropriate for the possible quantum states of our measuring instruments that will describe the interaction.

But we shall find that assuming an individual quantum system is actually in one of the possible eigenstates of a system greatly simplifies understanding two-particle entanglement (chapter 29).

This is also consistent with Einstein's objectively real view that a particle has a position, a continuous path, and various properties that are conserved as long as the particle suffers no interaction that could change any of those properties.

Einstein was right when he said that the wave function describes ensembles, that is, the statistical results for large numbers of systems.

All of quantum mechanics rests on the Schrödinger equation of motion that *deterministically* describes the time evolution of the *probabilistic* wave function, plus Dirac's three basic assumptions, the principle of superposition (of wave functions), the axiom of measurement (of expectation values for observables), and the projection postulate (the "collapse" of the wave function that introduces indeterminism or chance during interactions).

The most appropriate basis set is one in which the eigenfunction-eigenvalue pairs match up with the natural states of the measurement apparatus. In the case of polarizers, one basis is the two states of horizontal and vertical polarization.

Elements in the "transformation matrix" give us the probabilities of measuring the system and finding it in one of the possible quantum states or "eigenstates," each eigenstate corresponding to an "eigenvalue" for a dynamical operator like the energy, momentum, angular momentum, spin, polarization, etc.

---

7   Dirac, 1930, p.5

Diagonal (*n, n*) elements in the transformation matrix give us the eigenvalues for observables in quantum state *n*. Off-diagonal (*n, m*) matrix elements give us transition probabilities between quantum states *n* and *m*.

Notice the sequence - possibilities > probabilities > actuality: the wave function gives us the possibilities, for which we can calculate probabilities. Each experiment gives us one actuality. A very large number of identical experiments confirms our probabilistic predictions. Confirmations are always only statistics, of course.

For completeness, we offer a brief review of the fundamental principles of quantum mechanics, as developed by Paul Dirac.

The Schrödinger Equation.

The fundamental equation of motion in quantum mechanics is Erwin Schrödinger's famous wave equation that describes the evolution in time of his wave function ψ.

$$i\hbar \, \delta\psi / \delta t = H \, \psi \qquad (1)$$

Max Born interpreted the square of the absolute value of Schrödinger's wave function $|\psi_n|^2$ (or $< \psi_n \, | \, \psi_n >$ in Dirac notation) as providing the probability of finding a quantum system in a particular state *n*. This of course was Einstein's view for many years.

As long as this absolute value (in Dirac bra-ket notation) is finite,

$$< \psi_n \, | \, \psi_n > = \int \psi^* (q) \, \psi (q) \, dq < \infty, \qquad (2)$$

then ψ can be normalized to unity, so that the *probability* of finding a particle somewhere $< \psi \, | \, \psi > = 1$, which is necessary for its interpretation as a probability. The normalized wave function can then be used to calculate "observables" like the energy, momentum, etc. For example, the probable or expectation value for the position *r* of the system, in configuration space *q*, is

$$< \psi \, | \, r \, | \, \psi > = \int \psi^* (q) \, r \, \psi (q) \, dq. \qquad (3)$$

Dirac's Principle of Superposition.

The Schrödinger equation (1) is a linear equation. It has no quadratic or higher power terms, and this introduces a profound - and for many scientists and philosophers the most disturbing - feature of quantum mechanics, one that is impossible in classical physics, namely the principle of super-position of quantum states. If ψ_a and ψ_b are both solutions of equation (1), then an arbitrary linear combination of these,

$$| \psi > = c_a | \psi_a > + c_b | \psi_b >, \qquad (4)$$

with complex coefficients c_a and c_b, is also a solution.

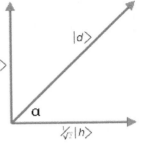

Together with statistical (probablistic) interpretation of the wave function, the principle of superposition accounts for the major mysteries of quantum theory, some of which we hope to resolve, or at least reduce, with an objective (observer-independent) explanation of irreversible information creation during quantum processes.

Observable information is critically necessary for measurements, though we note that observers can come along anytime after new information has been irreversibly recorded in the measuring apparatus as a consequence of the interaction with the quantum system. It is not the "conscious observer" standing by the apparatus that is responsible for the new information coming into existence.

The quantum (discrete) nature of physical systems results from there generally being a large number of solutions ψ_n (called eigenfunctions) of equation (1) in its time independent form, with energy eigenvalues E_n.

$$H \psi_n = E_n \psi_n, \qquad (5)$$

The discrete spectrum energy eigenvalues E_n limit interactions (for example, with photons) to specific energy differences $E_m - E_n$.

In the old quantum theory, Bohr postulated that electrons in atoms would be in "stationary states" of energy E_n, and that energy differences would be of the form $E_m - E_n = h\nu$, where ν is the frequency of the observed spectral line when an atom jumps from energy level E_m to E_n.

Einstein, in 1916, derived these two Bohr postulates from basic physical principles in his paper on the emission and absorption processes of atoms. What for Bohr were postulates or assumptions, Einstein grounded in quantum physics, though virtually no one appreciated his foundational work at the time, and few appreciate it today, his work mostly eclipsed by the Copenhagen physicists.

The eigenfunctions ψ_n are orthogonal to each other

$$< \psi_n \mid \psi_m > = \delta_{nm} \qquad (6)$$

where the "delta function"

$\delta_{nm} = 1$, if $n = m$, and $= 0$, if $n \neq m$. $\qquad$ (7)

Once they are normalized, the ψ_n form an orthonormal set of functions (or vectors) which can serve as a basis for the expansion of an arbitrary wave function φ

$$| \varphi > = \Sigma_0^\infty c_n | \psi_n >. \qquad (8)$$

The expansion coefficients are

$$c_n = < \psi_n \mid \varphi >. \qquad (9)$$

In the abstract Hilbert space, $< \psi_n \mid \varphi >$ is the "projection" of the vector φ onto the orthogonal axes of the ψ_n "basis" vector set.

Dirac's Axiom of Measurement.

The axiom of measurement depends on Heisenberg's idea of "observables," physical quantities that can be measured in experiments. A physical observable is represented as an operator, e.g., A, that is "Hermitean" (one that is "self-adjoint" - equal to its complex conjugate, $A^* = A$).

The diagonal n, n elements of the operator's matrix,

$$< \psi_n \mid A \mid \psi_n > = \int\int \psi^* (q) \, A \, (q) \, \psi \, (q) \, dq, \qquad (11)$$

are interpreted as giving the (probable) expectation value for A_n (when we make a measurement).

The off-diagonal n, m elements describe the uniquely quantum property of interference between wave functions and provide a measure of the probabilities for transitions between states n and m.

It is the intrinsic quantum probabilities that provide the ultimate source of indeterminism, and consequently of irreducible *irreversibility*, as we shall see.

Transitions between states are irreducibly random, like the decay of a radioactive nucleus (discovered by Rutherford in 1901) or the emission of a photon by an electron transitioning to a lower energy level in an atom (explained by Einstein in 1916).

The axiom of measurement is Dirac's formalization of Bohr's 1913 postulate that atomic electrons will be found in stationary states with energies E_n. In 1913, Bohr visualized them as orbiting the nucleus. Later, he said they could not be visualized, but chemists routinely visualize them as clouds of probability amplitude with easily calculated shapes that correctly predict chemical bonding.

The off-diagonal transition probabilities are the formalism of Bohr's "quantum jumps" between his stationary states, emitting or absorbing energy $h\nu = E_m - E_n$. Einstein explained clearly in 1916 that the jumps are accompanied by his discrete light quanta (photons), but Bohr continued to insist that the radiation was a classical continuous wave for another ten years, deliberately ignoring Einstein's foundational efforts in what Bohr might have felt was his own area of expertise (quantum mechanics).

The axiom of measurement asserts that a large number of measurements of the observable A, known to have eigenvalues An, will result in the number of measurements with value An, that is proportional to the probability of finding the system in eigenstate ψ_n. It is a statistical result that is incomplete, according to Einstein, because it contains only statistical information about an individual measurement. Quantum mechanics gives us only probabilities for finding individual systems in specific eigenstates.

Dirac's Projection Postulate.

Dirac's third novel concept of quantum theory is often considered the most radical. It has certainly produced some of the most radical ideas ever to appear in physics, in attempts by various "interpretations" of quantum mechanics to deny the "collapse of the wave function."

Dirac's projection postulate is actually very simple, and arguably intuitive as well. It says that when a measurement is made, the system of interest will be found in (will instantly "collapse" into) one of the possible eigenstates of the measured observable.

Now the proper choice of the "basis set" of eigenfunctions depends on the measurement apparatus. The natural basis set of vectors is usually one whose eigenvalues are the observables of our measurement system.

In Dirac's bra and ket notation, the orthogonal basis vectors in our example are $| v >$, the photon in a vertically polarized state, and $| h >$, the photon in a horizontally polarized state. These two states are eigenstates of our polarization measuring apparatus.

Given a quantum system in an initial state $|\varphi>$, according to equation 8, we can expand it in a linear combination of the eigenstates of our measurement apparatus, the $|\psi_n>$.

$$| \varphi > = \Sigma_0^\infty c_n | \psi_n >.$$

In the case of Dirac's polarized photons, the diagonal state $|d>$ is a linear combination of the horizontal and vertical states of the measurement apparatus, $|v>$ and $|h>$.

$$|d> = (1/\sqrt{2}) |v> + (1/\sqrt{2}) |h>. \qquad (12)$$

When we square the $(1/\sqrt{2})$ coefficients, we see there is a 50% chance of measuring the photon as either horizontal or vertically polarized.

According to Dirac's axiom of measurement, one of these possibilities is simply made actual, and it does so, said Max Born, in proportion to the absolute square of the complex probability amplitude wave function $|\psi_n|^2$.

In this way, ontological chance enters physics, and it is partly this fact of quantum randomness and indeterminism that bothered both Einstein ("God does not play dice") and Schrödinger (whose equation of motion for the wave function is deterministic).

But Dirac pointed out that not every measurement is indeterministic. Some measurements do not change the state.

When a photon is prepared in a vertically polarized state $|v>$, its interaction with a vertical polarizer is easy to visualize. We can picture the state vector of the whole photon simply passing through the polarizer unchanged (Pauli's measurement of the first kind).

The same is true of a photon prepared in a horizontally polarized state $|h>$ going through a horizontal polarizer. And the interaction of a horizontal photon with a vertical polarizer is easy to understand. The vertical polarizer will absorb the horizontal photon completely.

Pauli's Two Kinds of Measurement

In the case of a photon simply passing through a polarizer, no new information enters the universe. WOLFGANG PAULI called this a *measurement of the first kind*. Measuring a system that is known to be in a given quantum state may only confirm that it is in that state.

Today this is known as a *non-destructive* measurement.

> The method of measurement of the energy of the system discussed till now has the property that a repetition of measurement gives the same value for the quantity measured as in the first measurement...We shall call such measurements the *measurements of the first kind*. On the other hand it can also happen that the system is changed but in a controllable fashion by the measurement - even when, in the state before the measurement, the quantity measured had with certainty a definite value. In this method, the result of a repeated measurement is not the same as that of the first measurement. But still it may be that from the result of this measurement, an unambiguous conclusion can be drawn regarding the quantity being measured for the concerned system before the measurement. Such measurements, we call the measurements *of the second kind*.[8]

Measurements of the second kind are also known as a "state preparation." For example, we can take light of unknown polarization and pass it through a vertical polarizer. Any photon coming through has been *prepared in the vertical state*. All knowledge of the state before such a measurement is lost.

The new information created in a state preparation must be *irreversibly* recorded in the measurement apparatus, in order for there to be something the experimenter can observe. The recording increases the local negative entropy (information), so the apparatus most raise the global entropy, e.g., dissipating the heat generated in making the recording.

The diagonally polarized photon $|d>$, fully reveals the non-intuitive nature of quantum physics. We can visualize quantum indeterminacy, its statistical nature, and we can dramatically

---

8   Pauli, 1980, p.75

visualize the process of collapse, as a state vector aligned in one direction must rotate instantaneously into another vector direction.

As we saw above, the vector projection of $|d>$ onto $|v>$, with length $(1/\sqrt{2})$, when squared, gives us the probability 1/2 for photons to emerge from the vertical polarizer. But this is only a statistical statement about the expected probability for large numbers of identically prepared photons.

When we have only one photon at a time, we never get one-half of a photon coming through the polarizer. Critics of standard quantum theory, including Einstein, sometimes say that it tells us nothing about individual particles, only ensembles of identical experiments. There is truth in this, but nothing stops us from imagining the strange process of a single diagonally polarized photon interacting with the vertical polarizer.

There are two possibilities. We either get a whole photon coming through (which means that it "collapsed" into a vertical photon, or the diagonal vector was "reduced to" a vertical vector) or we get no photon at all. This is the entire meaning of "collapse." It is the same as an atom "jumping" discontinuously and suddenly from one energy level to another. It is the same as the photon in a two-slit experiment suddenly appearing at one spot on the photographic plate, where an instant earlier it might have appeared anywhere.

We can even visualize what happens when no photon appears. We can say that the diagonal photon was reduced to a horizontally polarized photon and was therefore completely absorbed.

How do we see the statistical nature and the indeterminacy?

First, statistically, in the case of many identical photons, we can say that half will pass through and half will be absorbed.

Secondly, the indeterminacy is simply that in the case of one photon, we have no ability to know which it will be. This is just as we cannot predict the time when a radioactive nucleus will decay, or the time and direction of an atom emitting a photon, as Einstein discovered in 1917, when we first learned that ontological chance is involved in quantum processes, especially in the interaction of matter and radiation.

This indeterminacy is a consequence of our diagonal photon state vector being "represented" (transformed) into a linear superposition of vertical and horizontal photon basis state vectors.

It is the principle of superposition together with the projection postulate that provides us with indeterminacy, statistics, and a way to "visualize" the collapse of a superposition of quantum states into one of the basis states.

Quantum mechanics is a probabilistic and statistical theory. The probabilities are theories about what experiments will show.

Theories are confirmed (statistically) when a very large number of experiments are performed with identical starting conditions.

Experiments provide the statistics (the frequency of outcomes) that confirm the predictions of quantum theory - with the highest accuracy of any physical theory ever invented!

But Dirac's principle of superposition of states, which gives us the probabilities of a system being found in different eigenstates, never means an *individual* system is in a combination of states!

Schrödinger's Cat (chapter 28) is always found to be dead or alive, not some bizarre combination of both.

And as Dirac made perfectly clear, we never find a photon split between a partial photon vertically polarized and another part horizontally polarized.

We always find the whole photon (or electron). And there is no reason that before the measurement, the particle is in some combination or superposition of states and lacks properties such as position, momentum, angular momentum, all of which are conserved quantities according to their conservation laws.

Thus Einstein's view of "objective reality," that particles have paths between measurements, is in complete agreement with Dirac's transformation theory.

We shall see in chapter 24, that the Copenhagen Interpretation denies Einstein's very simple and intuitive views of "reality."

Chapter 19

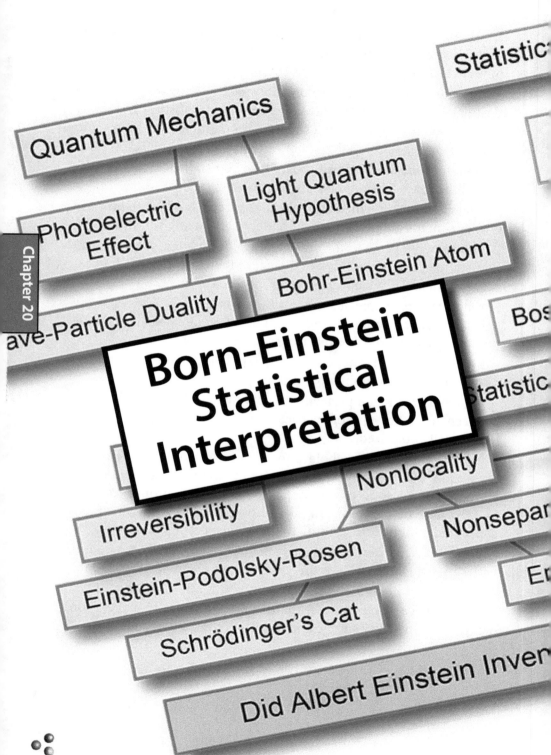

Statistic

Quantum Mechanics

Light Quantum
Hypothesis

Photoelectric
Effect

Bohr-Einstein Atom

ave-Particle Duality

Bos

**Born-Einstein
Statistical
Interpretation**

Statistic

Nonlocality

Irreversibility

Nonsepar

Einstein-Podolsky-Rosen

Er

Schrödinger's Cat

Did Albert Einstein Inven

Statistical Interpretation

It is often said that MAX BORN gave us the "statistical interpretation" of quantum mechanics that lies at the heart of NIELS BOHR's and WERNER HEISENBERG's principle of complementarity and the "Copenhagen Interpretation" of quantum mechanics.

But Born himself said many times he had only applied an idea of ALBERT EINSTEIN that had circulated privately for many years. To be sure, Born and Einstein quarreled for years over determinism and causality, but as we saw in chapter 11, it was Einstein who discovered "chance" in the interaction of matter and radiation, even if he considered it a "weakness in the theory."

As we showed in chapters 2 to 4, probability and statistics were very important in the two centuries before Born's work, but most physicists and philosophers saw the implied randomness to be "epistemic," the consequence of human ignorance. Random distributions of all kinds were thought to be completely deterministic at the particle level, with collisions between atoms following Newton's dynamical laws. LUDWIG BOLTZMANN's transport equation and H-Theorem showed that the increase of entropy is statistically irreversible at the macroscopic level, even if the motions of individual particles were time reversible.

Boltzmann did speculate that there might be some kind of molecular "chaos" or "disorder" that could cause particles traveling between collisions to lose the "correlations" or information about their past paths that would be needed for the paths to be time reversible and deterministic, but nothing came of this idea.

In his early career, ERWIN SCHRÖDINGER was a great exponent of fundamental chance in the universe. He followed his mentor FRANZ S. EXNER, who as a colleague of Boltzmann at the University of Vienna was a great promoter of statistical thinking.

In his inaugural lecture at Zurich in 1922, Schrödinger argued that available evidence can not justify our assumptions that physical laws are deterministic and strictly causal. His inaugural lecture was modeled on that of Exner in 1908.

Chapter 20

Exner's assertion amounts to this: It is quite possible that Nature's laws are of thoroughly statistical character. The demand for an absolute law in the background of the statistical law — a demand which at the present day almost everybody considers imperative — goes beyond the reach of experience. Such a dual foundation for the orderly course of events in Nature is in itself improbable. The burden of proof falls on those who champion absolute causality, and not on those who question it. For a doubtful attitude in this respect is to-day by far the more natural.[1]

Several years later, Schrödinger presented a paper on "Indeterminism in Physics" to the June, 1931 *Congress of A Society for Philosophical Instruction* in Berlin. He supported the idea of Boltzmann that "an actual continuum must consist of an infinite number of parts; but an infinite number is undefinable."

If nature is more complicated than a game of chess, a belief to which one tends to incline, then a physical system cannot be determined by a finite number of observations. But in practice a finite number of observations is all that we can make.

All that is left to determinism is to believe that an infinite accumulation of observations would in principle enable it completely to determine the system. Such was the standpoint and view of classical physics, which latter certainly had a right to see what it could make of it. But the opposite standpoint has an equal justification: we are not compelled to assume that an infinite number of observations, which cannot in any case be carried out in practice, would suffice to give us a complete determination.

In the history of science it is hard to find ears more likely to be sympathetic to a new idea than Schrödinger should have been to Max Born's suggestion that the square of the amplitude of Schrödinger's wave function $|\psi^2|$ should be interpreted statistically as the likelihood of finding a particle. And Schrödinger should have known Einstein thought quantum mechanics is statistical.

Yet Schrödinger objected strenuously, not so much to the probability and statistics as to the conviction of Born and his brilliant student Heisenberg that quantum phenomena, like

---

1   'What Is a Law of Nature?', *Science and the Human Temperament*, p.142.

quantum jumps between atomic energy levels, were only predictable *statistically*, and that there is a fundamental *indeterminacy* in the classical idea that particles have simultaneously knowable exact positions and velocities (momenta). Born, Heisenberg, and Bohr had declared classical determinism and causality untrue of the physical world.

It is likely that Schrödinger was ecstatic that his wave equation implied a deterministic physical theory. His wave function ψ evolves in time to give exact values for itself for all times and places. Perhaps Schrödinger thought that the waves themselves could provide a field theory of physics, much as fields in Newton's gravitational theory and in Maxwell's electromagnetic theory provide complete descriptions of nature. Schrödinger wondered whether nature might be only waves, no particles?

In July of 1926, Born used LOUIS DE BROGLIE's matter waves for electrons, as described by Schrödinger's wave equation, but he interpreted the wave as the *probability* of finding an electron going off in a specific collision direction, proportional to the square of the wave function ψ, now seen as a "probability amplitude."

Born's interpretation of the quantum mechanical wave function of a material particle as the probability (amplitude) of finding the material particle was a direct extension of Einstein's interpretation of light waves giving probability of finding photons.

To be sure, Einstein's interpretation may be considered only qualitative, where Born's was quantitative, since the new quantum mechanics now allowed exact calculations.

Nevertheless, Born initially gave full credit for the statistical interpretation to Einstein for the "ghost field" idea. Although the original idea is pure Einstein, it is widely referred to today as "Born's statistical interpretation," another example of others getting credit for a concept first seen by Einstein.

Born described his insights in 1926,

> Collision processes not only yield the most convincing experimental proof of the basic assumptions of quantum theory, but also seem suitable for explaining the physical meaning of the formal laws of the so-called "quantum

mechanics."... The matrix form of quantum mechanics that was founded by Heisenberg and developed by him and the author of this article starts from the thought that an exact representation of processes in space and time is quite impossible and that one must then content oneself with presenting the relations between the observed quantities, which can only be interpreted as properties of the motions in the limiting classical cases. On the other hand, Schrödinger (3) seems to have ascribed a reality of the same kind that light waves possessed to the waves that he regards as the carriers of atomic processes by using the de Broglie procedure; he attempts "to construct wave packets that have relatively small dimensions in all directions," and which can obviously represent the moving corpuscle directly.

Neither of these viewpoints seems satisfactory to me. Here, I would like to try to give a third interpretation and probe its utility in collision processes. I shall recall a remark that Einstein made about the behavior of the wave field and light quanta. He said that perhaps the waves only have to be wherever one needs to know the path of the corpuscular light quanta, and in that sense, he spoke of a "ghost field." It determines the probability that a light quantum - viz., the carrier of energy and impulse – follows a certain path; however, the field itself is ascribed no energy and no impulse.

One would do better to postpone these thoughts, when coupled directly to quantum mechanics, until the place of the electro-magnetic field in the formalism has been established. However, from the complete analogy between light quanta and electrons, one might consider formulating the laws of electron motion in a similar manner. This is closely related to regarding the de Broglie-Schrödinger waves as "ghost fields," or better yet, "guiding fields."

I would then like to pursue the following idea heuristically: The guiding field, which is represented by a scalar function ψ of the coordinates of all particles that are involved and time, propagates according to Schrödinger's differential equation. However, impulse and energy will be carried along as when corpuscles (i.e., electrons) are actually flying around. The paths of these corpuscles are determined only to the extent that they are constrained by the law of energy and impulse; moreover, only a probability that a certain path will be followed will be determined by the function ψ. One can perhaps summarize this, somewhat paradoxically, as: The motion of the particle follows the laws of probability, but the probability itself propagates in accord with causal laws.[2]

This last sentence is a remarkably concise description of the dualism in quantum mechanics, a strange mixture of indeterminism and determinism, of chance and necessity.

In his 1948 Waynflete lectures, Born elaborated on his understanding of chance,

> There is no doubt that the formalism of quantum mechanics and its statistical interpretation are extremely successful in ordering and predicting physical experiences. But can our desire of understanding, our wish to explain things, be satisfied by a theory which is frankly and shamelessly statistical and indeterministic? Can we be content with accepting chance, not cause, as the supreme law of the physical world?
>
> To this last question I answer that not causality, properly understood, is eliminated, but only a traditional interpretation of it, consisting in its identification with determinism. I have taken pains to show that these two concepts are not identical. Causality in my definition is the postulate that one physical situation depends on the other, and causal research means the discovery of such dependence. This is still true in quantum physics, though the objects of observation for which a dependence is claimed are different: they are the probabilities of elementary events, not those single events themselves.[3]

Ever since 1930, when Born's young graduate student Heisenberg had been selected for the Nobel Prize in physics although much of the theory was his own work, Born felt he had been treated unfairly.

He finally received recognition, with the Nobel Prize for physics in 1954, for his "statistical interpretation." But Born's voluminous correspondence with Einstein reveals that he had perhaps come to think that Einstein's supposed determinism meant Einstein did not believe in the statistical nature of quantum physics, so this idea may now rightfully belong to Born. He called it "his own" in the 1950's.

---

2   Born. 1926, p. 803.

3   Born, 1964, p.102

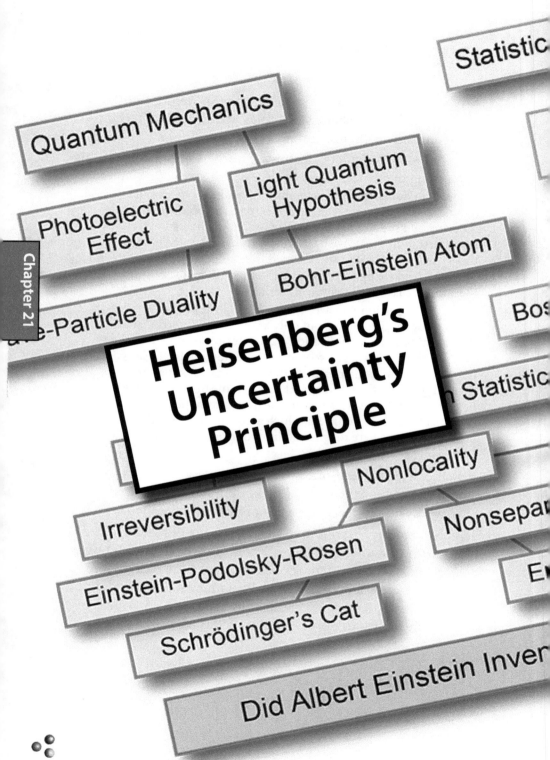

Statistic

Quantum Mechanics

Light Quantum
Hypothesis

Photoelectric
Effect

Bohr-Einstein Atom

e-Particle Duality

Bos

**Heisenberg's
Uncertainty
Principle**

Statistic

Nonlocality

Irreversibility

Nonsepar

Einstein-Podolsky-Rosen

E

Schrödinger's Cat

Did Albert Einstein Inver

Heisenberg's Uncertainty Principle

From the time in the 1950's I first started work on the problem of how information structures formed in the universe and the related problems of free will and creativity, down to the publication of my first book in 2011, *Free Will: The Scandal in Philosophy*, my source for the random element needed to generate alternative possibilities, without which no new information is possible, was WERNER HEISENBERG's uncertainty principle of 1927.

I wrote that "quantum physics in the twentieth century opened a crack in the wall of physical determinism."[1] My source was ARTHUR STANLEY EDDINGTON's great book, *The Nature of the Physical World*, the print version of his Gifford Lectures earlier in the year, with one great alteration.

In the delivered lectures, Eddington had described himself as unable "to form a satisfactory conception of any kind of law or causal sequence which shall be other than deterministic." A year later, in response to Heisenberg's uncertainty principle, Eddington revised his lectures for publication and dramatically announced "physics is no longer pledged to a scheme of deterministic law." He went even farther and enthusiastically identified indeterminism with freedom of the will. "We may note that science thereby withdraws its moral opposition to freewill."[2]

Eddington was the most prominent interpreter of the new physics to the English-speaking world. He confirmed Einstein's general theory of relativity with his eclipse observations in 1919, helping make Einstein a household word. And Eddington's praise of uncertainty contributed to making the young Heisenberg the symbolic head of the "founders" of the new quantum mechanics.

The Nobel Prizes of 1932/1933 for atomic physics were shared among Heisenberg, ERWIN SCHRÖDINGER, and PAUL DIRAC. Heisenberg's key contribution in his 1925 matrix mechanics was the discovery that position q and momentum p are complex conjugate quantities that do not commute. $pq \neq qp$!

---

1   Doyle, 2011, p.4.
2   Eddington, 1927, p.294-295

Dirac made this non-commutativity the fundamental fact of his 1926 transformation theory, in the form $pq - qp = -ih/2\pi = -i\hbar$. In 1927, Heisenberg proposed the idea that there is a limit to the accuracy with which one can make simultaneous measurements of the position and momentum, which he called a straightforward consequence of the commutativity rule as expressed by Dirac.

Heisenberg's Microscope

Heisenberg famously explained the joint uncertainty in position Δq and in momentum Δp in terms of measuring the properties of an electron under a microscope.

For example, let one illuminate the electron and observe it under a microscope. Then the highest attainable accuracy in the measurement of position is governed by the wavelength of the light. However, in principle one can build, say, a γ-ray microscope and with it carry out the determination of position with as much accuracy as one wants. In this measurement there is an important feature, the Compton effect. Every observation of scattered light coming from the electron presupposes a photoelectric effect (in the eye, on the photographic plate, in the photocell) and can therefore also be so interpreted that a light quantum hits the electron, is reflected or scattered, and then, once again bent by the lens of the microscope, produces the photoeffect. At the instant when position is determined— therefore, at the moment when the photon is scattered by the electron—the electron undergoes a discontinuous change in momentum. This change is the greater the smaller the wavelength of the light employed—that is, the more exact the determination of the position. At the instant at which the position of the electron is known, its momentum therefore can be known up to magnitudes which correspond to that

image detector

lens

scattered photon

incoming photon

recoil electron

Heisenberg's Microscope
Compare Compton Effect

discontinuous change. Thus, the more precisely the position is determined, the less precisely the momentum is known, and conversely. In this circumstance we see a direct physical interpretation of the equation $pq - qp = -i\hbar$. Let q_1 be the precision with which the value q is known (q_1 is, say, the mean error of q) therefore here the wavelength of the light. Let p_1 be the precision with which the value p is determinable; that is, here, the discontinuous change of p in the Compton effect. Then, according to the elementary laws of the Compton effect p_1 and q_1 stand in the relation

$$p_1\, q_1 \sim h. \qquad (1)$$

Here we can note that equation (1) is a precise expression for the facts which one earlier sought to describe by the division of phase space into cells of magnitude h.

...in all cases in which relations exist in classical theory between quantities which are really all exactly measurable, the corresponding exact relations also hold in quantum theory (laws of conservation of momentum and energy). Even in classical mechanics we could never practically know the present exactly, vitiating Laplace's demon. But what is wrong in the sharp formulation of the law of causality, "When we know the present precisely, we can predict the future," it is not the conclusion but the assumption that is false. Even in principle we cannot know the present in all detail. For that reason everything observed is a selection from a plenitude of possibilities and a limitation on what is possible in the future. As the statistical character of quantum theory is so closely linked to the inexactness of all perceptions, one might be led to the presumption that behind the perceived statistical world there still hides a "real" world in which causality holds. But such speculations seem to us, to say it explicitly, fruitless and senseless. Physics ought to describe only the correlation of observations. One can express the true state of affairs better in this way: Because all experiments are subject to the laws of quantum mechanics, and therefore to equation (1), it follows that quantum mechanics establishes the final failure of causality...one can say, if one will, with Dirac, that the statistics are brought in by our experiments. [3]

---

3   Heisenberg, 1927, p.64

Now this idea that it is our experiments that makes quantum mechanics statistical is very subtle. Bohr suggested Heisenberg use the word uncertainty (*Unsicherheit* in German) because it connotes an *epistemological* problem, knowledge of the world in our minds. A reluctant Heisenberg went along, but even the words he preferred, *Unbestimmtheit* or *Ungenauigkeit*, connote vagueness or indeterminacy as a property of our interaction with the world and not necessarily an *ontological* property of nature itself.

Einstein's objective reality agrees that the statistical nature of quantum mechanics lies in the results from many experiments, which only give us statistical data. But for Einstein there is an underlying reality of objects following continuous paths, conserving their fundamental properties when they are not acted upon.

Heisenberg had submitted his uncertainty paper for publication without first showing it to Bohr for his approval. When he did read it, Bohr demanded that Heisenberg withdraw the paper, so that it could be corrected. Heisenberg, quite upset, refused, but he did agree to add this paragraph in proof, admitting several errors.

After the conclusion of the foregoing paper, more recent investigations of Bohr have led to a point of view which permits an essential deepening and sharpening of the analysis of quantum-mechanical correlations attempted in this work. In this connection Bohr has brought to my attention that I have over-looked essential points in the course of several discussions in this paper. Above all, the uncertainty in our observation does not arise exclusively from the occurrence of discontinuities, but is tied directly to the demand that we ascribe equal validity to the quite different experiments which show up in the corpuscular theory on one hand, and in the wave theory on the other hand. In the use of an idealized gamma-ray microscope, for example, the necessary divergence of the bundle of rays must be taken into account. This has as one consequence that in the observation of the position of the electron the direction of the Compton recoil is only known with a spread which then leads to relation (1). Furthermore, it is not sufficiently stressed that the simple theory of the Compton effect, strictly speaking, only applies to free electrons. The consequent care needed in employing the uncertainty relation is, as Professor Bohr has explained, essential, among other things, for a comprehensive discussion of the transition from micro- to macromechanics. Finally, the discussion of resonance fluorescence is not entirely correct because the connection between the phase of the light and that of the electronic motion is not so simple as was assumed. I owe great thanks to Professor Bohr for sharing with me at an early stage the results of these more recent investigations of his—to appear soon in a paper on the conceptual structure of quantum theory—and for discussing them with me. [4]

As we shall see in chapter 24, a core tenet of the Copenhagen Interpretation is Heisenberg's idea that experiments bring particle properties into existence. Heisenberg described this as "the 'path' only comes into being because we observe it" (*Die "Bahn" entsteht erst dadurch, dass wir sie beobachten*).

Einstein, while disliking the statistical nature of quantum mechanics (which he himself discovered), nevertheless defended what he called the "objective" nature of reality, independent of the human mind or our experimental methods. He wanted to know whether a particle has a path *before* it is measured. He sarcastically asked (his biographer, Abraham Pais), is the moon only there when we are looking at it? Einstein (and we) use conservation principles to visualize the Compton Effect and Heisenberg's Microscope!

In the next chapter. we shall see that in his Como lecture later in 1927, Bohr further embarrassed and upset Heisenberg by publishing how position and momentum uncertainty can be explained completely using only properties of light waves, as in Schrödinger's wave mechanics. Bohr said that it actually has nothing to do with collisions disturbing the state of a particle! [5]

Perhaps as a consequence, from then on Heisenberg became quite deferential to Bohr. He traveled the world lecturing on the greatness of Bohr's "Copenhagen Interpretation." Despite this, Heisenberg continued to describe his uncertainty principle as a result of the Compton Effect. As a result, Heisenberg's microscope is still mistakenly taught as the reason for quantum uncertainty in many physics textbooks and popular science treatments.

Chapter 21

---

4   *ibid.*, p.83
5   See chapter 22

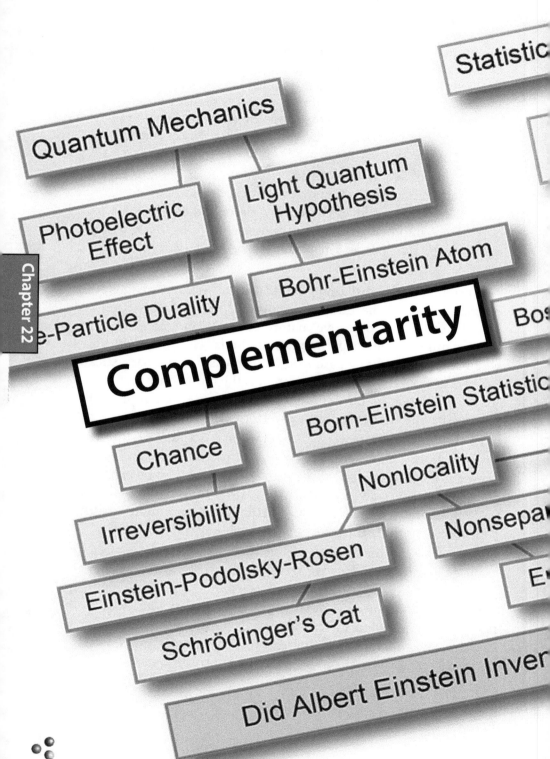

Bohr Complementarity

Among all the major scientists of the twentieth century, Niels Bohr may have most wanted to be considered a philosopher. Bohr introduced his concept of *complementarity* in a lecture at Lake Como in Italy in 1927, shortly before the fifth Solvay conference. It was developed in the same weeks as WERNER HEISENBERG was formulating his uncertainty principle. Complementarity, based largely on the wave-particle duality proposed by Einstein in 1909, lies at the core of the Copenhagen Interpretation of quantum mechanics.

Over the years, Bohr suggested somewhat extravagantly that complementarity could explain many great philosophical issues: it can illuminate the mind/body problem, it might provide for the difference between organic and inorganic matter, and it could underlie other great dualisms like subject/object, reason versus passion, and even free will versus causality and determinism.

Information philosophy identifies the wave function as pure abstract information, providing a theoretical prediction of the probability of finding particles, of matter or energy, at different positions in space and time. As such, it is similar in some sense to the idea of an *immaterial* mind in the material body. In this respect, Bohr was correct.

Like most educated persons of his time, Bohr knew of IMMANUEL KANT's noumenal/phenomemal dualism. He often spoke as if the goal of his complementarity was to reconcile opposites. He likened it to the eastern yin and yang, and his grave is marked with the yin/yang symbol.

Bohr was often criticized for suggesting that both A and Not-A could be the case. This was a characteristic of GEORG W. F. HEGEL's dialectical materialism. Had Bohr absorbed some Hegelian thinking? Another Hegelian trait was to speak indirectly and obscurely of the most important matters, and sadly this was Bohr's way, to the chagrin of many of his disciples. They sarcastically called his writing "obscure clarity." They hoped for clarity but got mostly fuzzy thinking when Bohr stepped outside of quantum mechanics.

Chapter 22

Bohr might very much have liked the current two-stage model for free will incorporating both randomness and an adequate statistical determinism. He might have seen it as a shining example of his complementarity.

As a philosopher, Bohr was a logical positivist, greatly influenced by ERNST MACH. Mach put severe epistemological limits on knowing the Kantian "things in themselves," just as Kant had put limits on reason. The British empiricist philosophers JOHN LOCKE and DAVID HUME had put the "primary" objects beyond the reach of our "secondary" sensory perceptions.

Bohr was an avid follower of the analytic philosophy of BERTRAND RUSSELL. He admired the *Principia Mathematica* of Russell and ALFRED NORTH WHITEHEAD.

Bohr seemed to deny the existence of Einstein's "objective reality," but clearly knew and said often that the physical world is largely independent of human observations. In classical physics, the physical world is assumed to be completely independent of the act of observing the world. Copenhageners were proud of their limited ability to know. Bohr said:

> There is no quantum world. There is only an abstract quantum physical description. It is wrong to think that the task of physics is to find out how nature is. Physics concerns what we can say about nature. [1]

Agreeing with Russell, LUDWIG WITTGENSTEIN, and other twentieth-century analytic language philosophers, Bohr emphasized the importance of conventional language as a tool for knowledge. Since language evolved to describe the familiar world of "classical" objects in space and time, Bohr and Heisenberg insisted that somewhere between the quantum world and the classical world there must come a point when our observations and measurements will be expressible in classical concepts. They argued that a measurement apparatus and a particular observation must be describable classically in order for it to be understood and for it to become knowledge in the mind of the observer. And controversially, they maintained that a measurement is not "complete" until it is knowledge in the mind

---

1   Quoted by Aage Petersen, *Bulletin of the Atomic Scientists.* Sep 1963, Vol. 19 Issue 7, p.12

of a "conscious observer." This is a step too far. The physical change in an information structure undergoing a measurement is complete when the new information is recorded physically, well before it is understood in any observer's mind.

Bohr was convinced that his complementarity implies that quantum mechanics is "complete." This was vigorously challenged by Einstein in his EPR paper of 1935.

Heisenberg's Microscope Revisited

As we saw in the last chapter, "Heisenberg's Microscope" showed that low-energy long-wavelength photons would not disturb an electron's momentum, but their long waves provided a blurry picture at best, so they lacked the resolving power to measure the position accurately. Conversely, if a high-energy, short wavelength photon is used (e.g., a gamma-ray), it might measure momentum, but the recoil of the electron ("Compton Effect") would be so large that its position becomes uncertain.

But in his Como Lecture, Bohr showed Heisenberg's disturbance of a *particle* is not the fundamental cause. He said that one can correct for the disturbance (the recoil) but can not eliminate the limits on resolving power of the measuring instrument, a consequence of the *wave* picture, not the particle picture.

Bohr cleverly derived Heisenberg's indeterminacy principle solely from space-time considerations about waves, greatly upsetting Heisenberg.

Adding to his embarassment, MAX BORN tells a story that Heisenberg could not answer his thesis examiner Willy Wien's question on resolving power and nearly failed the oral exam for his doctorate. [2]

Born says Heisenberg looked up the answers to all the questions he could not answer, and the optical formula for resolution became the basis for his famous example of the microscope a few years later.

So when Bohr pointed out the mistake in Heisenberg's first uncertainty paper draft suggesting that a "disturbance" was the source of the uncertainty. Heisenberg says he was "brought to tears."

---

2   Born, 1978, p.213

Bohr's Uncertainty Derivation

A "wave packet" with significant values in a spatially limited volume can be made from a superposition of plane waves with a range of frequencies.

Let Δt be the time it takes a wave packet to pass a certain point. Δv is the range of frequencies of the superposed waves.

In space instead of time, the wave packet is length Δx and the range of waves per centimeter is $\Delta \sigma$.

Bohr showed that the range of frequencies Δv needed so the wave packet is kept inside length of time Δt is related as

$\Delta v \, \Delta t = 1.$

A similar argument in space relates the physical size of a wave packet Δx to the variation in the number of waves per centimeter $\Delta \sigma$. σ is the so-called wave number $= 1/\lambda$ (λ is the wavelength):

$\Delta \sigma \, \Delta x = 1.$

If we multiply both sides of the above equations by Planck's constant h, and use the relation between energy and frequency $E = h v$ (and the similar relation between momentum and wavelength $p = h \sigma = h / \lambda$), the above become the Heisenberg indeterminacy relations:

$\Delta E \, \Delta t = h, \quad \Delta p \, \Delta x = h.$

This must surely have dazzled and perhaps deeply upset Heisenberg. Bohr had used only the space and time properties of waves to derive the physical limits of Heisenberg's uncertainty principle!

Bohr was obviously impressed by the new de Broglie - Schrödinger wave mechanics. His powerful use of Schrödinger's new wave mechanics frustrated Heisenberg, whose matrix mechanics was the first derivation of the new quantum principles, especially the non-commutativity of position and momentum operators.

Chapter 22

The equal embrace of particle and wave pictures was the core idea of Bohr's new complementarity, a position that Heisenberg defended vigorously in coming years, though without abandoning his microscope!

Bohr was pleased that Schrödinger's wave function provides a "natural" explanation for the "quantum numbers" of the "stationary states" in his quantum postulate. They are just the nodes in the wave function. On the other hand, Schrödinger himself hoped to replace particles and "unnatural" quantum jumps of Bohr's quantum postulate by resonances in his wave field. This led to many years of bitter disagreement between Bohr and Schrödinger.

Free Choice in Quantum Mechanics

Complementarity led Bohr and Heisenberg to a very important idea. Because there are always two complementary ways to approach any problem in quantum physics. They said that the result of an experiment depends on the "free choice" of the experimenter as to what to measure.

The quantum world of photons and electrons might look like waves or look like particles depending on what we look for, rather than what they "are" as "things in themselves." This is partly true.

In classical physics, simultaneous values exist for the position and momentum of elementary particles like electrons. In quantum physics, measuring one of these with high accuracy reduces the accuracy of the other, because of the uncertainty principle.

Indeed, in quantum mechanics, Bohr and Heisenberg claimed that neither of these properties could be said to exist until an experimenter freely decides to make a measurement.

Heisenberg says the property comes into existence as a result of the experiment. This is true, but only in a limited sense. If the experimenter decides to measure position, the result is a position. If momentum is measured, then the result is a momentum.

Einstein asked whether the particle has a position (and a path) before a particle is measured (his "objective reality"). He thought the idea that fundamental physical properties like momentum do not exist before a measurement is simply absurd.

Conservation laws allow us to *retrodict* those properties between successive measurements, as we shall see.

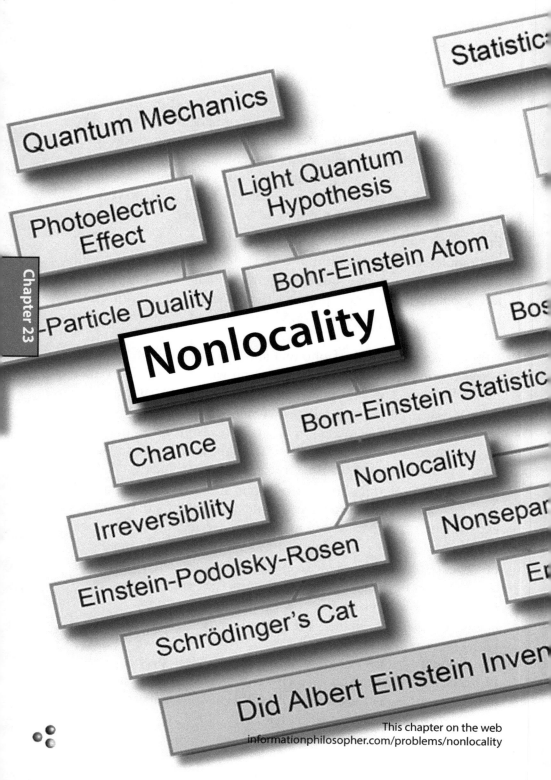

Quantum Mechanics

Statistic

Light Quantum
Hypothesis

Photoelectric
Effect

Bohr-Einstein Atom

-Particle Duality

Bos

Nonlocality

Born-Einstein Statistic

Chance

Nonlocality

Irreversibility

Nonsepar

Einstein-Podolsky-Rosen

Er

Schrödinger's Cat

Did Albert Einstein Inven

This chapter on the web
informationphilosopher.com/problems/nonlocality

Nonlocality at the Solvay Conference in 1927

Nonlocality is today strongly associated with the idea of entanglement (see chapter 29), but nonlocality was discovered as a property of a single quantum of light, whereas entanglement is a joint property of two quantum particles, depending on an even more subtle property called *nonseparability* (chapter 33).

Nonlocality is thought to be an essential element of light having wave and particle aspects, as Einstein described it first in 1909. But when understood as an "action-at-a-distance" faster than the speed of light, we shall show that this nonlocality *does not exist*.

We can visualize the wave function of quantum mechanics in the following way. It was Einstein who first said that the light wave tells us about probabilities of finding particles of light. Later MAX BORN made it quantitative. He identified the Schrödinger wave function Ψ as a probability amplitude whose squared modulus $|\Psi|^2$ gives the probability of finding a particle in a particular point.

We can think of Ψ as a "possibilities function," showing all the locations in space where there is a non-zero probability of finding a particle. The power of quantum mechanics is that we can *calculate* precisely the probability of finding the particle for each possibility.

Since WERNER HEISENBERG and PAUL DIRAC first discussed the "collapse" of the wave function (Dirac's projection postulate), it has been appropriate to say that "one of many possibilities has been made actual."

In the case of the photon, for example, it is localized when it has been scattered or absorbed by an electron. In the case of an electron, it might be a collision with another particle, or recombining with an ion to become bound in an atom, or absorbed into a metal and ejecting an electron as Einstein first explained.

The electron is actually never found at an infinitesimal point in four-dimensional space time, but remains "nonlocal" inside the minimal phase-space volume h^3 required by the uncertainty principle (for example, a particular electron orbital wave function and corresponding energy state).

Chapter 23

Einstein was first to have seen single-particle nonlocality, in 1905, when he tried to understand how a spherical wave of light that goes off in many directions can be wholly absorbed at a single location. In his famous paper on the photoelectric effect (for which he was awarded the Nobel Prize), Einstein hypothesized that light must be transmitted from one place to another as a *discrete* and *physically localized quantum* of energy.

Einstein did not then use the term nonlocal or "local reality," but we can trace his thoughts backwards from 1927 and 1935 to see that quantum nonlocality (and later *nonseparability*) were always major concerns for him, because they are not easily made consistent with a *continuous* field theory and they both *appear* to be inconsistent with his principle of relativity.

Einstein clearly described wave-particle duality as early as 1909, over a dozen years before the duality was made famous by Louis DE BROGLIE's thesis argued that clearly localized *material* particles also have a wavelike property. See chapter 9.

The fifth Solvay conference was titled "Electrons and Photons." It is no exaggeration to say that at that time, no physicist knew more than Einstein about electrons and photons. Yet he gave no major paper at the conference. He did give a short talk at a blackboard that prefigures his explosive EPR paper eight years later.

The fragments that remain of what Einstein actually said at the conference show a much deeper criticism of quantum mechanics. Einstein's nonlocality remarks were not a formal presentation and were not even reported in the conference proceedings. We know them only from brief notes on the general discussion and from what others tell us that Einstein said.

In his contribution to Paul Schilpp's volume on Einstein's work, NIELS BOHR said that Einstein went to the blackboard and drew a diagram which Bohr reconstructed in 1949:

> At the general discussion in Como, we all missed the presence of Einstein, but soon after, in October 1927, I had the opportunity to meet him in Brussels at the Fifth Physical Conference of the Solvay Institute, which was devoted to the theme "Electrons and Photons." At the Solvay meetings,

Einstein had from their beginning been a most prominent figure, and several of us came to the conference with great anticipations to learn his reaction to the latest stage of the development which, to our view, went far in clarifying the problems which he had himself from the outset elicited so ingeniously. During the discussions, where the whole subject was reviewed by contributions from many sides and where also the arguments mentioned in the preceding pages were again presented, Einstein expressed, however, a deep concern over the extent to which causal account in space and time was abandoned in quantum mechanics.

To illustrate his attitude, Einstein referred at one of the sessions to the simple example, illustrated by Fig. 1, of a particle (electron or photon) penetrating through a hole or a narrow slit in a diaphragm placed at some distance before a photographic plate.

On account of the diffraction of the wave connected with the motion of the particle and indicated in the figure by the thin lines, it is under such conditions not possible to predict with certainty at what point the electron will arrive at the photographic plate, but only to calculate the probability that, in an experiment, the electron will be found within any given region of the plate.

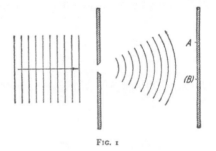

FIG. 1

The apparent difficulty, in this description, which Einstein felt so acutely, is the fact that, if in the experiment the electron is recorded at one point A of the plate, then it is out of the question of ever observing an effect of this electron at another point (B), although the laws of ordinary wave propagation offer no room for a correlation between two such events.[1]

The "nonlocal" effect at point B is the probability of an electron being found at point B going to zero instantly (as if an "action at a distance") when an electron is localized at point A

___

1   Schilpp, 1949, p. 211-213

And here are the notes on Einstein's actual remarks:[2]

MR EINSTEIN. - Despite being conscious of the fact that I have not entered deeply enough into the essence of quantum mechanics, nevertheless I want to present here some general remarks.

One can take two positions towards the theory with respect to its postulated domain of validity, which I wish to characterise with the aid of a simple example.

Let S be a screen provided with a small opening O, and P a hemispherical photographic film of large radius. Electrons impinge on S in the direction of the arrows. Some of these go through O, and because of the smallness of O and the speed of the particles, are dispersed uniformly over the directions of the hemisphere, and act on the film.

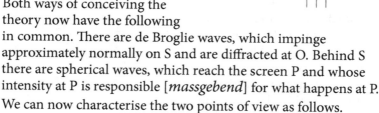

Both ways of conceiving the theory now have the following in common. There are de Broglie waves, which impinge approximately normally on S and are diffracted at O. Behind S there are spherical waves, which reach the screen P and whose intensity at P is responsible [*massgebend*] for what happens at P. We can now characterise the two points of view as follows.

1. Conception I. - The de Broglie-Schrödinger waves do not correspond to a single electron, but to a cloud of electrons extended in space. The theory gives no information about individual processes, but only about the ensemble of an infinity of elementary processes.

The waves give the probability or possibilities for a single electron being found at different locations in an ensemble of identical experiments. The waves "guide" the electrons to their positions, as will be seen in the two-slit experiment.

2. Conception II. - The theory claims to be a complete theory of individual processes. Each particle directed towards the screen, as far as can be determined by its position and speed,

The theory is not complete in this sense. It is a theory that makes probabilistic predictions that are confirmed perfectly by the statistics of many experiments.

2   Bacciagaluppi and Valentini, p.440

is described by a packet of de Broglie-Schrödinger waves of short wavelength and small angular width. This wave packet is diffracted and, after diffraction, partly reaches the film P in a state of resolution [*un etat de resolution*].

According to the first, purely statistical, point of view $|\psi|^2$ expresses the probability that there exists at the point considered a particular particle of the cloud, for example at a given point on the screen.

According to the second, $|\psi|^2$ expresses the probability that at a given instant the same particle is present at a given point (for example on the screen). Here, the theory refers to an individual process and claims to describe everything that is governed by laws.

By the same particle, Einstein means that the one individual particle has a possibility of being at more than one (indeed many) locations on the screen. This is so.

The second conception goes further than the first, in the sense that all the information resulting from I results also from the theory by virtue of II, but the converse is not true. It is only by virtue of II that the theory contains the consequence that the conservation laws are valid for the elementary process; it is only from II that the theory can derive the result of the experiment of Geiger and Bothe, and can explain the fact that in the Wilson chamber the droplets stemming from an α-particle are situated very nearly on continuous lines.

But on the other hand, I have objections to make to conception II. The scattered wave directed towards P does not show any privileged direction. If $|\psi|^2$ were simply regarded as the probability that at a certain point a given particle is found at a given time, it could happen that the same

Einstein is right that the one elementary process has a possibility of action elsewhere, but that could not mean producing an actual second particle. That would contradict conservation laws.

elementary process produces an action in two or several places on the screen. But the interpretation, according to which $|\psi|^2$ expresses the probability that this particle is found at a given point, assumes an entirely peculiar mechanism of action at a distance, which prevents the wave continuously distributed in space from producing an action in two places on the screen.

The "mechanism" of action-at-a-distance is simply the disappearance of possibilities elsewhere when a particle is actualized (localized) somewhere

In my opinion, one can remove this objection only in the following way, that one does not describe the process solely by the Schrödinger wave, but that at the same time one localises the particle during the propagation. I think that Mr de Broglie is right to search in this direction. If one works solely with the Schrödinger waves, interpretation II of $|\psi|^2$ implies to my mind a contradiction with the postulate of relativity.

Here Einstein's "objective reality" pictures a localized particle propagating under the guidance of Schrödinger's wave function. De Broglie's idea will be developed 25 years later by David Bohm, who will add an explicit potential traveling faster than the speed of light, which Einstein will reject.

I should also like to point out briefly two arguments which seem to me to speak against the point of view II. This [view] is essentially tied to a multi-dimensional representation (configuration space), since only this mode of representation makes possible the interpretation of $|\psi|^2$ peculiar to conception II. Now, it seems to me that objections of principle are opposed to this multi-dimensional representation. In this

The permutation of two identical particles does not produce two different points in multidimensional (configuration space). For example, interchange of the two electrons in the filled first electron shell, $1s^2$, just produces a change of sign for the antisymmetric two-particle wave function, no difference for $|\psi|^2$.

representation, indeed, two configurations of a system that are distinguished only by the permutation of two particles of the same species are represented by two different points (in configuration space), which is not in accord with the new results in statistics. Furthermore, the feature of forces of acting only at small spatial distances finds a less natural expression in configuration space than in the space of three or four dimensions. [3]

Bohr's reaction to Einstein's presentation has been preserved. He didn't understand a word! He ingenuously claims he does not know what quantum mechanics is. His response is vague and ends with simple platitudes.

MR BOHR. I feel myself in a very difficult position because I don't understand what precisely is the point which Einstein wants to [make]. No doubt it is my fault.

As regards general problem I feel its difficulties. I would put [the] problem in [an]other way. I do not know what quantum mechanics is. I think we are dealing with some mathematical methods which are adequate for description of our experiments Using a rigorous wave theory we are claiming something which

3   Bacciagaluppi and Valentini, pp.440-442

the theory cannot possibly give. [We must realise] that we are
away from that state where we could hope of describing things
on classical theories. [I] Understand [the] same view is held
by Born and Heisenberg. I think that we actually just try to
meet, as in all other theories, some requirements of nature, but
[the} difficulty is that we must use words which remind [us]
of older theories. The whole foundation for causal spacetime
description is taken away by quantum theory, for it is based
on [the] assumption of observations without interference. ...
excluding interference means exclusion of experiment and the
whole meaning of space and time observation ... because we
[have] interaction [between object and measuring instrument]
and thereby we put us on a quite different standpoint than
we thought we could take in classical theories. If we speak of
observations we play with a statistical problem There are certain
features complementary to the wave pictures (existence of
individuals). ...

The saying that spacetime is an abstraction might seem a
philosophical triviality but nature reminds us that we are dealing
with something of practical interest. Depends on how I consider
theory. I may not have understood, but I think the whole thing
lies [therein that the] theory is nothing else [but] a tool for
meeting our requirements and I think it does. [4]

Twenty-two years later, in Bohr's contribution to the Schilpp
volume, he had no better response to Einstein's 1927 concerns. Bohr
chose to retell the story of how he and Heisenberg refuted every
attempt by Einstein to attack the uncertainty principle.

Although Bohr seems to have missed Einstein's point completely,
Heisenberg at least came to understand it. In his 1930 lectures at
the University of Chicago, Heisenberg presented a critique of both
particle and wave pictures, including a new example of Einstein's
nonlocal action-at-a-distance, using reflected and transmitted
waves at a mirror surface that Einstein had developed since 1927.

Heisenberg wrote:
In relation to these considerations, one other idealized
experiment (due to Einstein) may be considered. We imagine
a photon which is represented by a wave packet built up out
of Maxwell waves. It will thus have a certain spatial extension

4   Bacciagaluppi and Valentini, pp, 442-443

and also a certain range of frequency. By reflection at a semi-transparent mirror, it is possible to decompose it into two parts, a reflected and a transmitted packet. There is then a definite probability for finding the photon either in one part or in the other part of the divided wave packet. After a sufficient time the two parts will be separated by any distance desired; now if an experiment yields the result that the photon is, say, in the reflected part of the packet, then the probability of finding the photon in the other part of the packet immediately becomes zero. The experiment at the position of the reflected packet thus exerts a kind of action (reduction of the wave packet) at the distant point occupied by the transmitted packet, and one sees that this action is propagated with a velocity greater than that of light. However, it is also obvious that this kind of action can never be utilized for the transmission of signals so that it is not in conflict with the postulates of the theory of relativity.[5]

Heisenberg has seen that the point of "Einstein's experiment" was nonlocality, not an attack on his uncertainty principle. We shall see that for the next ten years at least, and in many cases for the rest of Einstein's life, followers of the Copenhagen Interpretation were convinced that Einstein was stuck in the past, primarily interested in denying their work and restoring determinism to physics.

If Heisenberg had read (or reread) Einstein's 1905 article on the light-quantum hypothesis at this time, he would have surely seen that Einstein's light wave had "immediately become zero" everywhere when all its energy is absorbed in the metal and an electron is ejected by the photoelectric effect.

It is only Einstein's mistaken assumption that a light wave consists of some form of energy distributed everywhere (a cloud of electrons) that there is a conflict with special relativity. But there is also a worrisome simultaneity of events in a spacelike separation.

Once we see the wave as just a mathematical abstract function that gives the probability of finding a particle of light, the conflict with relativity disappears. When a particle is found in one place, the probabilities of it being elsewhere simply disappear.

There is nothing happening faster than light in the sense of material or energy coming instantly from all directions to appear at a single point. Nonlocality is just the *appearance* of something moving faster than light speed. There is no "action-at-a-distance."

5   Heisenberg, 1930, p.39

Chapter 23

If nonlocality is defined as an "action" by one particle on another in a spacelike separation ("at a distance") at speeds faster than light, then nonlocality simply does not exist.

"Collapse" of the Wave Function

As Einstein's blackboard drawing at the Solvay Conference shows us, the wave function propagates like a light wave in all directions, but when the particle appears, it is found at a single point.

Using Einstein's idea of "objective reality," without any interactions that could change the momentum, the particle must have traveled in a straight line from the origin to the point where it is found.

And although we cannot know the actual path taken by any particle, Einstein strongly believed that such paths exist in his "local" and "objective reality."

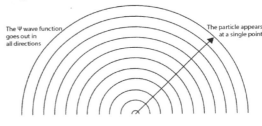

The Ψ wave function goes out in all directions

The particle appears at a single point

Einstein tells us the wave represents the probability of finding the particle. (Today it is the absolute square of the complex wave function $|\Psi|^2$ that gives us the probability.) All directions are equally probable until the moment when the particle is found somewhere. At that moment, the probability of its being elsewhere goes to zero.

This has been interpreted as a "collapse." If the wave had been carrying energy in all directions, or matter as Schrödinger thought, energy and matter would indeed have had to "collapse" to the point.

But nothing moves in this picture. It is just that the probability wave disappears when the particle appears. The use of the word "collapse," with its connotation of objects falling together, was an unfortunate choice.

Everything physical that is happening in this picture is happening locally! There is nothing nonlocal going on. But then why was Einstein worried? What did he see in 1927?

He saw events at two points (A and B in his drawing) in a spacelike separation occurring "simultaneously," a concept that his new special theory of relativity says is *impossible* in any absolute sense.

A related nonlocality or "impossible simultaneity" is involved in the mystery of entanglement. See chapters 26 to 29.

Chapter 23

The Two-Slit Experiment

Although Einstein's presentation at the fifth Solvay conference was an unprepared modest talk at the blackboard, his debates with Bohr at morning breakfast and evening dinner have become world famous, thanks to Bohr and his associates bragging about how they won every point against Einstein.

It is not obvious that Bohr understood what exactly Einstein waas debating about, as we saw in his remarks after Einstein's talk. Bohr said he was defending against Einstein's attack on the uncertainty principle.   And uncertainty did come up, when Einstein tried to defend his "objective reality" view that the electron (or photon) must go through just one slit in the famous two-slit experiment.

Bohr described their debate with another figure.

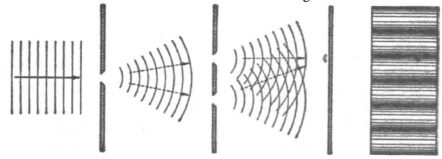

FIG. 3

He said,

> as indicated by the broken arrows, the momentum transferred to the first diaphragm ought to be different if the electron was assumed to pass through the upper or the lower slit in the second diaphragm, Einstein suggested that a control of the momentum transfer would permit a closer analysis of the phenomenon and, in particular, to decide through which of the two slits the electron had passed before arriving at the plate. [6]

Note that Einstein was hoping to establish the path of the particle, Bohr' was touting his idea of *complementarity*, which says we can *either* trace the path of a particle *or* observe interference effects, but not both at the same time.

---

6   Schilpp, 1949, p.216-217

The Copenhagen Interpretation (see next chapter) maintains that it is impossible to acquire any information about particle paths between measurements. This is true. Without measurements we know nothing. But Copenhagen, especially Heisenberg, insisted that the 'path' only comes into being because we observe it.

This leads to the anthropomorphic view that particles have no definite properties until they are measured. Einstein's view is that just becuse we don't know what is going on from moment to moment, it does not mean that properties are not being conserved. The moon is there even when we are not looking, etc.

We will return to the "one deep mystery" in the two-slit experiment in chapter 33.

Nature's Choice and the Experimenter's Choice

In the same session at Solvay where Einstein raised objections to the Copenhagen Interpretation, Bohr described a discussion about randomness in quantum events and the "free choice" of an experimenter as to what to measure. In the latter case, Heisenberg is correct. The measurement does define the properties seen.

> On that occasion an interesting discussion arose also about how to speak of the appearance of phenomena for which only predictions of statistical character can be made. The question was whether, as to the occurrence of individual effects, we should adopt a terminology proposed by Dirac, that we were concerned with a choice on the part of "nature" or, as suggested by Heisenberg, we should say that we have to do with a choice on the part of the "observer" constructing the measuring instruments and reading their recording. Any such terminology would, however, appear dubious since, on the one hand, it is hardly reasonable to endow nature with volition in the ordinary sense, while, on the other hand, it is certainly not possible for the observer to influence the events which may appear under the conditions he has arranged. To my mind, there is no other alternative than to admit that, in this field of experience, we are dealing with individual phenomena and that our possibilities of handling the measuring instruments allow us only to make a choice between the different complementary types of phenomena we want to study.[7]

_____

7   *ibid.*, p.223

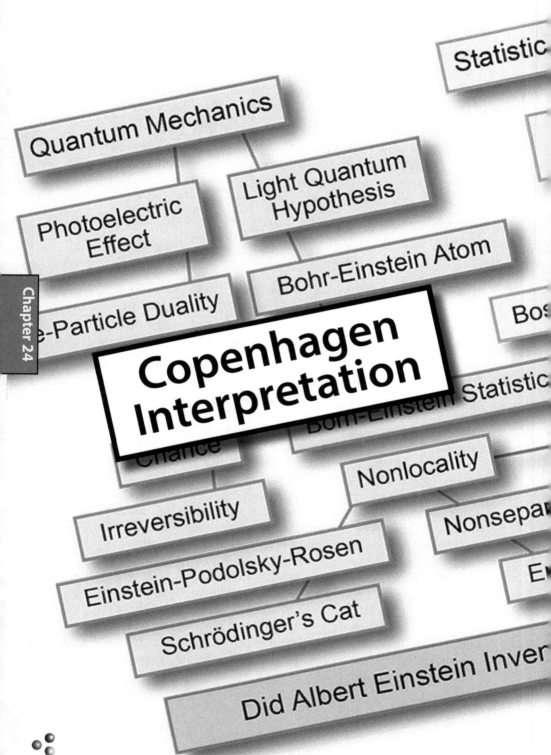

Statistic

Quantum Mechanics

Light Quantum Hypothesis

Photoelectric Effect

Bohr-Einstein Atom

Bos

e-Particle Duality

Copenhagen Interpretation

Born-Einstein Statistic

Chance

Nonlocality

Irreversibility

Nonsepa

Einstein-Podolsky-Rosen

E

Schrödinger's Cat

Did Albert Einstein Inver

Copenhagen Interpretation

The idea that there was a Copenhagen way of thinking was christened as the "*Kopenhagener Geist der Quantentheorie*" by WERNER HEISENBERG in his 1930 textbook *The Physical Principles of Quantum Theory*, based on his 1929 lectures in Chicago (given at the invitation of ARTHUR HOLLY COMPTON).

The basic ideas of Copenhagen thinking were presented by NIELS BOHR and Heisenberg at the 1927 Solvay conference on physics entitled "Electrons and Photons."

It is a sad fact that ALBERT EINSTEIN, who had discovered more than any other scientist on the quantum interaction of electrons and photons, was largely ignored or misunderstood when he clearly described nonlocality at the 1927 conference. As we saw in the previous chapter, Bohr said he could not understand what Einstein was talking about.

At the Solvay conference, Bohr and Heisenberg consolidated their Copenhagen view as a "complete" picture of quantum physics, despite the fact that they could not, or would not, visualize or otherwise explain exactly what is going on in the microscopic world of "quantum reality." Electron paths (especially orbits) that cannot be observed, they said, simply do not exist!

Bohr and Heisenberg opposed Einstein's concept of an underlying "objective reality," but they clearly knew and said that the physical world is largely independent of human observations. In classical physics, the physical world is assumed to be completely independent of the act of observing the world.

In quantum physics however, Heisenberg said that the result of an experiment depends on the "free choice" of the experimenter as to what to measure. The quantum world of photons and electrons might look like waves or look like particles depending on what we look for, rather than what they "are" as "things in themselves."

Copenhageners were proud of their limited ability to know what is going on in the microscopic world.

Chapter 24

According to his friend Aage Petersen, Bohr said:

> There is no quantum world. There is only an abstract quantum physical description. It is wrong to think that the task of physics is to find out how nature is. Physics concerns what we can say about nature. [1]

Bohr thus put severe *epistemological* limits on knowing the "things in themselves," just as IMMANUEL KANT had put limits on reason in the phenomenal world. The British empiricist philosophers JOHN LOCKE and DAVID HUME had put the "primary" objects beyond the reach of our "secondary" sensory perceptions. In this respect, Bohr shared the positivist views of many other empirical scientists and philosophers, ERNST MACH for example.

Twentieth-century analytic language philosophers like BERTRAND RUSSELL and LUDWIG WITTGENSTEIN thought that philosophy (and even physics) could not solve some basic problems, but only "dis-solve" them by showing them to be conceptual errors resulting from the misuse of language.

Neither Bohr nor Heisenberg thought that macroscopic objects actually *are* classical. They both saw them as composed of microscopic quantum objects. The information interpretation of quantum mechanics says there is only one world, the quantum world. Averaging over large numbers of microscopic quantum objects explains why macroscopic objects *appear* to be classical.

On the other hand, Bohr and Heisenberg insisted that the language of classical physics is essential as a tool for knowledge.

Heisenberg wrote:

> The Copenhagen interpretation of quantum theory starts from a paradox. Any experiment in physics, whether it refers to the phenomena of daily life or to atomic events, is to be described in the terms of classical physics. The concepts of classical physics form the language by which we describe the arrangement of our experiments and state the results. We cannot and should not replace these concepts by any others. Still the application of these concepts is limited by the relations of uncertainty. We must keep in mind this limited range of applicability of the classical concepts while using them, but we cannot and should not try to improve them. [2]

---

1   *Bulletin of the Atomic Scientists.* Sep 1963, Vol. 19 Issue 7, p.12

2   Heisenberg, 1955, p. 44

Einstein wanted us to get beyond questions of logic and language to get to an "objective reality" he saw as independent of the mind of man. Logic alone tells us nothing of the physical world, he said.

But since language has evolved to describe the familiar world of "classical" objects in space and time, Bohr and Heisenberg insisted that somewhere between the quantum world and the classical world there must come a point where our observations and measurements can be expressible in classical concepts. They argued that a measurement apparatus and a particular observation must be describable classically in order for it to be understood and become knowledge in the mind of the observer.

The exact location of that transition from the quantum to the classically describable world was arbitrary, said Heisenberg. He called it a "cut" (*Schnitt*). Heisenberg's and especially JOHN VON NEUMANN's and EUGENE WIGNER's insistence on a critical role for a "conscious observer" has led to a great deal of nonsense being associated with the Copenhagen Interpretation and in the philosophy of quantum physics. Heisenberg may only have been trying to explain how knowledge reaches the observer's mind. But for von Neumann and Wigner, the mind was actually considered a *causal factor* in the behavior of the quantum system. It is not.

Today, a large number of panpsychists, some philosophers, some scientists, still believe that the mind of a conscious observer is needed to cause the "collapse of the wave function." We explore von Neumann's "psycho-physical parallelism" in the next chapter.

In the mid-1950's, Heisenberg reacted to DAVID BOHM's 1952 "pilot-wave" interpretation of quantum mechanics by calling his work with Bohr the "Copenhagen Interpretation" and indeed insisted it is the only correct interpretation of quantum mechanics. A significant fraction of working quantum physicists today say they agree with Heisenberg, though few have ever looked carefully into the fundamental assumptions of the Copenhagen Interpretation.

We'll see that much of the Copenhagen interpretation is standard quantum physics and correct. But it also contains a lot of nonsense that has made understanding quantum physics difficult and spawned several quantum mysteries that we hope to resolve.

Chapter 24

What Exactly Is in the Copenhagen Interpretation?

There are several major components to the Copenhagen Interpretation, which most historians and philosophers of science agree on:

No Observer-Independent Quantum Reality. The most radical concept of the Copenhagen school is that because the wave function gives us only probabilities about quantum properties, that these properties do not exist in the sense of Einstein's "objective reality."

No Path? Bohr, Heisenberg, and others said we cannot describe a particle as having a path, or a definite position before a measurement. Indeed, it is said a particle can be in two places at once, like going through the two slits in the two-slit experiment.

But just because we cannot know the path does not mean it cannot exist. Einstein's "objective reality" hoped for a deeper level of physics in which particles do have paths (even if we cannot know them) and, in particular, the paths obey conservation principles.

Conscious Observer. This is the claim that quantum systems cannot change their states without an observation being made by a conscious observer. Does the collapse only occur when an observer "looks at" the system? How exactly does the mind of the observer have causal power over the physical world? (the mind-body problem). JOHN BELL asked sarcastically, "does the observer need a Ph.D.?"

Einstein objected to the absurd idea that his bed had diffused throughout the room and only gathered itself back together when he opened the bedroom door and looked in. Does the moon only exist when somoone is looking at it?, he asked.

JOHN VON NEUMANN and EUGENE WIGNER seemed to believe that the mind of the observer was essential, but it is not found in the original work of Bohr and Heisenberg, so should perhaps not be a part of the Copenhagen Interpretation? It has no place in standard quantum physics today.

Wave-particle duality. Einstein's 1909 insight into this dual aspect of quantum mechanics led to Bohr's deep philosophical notion of *complementarity*, though Bohr did not mention Einstein.

Bohr wanted a synthesis of the particle-matrix mechanics theory of Heisenberg, MAX BORN, and PASCUAL JORDAN, with the wave mechanical theory of LOUIS DE BROGLIE and ERWIN SCHRÖDINGER,. Wave theory became critical to Bohr's concept of complementarity, which we sw in chapter 22.

Heisenberg had to have his arm twisted by Bohr in 1927 to accept the equal importance of the wave description.

Copenhagen says quantum objects *are both* waves and particles, that what you see depends on how you look at them. In Einstein's "objective reality," physical objects are particles. Waves are mathematical theories about their behavior, giving us the *probabilities* of where they will be found, and with what properties.

No Visualizability? Bohr and Heisenberg both thought we could not produce models of what is going on at the quantum level. Bohr thought that since the wave function cannot be observed we can't say anything about it. Heisenberg said it was a probability and the basis for the statistical nature of quantum mechanics.

Whenever we draw a diagram of waves impinging on the two-slits, we are in fact visualizing the wave function as possible locations for a particle, with calculable probabilities for each possible location.

The Quantum Postulates. Bohr postulated that quantum systems (beginning with his "Bohr atom" in 1913) have "stationary states" which make discontinuous "quantum jumps" between the states with the emission or absorption of radiation. Until at least 1925 Bohr insisted the radiation itself is *continuous*. Einstein had said radiation is a *discrete* localized "light quantum" (later called a photon) as early as 1905.

Ironically, ignorant of the history (dominated by Bohr's account), most of today's physics textbooks teach the "Bohr atom" as emitting or absorbing photons - Einstein's light quanta!

Indeterminacy principle. Heisenberg sometimes called it his "uncertainty" principle, which implies human ignorance, making it an epistemological (knowledge) problem rather than an ontological (reality) problem. Indeterminacy is another example of complementarity, between the non-commuting conjugate variables

Chapter 24

momentum and position, for example, $\Delta p \, \Delta x \geq h$. Energy and time, as well as action and the angle variables, are also complementary.

Completeness. Copenhageners claim that Schrödinger's wave function ψ provides a "complete" description of a quantum system, despite the fact that conjugate variables like position and momentum cannot both be known with arbitrary accuracy, as they can in classical systems. There is less information in the quantum world than classical physics requires. The wave function ψ evolves according to the unitary deterministic Schrödinger equation of motion, conserving that information. When one possibility discontinuously becomes actual, new information may be irreversibly created and recorded by a measurement apparatus.

Einstein, however, maintained that quantum mechanics is *incomplete*, because it provides only statistical information derived from ensembles of quantum systems.

Correspondence principle. Bohr maintained that in the limit of large quantum numbers, the atomic structure of quantum systems approaches the behavior of classical systems. Bohr and Heisenberg both described this case as when Planck's quantum of action h can be neglected. They mistakenly described this as $h \to 0$.

Planck's h is a constant of nature, like the velocity of light. The quantum-to-classical transition is when the action of a macroscopic object is large compared to h. Bohr compared it to non-relativistic physics when the velocity v is small compared to the velocity of light. It is not an apt comparison because h never becomes small. It is when the number of quantum particles increases (as mass increases) that large macroscopic objects behave like classical objects. Position and velocity become arbitrarily accurate as $h/m \to 0$.

$\Delta v \, \Delta x \geq h/m$.

The correspondence between classical and quantum physics occurs for large numbers of particles that can be averaged over and for large quantum numbers. This is known as the quantum-to-classical transition.

Standard Quantum Physics. PAUL DIRAC formalized quantum mechanics with three fundamental concepts, all very familiar and accepted by Bohr, Heisenberg, and the other Copenhageners:

Axiom of measurement. Bohr's stationary quantum states have eigenvalues with corresponding eigenfunctions (the eigenvalue-eigenstate link).

Superposition principle. According to Dirac's transformation theory, ψ can be represented as a linear combination of vectors that are a proper basis for the combined target quantum system and the measurement apparatus.

Projection postulate. The collapse of the wave function ψ, which is irreversible, upon interacting with the measurement apparatus and creating new information.

Irreversibility. Without irreversible recording of information in the measuring apparatus (a pointer reading, blackened photographic plate, Geiger counter firing, etc.), there would be nothing for observers to see and to know.

All the founders of quantum mechanics mention the need for irreversibility. The need for entropy transfer to stabilize irreversibly recorded information so it could be observed was first shown by LEO SZILARD in 1929, later by LEON BRILLOUIN and ROLF LANDAUER.

Classical apparatus. Bohr's requirement that the macroscopic measurement apparatus be described in ordinary "classical" language is a third kind of "complementarity," between the microscopic quantum system and the macroscopic "classical apparatus."

But Born and Heisenberg never actually said the measuring apparatus is "classical." They knew that everything is fundamentally a quantum system.

Statistical Interpretation (probability and acausality). Born interpreted the squared modulus of Schrödinger's complex wave function as the probability of finding a particle. Einstein's "ghost field" or "guiding field," de Broglie's pilot or guide wave, and Schrödinger's wave function as the distribution of the electric charge density were similar views in earlier years.

All the predicted properties of physical systems and the "laws of nature" are only probabilistic (acausal). All the results of physical experiments are purely statistical information.

Theories give us probabilities. Experiments give us statistics.

Large numbers of identical experiments provide the statistical evidence for the theoretical probabilities predicted by quantum mechanics. We know nothing about paths of individual particles.

Chapter 24

Bohr's emphasis on epistemological questions suggests he thought that the statistical uncertainty may only be in our knowledge. It may not describe nature itself. Or at least Bohr thought that we can not describe a "reality" for quantum objects, certainly not with classical concepts and language. But we shall see that the concept of an abstract and immaterial wave function (ψ as pure information moving through space, determined by boundary conditions) makes quantum phenomena "visualizable."

Ontological acausality, chance, and a probabilistic or statistical nature were first seen by Einstein in 1916, as Born acknowledged. He knew that "his statistical interpretation" was based entirely on the work of Einstein, who generously gave Born credit, partly because of his doubts about any theory in which "God plays dice!"

Two-slit experiment. A "gedanken" experiment in the 1920's, but a real experiment today, exhibits the combination of wave and particle properties.

Note that what the two-slit experiment really shows is

- first, the wave function deterministically and *continuously* exploring all the possibilities for interaction, its values determined by the boundary conditions of the experiment.

- second, the particle randomly and *discontinuously* chooses one of those possibilities to become actual. In Einstein's "objective reality" view, the particle goes through one slit, and the wave function, being different when two slits are open, guides the particle to display the two-slit interference pattern.

Measurement problem. There are actually at least three definitions of the measurement problem not normally associated with the Copenhagen Interpretation..

1) The claim that the two dynamical laws, unitary deterministic time evolution according to the Schrödinger equation and indeterministic collapse according to Dirac's projection postulate are logically inconsistent. They cannot both be true, it's claimed.

The proper interpretation is simply that the two laws apply at different times in the evolution of a quantum object, one for possibilities, the other for an actuality (as Heisenberg knew):

- first, the unitary deterministic evolution moves through space exploring all the possibilities for interaction, or may simply be defined at all positions by the boundary conditions of an experiment.
- second, the indeterministic collapse randomly (acausally) selects one of those possibilities to become actual.

2) The original concern that the "collapse dynamics" (von Neumann Process 1) is not part of the formalism (von Neumann Process 2) but an *ad hoc* element, with no rules for when to apply it.

If there was a deterministic law that predicted a collapse, or the decay of a radioactive nucleus, it would not be quantum mechanics!

3) Decoherence theorists (chapter 34) define the measurement problem as the failure to observe macroscopic superpositions, for example, Schrödinger's Cat (chapter 28).

Opposition to the Copenhagen Interpretation

Einstein, de Broglie, and especially Schrödinger insisted on a more "complete" picture, not merely what can be said, but what we can "see," a visualization (*Anschaulichkeit*) of the microscopic world. But de Broglie and Schrödinger's emphasis on the wave picture made it difficult to understand material particles and their "quantum jumps." Indeed, Schrödinger and more recent physicists like John Bell and the decoherence theorists H. D. ZEH and WOJCIECH ZUREK deny the existence of particles and the collapse of the wave function.

Perhaps the main claim of those today denying the Copenhagen Interpretation (as well as standard quantum mechanics) is that "there are no quantum jumps." Decoherence theorists and others favoring HUGH EVERETT's Many-Worlds Interpretation reject Dirac's projection postulate, a cornerstone of quantum theory.

Heisenberg had initially insisted on his own "matrix mechanics" of particles and their discrete, discontinuous, indeterministic behavior, the "quantum postulate" of unpredictable events that undermine the classical physics of causality. But Bohr told Heisenberg that his matrix mechanics was too narrow a view of the problem. The "complementary" wave picture must be included, Bohr insisted. This greatly disappointed Heisenberg and almost ruptured their relationship. But Heisenberg came to accept the criticism and he eventually endorsed all of Bohr's deeply philosophical view that quantum reality is unvisualizable.

In his September Como Lecture, a month before the 1927 Solvay conference, Bohr introduced his theory of "complementarity" as a "complete" theory. It combines the contradictory notions of wave and particle. Since both are required, they complement (and "complete") one another, he thought.

Although Bohr is often credited with integrating the dualism of waves and particles, it was Einstein who predicted a "fusion" of these would be necessary as early as 1909. But in doing so, Bohr obfuscated further what was already a mysterious picture. How could something possibly be both a discrete particle and a continuous wave? Did Bohr endorse the continuous deterministic wave-mechanical views of Schrödinger? Not exactly, but that Bohr accepted Schrödinger's wave mechanics as equal to and complementing his matrix mechanics was most upsetting to Heisenberg.

Bohr had astonished Heisenberg by deriving (in Bohr's Como Lecture) the uncertainty principle from the space-time wave picture alone, with no reference to the causal dynamics of Heisenberg's picture! After this, Heisenberg did the same derivation in his 1930 text and subsequently completely accepted complementarity. Heisenberg spent the next several years widely promoting Bohr's views to scientists and philosophers around the world.

Bohr said these contradictory pictures were "complementary" and that both were needed for a "complete" picture. He vigorously denied Einstein's claim that quantum mechanics is "incomplete," despite Bohr's acceptance of the fact that simultaneous knowledge of exact position and momentum is impossible. Classical physics has twice the number of precisely knowable variables (and thus twice the information) as quantum physics. In this sense, classical physics seems more "complete," quantum physics "incomplete."

Many critics of Copenhagen thought that Bohr deliberately embraced logically contradictory notions - of *continuous* deterministic waves and *discrete* indeterministic particles - perhaps as evidence of the Kantian "antinomies" that put limits on reason and human knowledge. These "contradictions" only strengthened Bohr's epistemological resolve and his insistence that physics requires a subjective view unable to reach Einstein's "objective reality" - the Kantian "things in themselves."

Subject and object were prominent examples of Bohr's complementarity. As Heisenberg described it in his 1955 explanation of the Copenhagen Interpretation

> This again emphasizes a subjective element in the description of atomic events, since the measuring device has been constructed by the observer, and we have to remember that what we observe is not nature in itself but nature exposed to our method of questioning. [3]

Some critics object to the idea that the "free choice" of the experimenter determines what properties appear, but this is correct. If we measure the z-component of spin, we get a definite answer for z, and know nothing about x- or y-components.

Key objections to the Copenhagen Interpretation include:

- The many unreasonable philosophical claims for "complementarity," e.g., that it solves the mind-body problem?
- The basic "subjectivity" of the Copenhagen interpretation. It deals with epistemological knowledge of things, rather than the objectively real "things themselves."
- Bohr's strong claim that there is no quantum world, or at least that we can know nothing about it.
- The idea that nothing exists until an observer measures it.

There is in fact only one world. It is a quantum world. Ontologically it is *indeterministic*, but epistemically, common sense and everyday experience inclines us to see it as only *adequately deterministic*.

Bohr and Heisenberg's Copenhagen Interpretation insists we use classical (deterministic?) concepts and everyday language to communicate our knowledge about quantum processes.

This may be a desirable goal when we begin to teach lay persons about the mysteries of quantum mechanics, but there comes a time when our deeper goal is for them to learn about the nature of the "objective reality" that Einstein wanted us to see.

---

3   Heisenberg, 1955, p. 58

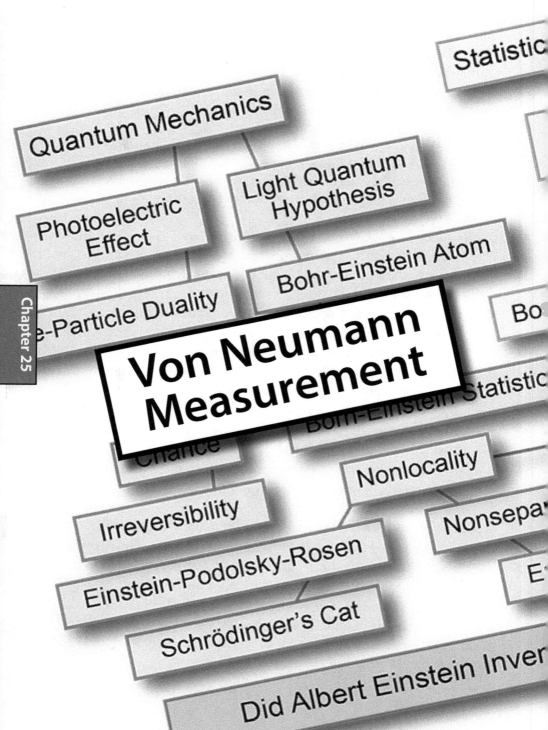

Statistic

Quantum Mechanics

Light Quantum
Hypothesis

Photoelectric
Effect

Bohr-Einstein Atom

e-Particle Duality   Bo

Von Neumann
Measurement

Chance   Born-Einstein Statistic

Nonlocality

Irreversibility

Nonsepa

Einstein-Podolsky-Rosen

E

Schrödinger's Cat

Did Albert Einstein Inver

Von Neumann Measurement

In his 1932 *Mathematical Foundations of Quantum Mechanics* (in German, English edition 1955), JOHN VON NEUMANN explained that two fundamentally different processes are going on in quantum mechanics (in a temporal sequence for a given particle - not happening at the same time).

Process 1. A non-causal process, in which the measured electron jumps randomly into one of the possible physical states (eigenstates) of the measuring apparatus plus electron.

The probability for each eigenstate is given by the square of the coefficients c_n of the expansion of the original system state (wave function ψ) in an infinite set of wave functions φ that represent the eigenfunctions of the measuring apparatus plus electron.

The coefficients $c_n = <\varphi_n | \psi>$.

As we saw in chapter 19, this is PAUL DIRAC's *principle of superposition*. c_n^2 is the probability that the electron will be found in the nth eigenstate. This is Dirac's *projection postulate*. When measured it is found to have the eigenvalue corresponding to that eigenstate. This is Dirac's *axiom of measurement*.

This is as close as we get to a description of the motion of the particle aspect of a quantum system. According to von Neumann, the particle simply shows up somewhere as a result of a measurement. Exact predictions for an individual particle are not possible,. This is why Einstein called quantum mechanics *incomplete*.

Information physics says that for a particle to show up, a new stable information structure must be created, information that may be observed only *after* it has been created (recorded).

Process 2. A causal process, in which the electron wave function ψ evolves deterministically according to ERWIN SCHRÖDINGER's wave equation of motion,

$(ih/2\pi)\, \partial\psi/\partial t = H\psi.$

This evolution describes only the motion of the probability amplitude wave ψ between measurements. The individual particle

Chapter 25

path itself can not be observed. It it were, new information from the measurement would require a new wave function.

MAX BORN had concisely described these two processes years earlier. "The motion of the particle follows the laws of probability, but the probability itself propagates in accord with causal laws." [1]

Von Neumann claimed there is a major difference between these two processes. Process 1 is thermodynamically *irreversible*. Process 2 is reversible. But only when it describes a time during which the particle has no known interactions. Any interactions destroy the "coherence" of the wave functions.

Information physics establishes that *indeterministic* process 1 may create stable new information. An irreversible process 1 is always involved when new information is created. In chapter 12, we showed that the irreversibility of microscopic processes depends on the interaction between matter and radiation.

Process 2 is *deterministic* and information preserving or conserving. But process 2 is an idealization. It assumes that deterministic laws of motion exist. These are differential equations describing continuous quantities. As Born emphasized, continuous quantities evolving deterministically are only probabilities!

Process 1 has come to be called the "collapse of the wave function" or the "reduction of the wave packet." It gave rise to the so-called "problem of measurement," because its randomness prevents it from being a part of the deterministic mathematics of process 2. According to von Neumann, the particle simply shows up somewhere as a result of a measurement. Einstein described these very processes in his 1905 work on the photoelectric effect.

Information physics says that the particle "shows up" only when a new stable information structure is created, information that subsequently can be observed. We might then add an additional condition to process 1.

Process 1b. Note that the information created in Von Neumann's Process 1 will only be stable if an amount of positive entropy greater than the negative entropy in the new information structure is transported away, in order to satisfy the second law of thermodynamics.

---

1   "Quantum mechanics of collision processes," *Zeit. f. Phys.* 1926, p.804

The Measurement Problem

The original problem, said to be a consequence of NIELS BOHR's "Copenhagen Interpretation" of quantum mechanics, was to explain how our measuring instruments, which are usually macroscopic objects and treatable with classical physics, can give us information about the microscopic world of atoms and sub-atomic particles like electrons and photons.

Bohr's idea of "complementarity" insisted that a specific experiment could reveal only partial information - for example, a particle's position. "Exhaustive" information requires complementary experiments, for example to also determine a particle's momentum (within the limits of WERNER HEISENBERG's indeterminacy principle).

Von Neumann's measurement problem is the logical contradiction between his two processes describing the time evolution of quantum systems; the unitary, continuous, deterministic, and information-conserving Schrödinger equation versus the non-unitary, discontinuous, indeterministic and information-creating collapse of the wave function.

The mathematical formalism of quantum mechanics provides no way to predict when the wave function stops evolving in a unitary fashion and collapses. Experimentally and practically, however, we can say that this occurs when the microscopic system interacts with a measuring apparatus. The Russian physicists Lev Landau and Evgeny Lifshitz described it in their 1958 textbook Quantum Mechanics"

> The possibility of a quantitative description of the motion of an electron requires the presence also of physical objects which obey classical mechanics to a sufficient degree of accuracy. If an electron interacts with such a "classical object", the state of the latter is, generally speaking, altered. The nature and magnitude of this change depend on the state of the electron, and therefore may serve to characterise it quantitatively...
>
> We have defined "apparatus" as a physical object which is governed, with sufficient accuracy, by classical mechanics.

Such, for instance, is a body of large enough mass...

Thus quantum mechanics occupies a very unusual place among physical theories: it contains classical mechanics as a limiting case [correspondence principle], yet at the same time it requires this limiting case for its own formulation. [2]

The Measurement Apparatus

The apparatus must allow different components of the wave function to evolve along distinguishable paths into different regions of space, where the different regions correspond to (are correlated with) the physical properties we want to measure. We then can locate a detector in these different regions of space to catch particles travelling a particular path.

We do not say that the system is on a particular path in this first step. That would cause the probability amplitude wave function to collapse. This first step is reversible, at least in principle. It is deterministic and an example of von Neumann process 2.

Let's consider the separation of a beam of photons into horizontally and vertically polarized photons by a birefringent crystal.

We need a beam of photons (and the ability to reduce the intensity to a single photon at a time). Vertically polarized photons pass straight 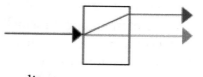 through the crystal. They are called the ordinary ray.

Horizontally polarized photons, however, are deflected at an angle up through the crystal, then exit the crystal back at the original angle. They are called the extraordinary ray.

Note that this first part of our apparatus accomplishes the separation of our two states into distinct physical regions.

We have not actually measured yet, so a single photon passing through our measurement apparatus is described as in a linear combination (a superposition) of horizontal and vertical polarization states,

$$|\psi> = (1/\sqrt{2}) |h> + (1/\sqrt{2}) |v> \qquad (1)$$

2   *Quantum Mechanics*, Lev Landau and Evgeny Lifshitz, pp.2-3

To show that von Neumann's process 2 is reversible, we can add a second birefringent crystal upside down from the first, but inline with the superposition of physically separated states,

Since we have not made a measurement and do not know the path of the photon, the phase information in the (generally complex) coefficients of equation (1) has been preserved, so when they combine in the second crystal, they emerge in a state identical to that before entering the first crystal (final arrow).

We can now create an information-creating, irreversible example of process 1. Suppose we insert something between the two crystals that is capable of a measurement to produce observable information. We need detectors, for example two charge-coupled devices that locate the photon in one of the two rays.

We can write a quantum description of the CCDs, one measuring horizontal photons, $|A_h>$ (the upper extraordinary ray), and the other measuring vertical photons, $|A_v>$ (passing straight through).

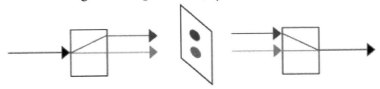

We treat the detection systems quantum mechanically, and say that each detector has two eigenstates, e.g., $|A_{h0}>$, corresponding to its initial state and correlated with no photons, and the final state $|A_{h1}>$, in which it has detected a horizontal photon.

When we actually detect the photon, say in a horizontal polarization state with statistical probability 1/2, there are two "collapses" or "quantum jumps" that occur.

The first is the jump of the probability amplitude wave function $|\psi>$ of the photon in equation (1) into the horizontal state $|h>$.

The second is the quantum jump of the horizontal detector from $|A_{h0}>$ to $|A_{h1}>$. These two happen together, as the quantum states have become correlated with the states of the sensitive detectors in the classical apparatus.

One can say that the photon has become entangled with the sensitive horizontal detector area, so that the wave function describing their interaction is a superposition of photon and apparatus states that cannot be observed independently.

$$|\psi> + |A_{h0}> \quad => \quad |\psi, A_{h0}> \quad => \quad |h, A_{h1}>$$

These jumps destroy (unobservable) phase information, raise the (Boltzmann) entropy of the apparatus, and increase visible information (Shannon entropy) in the form of the visible spot. The entropy increase takes the form of a large chemical energy release when the photographic spot is developed (or a cascade of electrons in a CCD).

Note that the birefringent crystal and the parts of the macroscopic apparatus other than the sensitive detectors are treated classically.

We see that our example agrees with von Neumann. A measurement which finds the photon in a specific polarization state is thermodynamically irreversible, whereas the deterministic evolution described by Schrödinger's equation is time reversible and can be reversed experimentally, provided no decohering interaction occurs.

We thus establish a clear connection between a measurement, which increases the information by some number of bits (negative Shannon entropy), and the compensating increase in the (positive Boltzmann) entropy of the macroscopic apparatus, needed to satisfy the second law of thermodynamics.

Note that the Boltzmann entropy can be radiated away (ultimately into the night sky to the cosmic microwave background) only because the expansion of the universe, discovered by Einstein, provides a sink for the positive entropy.

The *Schnitt* and Conscious Observer

Von Neumann developed WERNER HEISENBERG's idea that the collapse of the wave function requires a "cut" (*Schnitt* in German) between the microscopic quantum system and the observer. He said it did not matter where this cut was placed, because the mathematics would produce the same experimental results.

There has been a lot of controversy and confusion about this cut. EUGENE WIGNER placed it outside a room which includes the measuring apparatus and an observer A, and just before observer B makes a measurement of the physical state of the room, which is imagined to evolve deterministically according to process 2 and the Schrödinger equation.

Von Neumann contributed a lot to this confusion in his discussion of subjective perceptions and "psycho-physical parallelism." He wrote:

> [I]t is a fundamental requirement of the scientific viewpoint -- the so-called principle of the psycho-physical parallelism -- that it must be possible so to describe the extra-physical process of the subjective perception as if it were in reality in the physical world -- i.e., to assign to its parts equivalent physical processes in the objective environment, in ordinary space.

> In a simple example, these concepts might be applied about as follows: We wish to measure a temperature. If we want, we can pursue this process numerically until we have the temperature of the environment of the mercury container of the thermometer, and then say: this temperature is measured by the thermometer. But we can carry the calculation further, and from the properties of the mercury, which can be explained in kinetic and molecular terms, we can calculate its heating, expansion, and the resultant length of the mercury column, and then say: this length is seen by the observer.

> Going still further, and taking the light source into consideration, we could find out the reflection of the light quanta on the opaque mercury column, and the path of the remaining light quanta into the eye of the observer, their refraction in the eye lens, and the formation of an image on the retina, and then we would say: this image is registered by the retina of the observer.

> And were our physiological knowledge more precise than it is today, we could go still further, tracing the chemical reactions which produce the impression of this image on the retina, in the optic nerve tract and in the brain, and then in the end say: these chemical changes of his brain cells are perceived by the observer. But in any case, no matter how far we calculate -- to the mercury vessel, to the scale of the thermometer, to the retina, or into the

brain, at some time we must say: and this is perceived by the observer. That is, we must always divide the world into two parts, the one being the observed system, the other the observer...

The boundary between the two is arbitrary to a very large extent... That this boundary can be pushed arbitrarily deeply into the interior of the body of the actual observer is the content of the principle of the psycho-physical parallelism -- but this does not change the fact that in each method of description the boundary must be put somewhere, if the method is not to proceed vacuously, i.e., if a comparison with experiment is to be possible. Indeed experience only makes statements of this type: an observer has made a certain (subjective) observation; and never any like this: a physical quantity has a certain value.

Now quantum mechanics describes the events which occur in the observed portions of the world, so long as they do not interact with the observing portion, with the aid of the process 2, but as soon as such an interaction occurs, i.e., a measurement, it requires the application of process 1. The dual form is therefore justified. However, the danger lies in the fact that the principle of the psycho-physical parallelism is violated, so long as it is not shown that the boundary between the observed system and the observer can be displaced arbitrarily in the sense given above. [3]

Information physics places the von Neumann/Heisenberg cut or boundary at the place and time of information creation. It is only *after* information is created that an observer could make an observation. Beforehand, there is no information to be observed.

Just as the new information recorded in the measurement apparatus cannot subsist unless a compensating amount of entropy is transferred away from the new information, something similar to Process 1b must happen in the mind of an observer if the new information is to constitute an "observation."

It is only in cases where information persists long enough for a human being to observe it that we can properly describe the observation as a "measurement" and the human being as an "observer." So, following von Neumann's "process" terminology, we can complete his theory of the measuring process by adding an anthropomorphic third process...

3   *The Mathematical Foundations of Quantum Mechanics*, pp. 418-21

Process 3 - a conscious observer recording new information in a mind. This is only possible if there are two local reductions in the entropy (the first in the measurement apparatus, the second in the mind), both balanced by even greater increases in positive entropy that must be transported away from the apparatus and the mind, so the overall increase in entropy can satisfy the second law of thermodynamics.

For some physicists, it is the wave-function collapse that gives rise to the "problem" of measurement because its randomness prevents us from including it in the mathematical formalism of the deterministic Schrödinger equation in process 2.

Information creation occurs as a result of the interaction between the indeterministic microscopic system and the adequately deterministic measuring apparatus. It is a severe case of anthropomorphism to think it requires the consciousness of an observer for the wave function itself to collapse.

The collapse of a wave function and information creation has been going on in the universe for billions of years before human consciousness emerged. The cosmic information-creating process requires no conscious observer. *The universe is its own observer.*

It is enough that the new information created is observable and stable, so that a human observer can look at it in the future. Information physics is thus subtly involved in the question of what humans can know (epistemology).

Many scientists and philosophers deny von Neumann's process 1, the collapse of the wave function (also PAUL DIRAC's projection postulate), claiming that the Schrödinger equation is all that is needed to describe a "unitary," information-conserving evolution of the "wave function of the universe." But in such a universe, nothing ever happens.

Information physics solves the problem of measurement by identifying the moment and place of the collapse of the wave function with the creation of a potentially observable information structure. Some interactions between matter and radiation create irreversible collapses but do not produce information structures that last long enough to be observed. These can never be the basis of measurements of "observables" by physicists.

Chapter 25

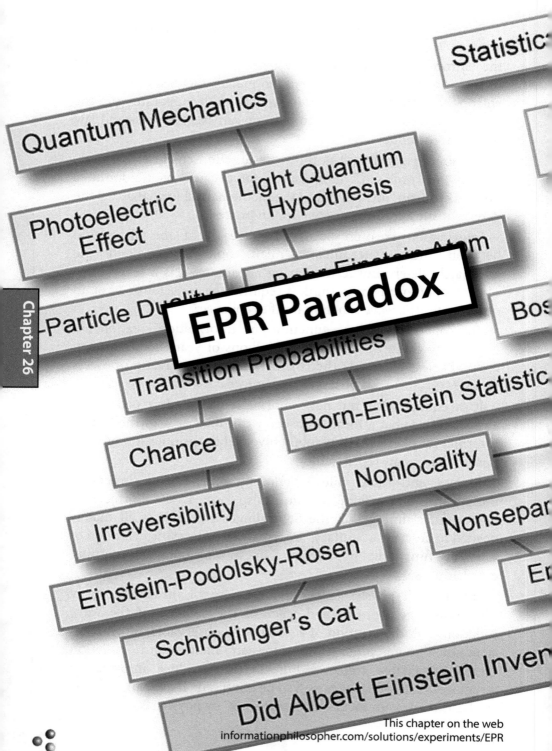

Statistic

Quantum Mechanics

Light Quantum
Hypothesis

Photoelectric
Effect

Bohr-Einstein Atom

Particle Duality

EPR Paradox

Bos

Transition Probabilities

Born-Einstein Statistic

Chance

Nonlocality

Irreversibility

Nonsepar

Einstein-Podolsky-Rosen

Er

Schrödinger's Cat

Did Albert Einstein Inven

This chapter on the web
informationphilosopher.com/solutions/experiments/EPR

Einstein-Podolsky-Rosen

The 1935 paper, "Can Quantum-Mechanical Description of Physical Reality Be Considered Complete?" by ALBERT EINSTEIN, BORIS PODOLSKY, and NATHAN ROSEN (and known by their initials as EPR) was originally proposed to exhibit internal contradictions in the new quantum physics.

Einstein's greatest scientific biographer, Abraham Pais, concluded in 1982 that the EPR paper "had not affected subsequent developments in physics, and it is doubtful that it ever will."[1]

This may have been the worst scientific prediction ever made, as EPR is identified today as the basis for the "second revolution in quantum mechanics." EPR has led us to exponentially more powerful quantum computing, ultra-secure quantum cryptography and quantum communications, and the *entangled* states that offer the exotic possibility of quantum teleportation.

Although many thousands of articles have been written analyzing the EPR paper, it is fair to say that no one has ever explained exactly what Einstein was worried about. The first and most famous reply was that of NIELS BOHR, who did not have a clue. Bohr just repeated his defense of the uncertainty principle and his philosophical notion of *complementarity*.

The EPR paper was obscure even to Einstein. It was written in English, which Einstein was just beginning to learn, by Podolsky, whose native language was Russian, and by Rosen, whose main contribution was an attack on the uncertainty principle, where Einstein had himself accepted uncertainty five years earlier.

For Einstein, uncertainty can be seen as a consequence of the *statistical* nature of quantum mechanics. Bohr and WERNER HEISENBERG had considered the possibility that uncertainty might be an *epistemological* limit on our knowledge due to the limiting resolving power of our measuring instruments.

In earlier times Einstein argued that an individual particle might "objectively" have simultaneous values for position and momentum even if quantum measurements, being statistical, can only estimate values as averages over many measurements. The

---

1   Pais, 1982, p.456

statistical deviations Δp and Δx around the mean values give us the uncertainty principle $\Delta p \Delta x = h/2\pi$.

In the EPR paper, Einstein argued that its statistical character makes quantum mechanics an *incomplete* theory relative to "*objectively real*" classical mechanics, where the outcome of a measurement is *independent* of the observer.

The EPR authors hoped to show that quantum theory could not describe certain "*elements of reality*" and thus was either *incomplete* or, as they may have hoped, demonstrably incorrect.

> the following requirement for a complete theory seems to be a necessary one: every element of the physical reality must have a counterpart in the physical theory. We shall call this the condition of completeness.
>
> We shall be satisfied with the following criterion, which we regard as reasonable. *If, without in any way disturbing a system, we can predict with certainty {i.e., with probability equal to unity) the value of a physical quantity, then there exists an element of physical reality corresponding to this physical quantity.*[2]

Using Heisenberg's uncertainty principle, the EPR authors wrote, "when the momentum of a particle is known, its coordinate has no physical reality." But if both momentum and position had simultaneous reality—and thus definite values—"these values would enter into the complete description, according to the condition of completeness."[3]

NIELS BOHR and his Copenhageners took this "incompleteness" as just one more of Einstein's attacks on quantum mechanics, especially its uncertainty principle.

Einstein shortly later gave an "objectively real" example of incompleteness that even a third grader can understand. Imagine you have two boxes, in one of which there is a ball. The other is empty. An *incomplete* statistical theory like quantum mechanics says, "the probability is one-half that the ball is in the first box." An example of a *complete* theory is "the ball *is* in the first box."[4]

---

2   Einstein, Podolsky, Rosen, 1935, p.777
3   *ibid.* p.778
4   June 19, 1935 letter to Schrödinger. See also Fine, 1996, p.36 and p.69.

Here Einstein is criticizing the Copenhagen Interpretation's use of PAUL DIRAC's principle of superposition, which we saw in chapter 19 is easily misinterpreted. Dirac suggests that we might *speak as if* a single particle is partly in each of the two states, that the ball above is "distributed" over Einstein's two boxes.

Dirac's "manner of speaking" gives the false impression that the single ball can actually be in the two boxes at the same time. This is seriously misleading. Dirac expressed the concern that some would be misled - don't "give too much meaning to it," he said.

Two Places or Paths at the Same Time?

Einstein's Boxes were his criticism of the most outlandish claim of the "orthodox" Copenhagen Interpretation, that particles can be in two places at the same time and move simultaneously along different paths. The square of the wave function Ψ^2 gives us the probability of finding a particle in different places. Specifically, this means that when we do many identical experiments, we find the statistics of many different places and paths agrees perfectly with the probabilities. But in each individual experiment, we always find the whole particle in a single place!

Einstein's Boxes example also criticizes the idea that particles do not even exist until they are measured by some observer. Einstein said, sarcastically, "Before I open them, the ball is not in *one* of the two boxes. Being in a definite box only comes about when I lift the covers."[5] Einstein used his conservation principles to argue that a particle can not go in and out of existence, split into two, or jump around arbitrarily violating conservation of momentum.

A third tenet of the Copenhagen Interpretation that Einstein criticized is that the properties of a particle are not determined in advance of measurement. Properties are sometimes random or indeterministic, and in some sense determined by the observer, where for Einstein real objects have properties *independent of the observer*. Where his first two criticisms above were accurate, and flaws in the standard interpretation of quantum mechanics, this criticism was in part one of Einstein's mistakes.

---

5   Fine, 1996, p.69.

Einstein's fourth and most revolutionary criticism leads directly to entanglement and the "second revolution" in quantum mechanics. This is what he described as *nonlocality* and *nonseparability*.

Einstein's fundamental concern in the EPR paper was not *incompleteness*, which caught Bohr's attention.. It was *nonlocality*, which had been on Einstein's mind for many years, but Bohr never understood what Einstein was talking about, as we saw in chapter 23. Nonlocality challenged Einstein's special relativity and his claims about the *impossibility of simultaneity*.

Two years before EPR, and just before Einstein left Europe forever in 1933, he attended a lecture on quantum electrodynamics by LEON ROSENFELD.[6] Keep in mind that Rosenfeld was perhaps the most dogged defender of the Copenhagen Interpretation. After the talk, Einstein asked Rosenfeld, "What do you think of this situation?"

> Suppose two particles are set in motion towards each other with the same, very large, momentum, and they interact with each other for a very short time when they pass at known positions. Consider now an observer who gets hold of one of the particles, far away from the region of interaction, and measures its momentum: then, from the conditions of the experiment, he will obviously be able to deduce the momentum of the other particle. If, however, he chooses to measure the position of the first particle, he will be able tell where the other particle is.

We can diagram a simple case of Einstein's question as follows.

Two particles moving with equal and opposite momentum leave the circle of interaction (later "entanglement") in the center. Given the position of one particle, the position of the second particle must be exactly the same distance on the other side of the center.

Measuring one particle tells you something about the other particle, now assumed to be at a large spacelike separation. Does that knowledge require information to travel faster than light? No.

---

6   Lahti and Mittelstaedt, 1985, p.136

Einstein asked Rosenfeld, "How can the final state of the second particle be influenced by a measurement performed on the first after all interaction has ceased between them?" This was the germ of the EPR paradox, and ultimately the problem of two-particle entanglement.

Why does Einstein question Rosenfeld and describe this as an "influence," suggesting an "action-at-a-distance?"

It might be paradoxical in the context of Rosenfeld's Copenhagen Interpretation, since the second particle is not itself measured and yet we know something about its properties, which Copenhagen says we cannot know without an explicit measurement..

The second particle must have knowable properties. When we measure the first particle, we learn its momentum. By conservation laws, we know the second particle's equal and opposite momentum, and this means that we can know its position. How does Rosenfeld explain this? We do not know his answer.

Nonlocality in 1905 and 1927 involved only one particle and the mysterious influence of the probability wave. But in the EPR paper Einstein has shown nonlocal effects between *two* separated particles.

Einstein's basic concern was that particles now very far apart may still share some common information, so that looking at one tells us something about the other. And it tells us instantly, faster than the speed of light.

He later called nonlocality *"spukhaft Fernwirkung"* or *"spooky action-at-a-distance."*[7] But calculating and predicting the position and momentum of a distant particle based on conservation principles is better described as *"knowledge-at-a-distance."*

There is no "action," in the sense of one particle changing the properties of the other.

But Einstein's idea of a measurement in one place "influencing" measurements far away challenged what he thought of as "local reality." These "influences" *appear* to be nonlocal.

What is it Einstein saw? What was Einstein worried about? We have been arguing that it challenged the *impossibility of simultaneity* implied by his theory of special relativity.

---

7   Born, 1971, p.155

Note that Einstein knew nothing of the simultaneous spin or polarization measurements by Alice and Bob that constitute modern entanglement experiments. But Einstein's insight into the guiding field of the probability wave function can be applied to both entanglement and the two-slit experiment, in which case it might solve two mysteries with one explanation.

It will show Einstein was wrong about the "impossibility" of simultaneity, but like many of his mistakes, gives us a deep truth.

Is Quantum Mechanics Complete or Incomplete?

NIELS BOHR had strong reasons, mostly philosophical, for defending completeness. For one thing, his idea of complementarity claimed to have found the two complementary sides of all dualisms that combine to explain the wholeness of the universe.

But also, Bohr was a great admirer of the *Principia Mathematica* of BERTRAND RUSSELL and ALFRED NORTH WHITEHEAD, which claimed to be a "complete" system of propositional logic. This claim was challenged by GOTTLOB FREGE's linguistic puzzles about sense and reference[8] and by Russell's own famous "paradox." But even more devastating was KURT GÖDEL's 1931 theorems about inconsistency and incompleteness in mathematics.

Gödel visited the Institute for Advanced Study in 1933 and developed a lifelong friendship with Einstein. In 1934 Gödel gave a lecture series on undecidable propositions. Einstein, and probably Podolsky and Rosen, attended. Incompleteness, in the form of limits on knowledge, was in the air.

Heisenberg's uncertainty principle can be understood as an epistemological limit, where Einstein's goal was an ontological understanding of the objectively real. Any measurement apparatus uses an electromagnetic interaction to locate a material particle, so it is limited by the finite wavelength of the light used to "see" the particle. In his 1927 Como lecture, Bohr embarrassed Heisenberg by deriving his uncertainty principle on the basis of light waves alone, which limit the so-called "resolving power" of any instrument.

---

8   Doyle, 2016b, p.241

Einstein may well have continued to believe that a real particle actually has precise properties like position and momentum, but that quantum measurements are simply unable to determine them. Heisenberg also called his principle *indeterminacy*.

What Einstein wanted to "complete" quantum mechanics was more information about the paths and properties of individual systems between measurements. The Copenhagen Interpretation dogmatically insisted that nothing can be known about quantum particles and their paths until they are measured.

That its position cannot be known can not justify the claim that a particle can therefore be anywhere, or have no position. For example, that it can be in multiple places at the same time, as the principle of superposition of probabilities mistakenly suggests. This was explained by PAUL DIRAC as just a "manner of speaking."

As we saw in chapter 19, Einstein perfectly understood Dirac's superposition principle as our inability to say whether a particular photon will pass a polarizer or not, although we can predict the *statistics* of photons passing through with high accuracy.

Einstein might have seen this randomness as connected to his 1916 discovery of ontological chance, and so might not have liked it.

Dirac called this inability to predict a path "Nature's choice." It is randomness or chance beyond the control of an experimenter.

By contrast to Dirac, Heisenberg insisted on what he and Bohr called the "free choice" of the experimenter, for example whether to measure for the position or the momentum of a particle. Einstein might well have endorsed this freedom as supporting his belief in the "free creations of the human mind."

In the EPR paper, the authors mention that we can freely choose to measure the first particle's momentum or its position.

Copenhagen is correct that we cannot know the instantaneous details of a particle's path and properties without continuous measurements during its travel, but we can use conservation laws and symmetry to learn something about a path *after the fact* of a measurement.

Chapter 26

Back to EPR, after the measurement on the first particle, conservation laws give us "knowledge-at-a-distance" about the second particle. With this knowledge, we can retrospectively construct the path of the second particle.

Because of its perceived "incompleteness," Einstein mistakenly suggested that "additional variables" might be needed in quantum mechanics. In chapter 30, we will see that in 1952 DAVID BOHM added a faster-than-light vector potential to make what Einstein thought were nonlocal events possible and to restore classical physical determinism to quantum mechanics.

Bohm also proposed an improved EPR experiment using *discrete* electron spins rather than *continuous* momentum values. Today the Bohm version has become the standard presentation of the EPR experiment, using either spin-1/2 material particles or spin-1 light particles (photons). The spatial components of spin values that are observed provide canonical examples of both Heisenberg's "free choice of the experimenter" and Dirac's "Nature's choice," neither of which was a part of Einstein's original concerns.

If we freely choose to measure electron spin in the z-direction, our choice brings the z-direction components into existence. The x- and y-components are indeterminate. Heisenberg was right. The experimenter has a "free choice."

But the particular value of the z-component is random, either +1/2 or -1/2. So Dirac was also right. This is "Nature's choice." Now this randomness is sometimes criticized as rendering all events *indeterministic* and the results of mere chance. It is said to threaten reason itself.

If events are really uncaused, some fear that scientific explanations would be impossible. In 1927, Heisenberg said that his quantum mechanics had introduced *acausality* into nature. He thought it might contribute to human freedom. But he did not seem to know that in 1916 Einstein discovered ontological chance when matter and radiation interact. Einstein's ontological chance is physically and metaphysically much deeper than Heisenberg's epistemological uncertainty.

EPR in the 21st Century.

The next six chapters describe how Einstein's radical ideas about nonlocality and nonseparability morph into the "second revolution" in quantum mechanics.

It is a story of twists and turns, which began with Einstein seeing "action-at-a-distance" between the continuous light wave spread out everywhere and the discrete light quantum detected at a particular spot on a screen (chapter 23).

In the EPR article, Einstein insisted this "action-at-a-distance" must be impossible once the particles separate far enough so they no longer can interact.

In later 1935, ERWIN SCHRÖDINGER reacted to Einstein's separability principle by saying that the "entangled" particles could not be separated as long as they did not interact with other particles (see chapters 27 and 28).

In 1952 Bohm proposed a new test of nonseparability could be done using electron spins. Bohm argued for a return to deterministic physics, which he thought Einstein wanted.

Twelve years later, JOHN BELL developed a theorem to distinguish between standard quantum mechanics, including Schrödinger's entanglement, and what Bell thought was Einstein's idea of a realistic physics and Bohm's determinism.

A few young physicists hoping for a new foundation for quantum mechanics set out to test Bell's theorem experimentally, motivated by the chance their work would invalidate quantum mechanics.

Instead, they found the predictions of quantum mechanics were confirmed, including Einstein's concern that widely separated events could simultaneously acquire new properties.

A pair of entangled particles is now the basis for what is called a "qubit," the elementary piece of data in quantum computing. These two particles are called an "EPR pair," after Einstein, or they are said to be in a "Bell state," after John Bell.

And so Einstein's insight and imagination, even when wrong, continue to this day to produce new science and technology.

This chapter on the web
informationphilosopher.com/problems/nonseparability

Nonseparability

Entangled particles are described by a single two-particle wave function ψ_{12} that cannot be separated into a product of single-particle wave functions ψ_1 and ψ_2 without a measurement or external interaction that "decoheres" or "disentangles" them.

The question for ALBERT EINSTEIN and ERWIN SCHRÖDINGER was how long the particles could retain any correlation as they traveled a great distance apart. Once disentangled, or "decohered," the two-particle wave function Ψ_{12} can be described as the product of two single-particle wave functions Ψ_1 and Ψ_2 and there will no longer be any quantum interference between them. But entangled particles, it turns out, do not decohere spontaneously. They cannot decohere without an external interaction (like a measurement).

Einstein had objected to *nonlocal* phenomena as early as the Solvay Conference of 1927, when he criticized the collapse of the single-particle wave function as involving instantaneous "action-at-a-distance" that looks like the spherical outgoing wave acting at more than one place on the screen. He had seen single-particle nonlocality as early as his light-quantum hypothesis paper of 1905, as we saw in chapter 23. But we showed that the collapse of the mathematical probabilities $|\Psi|^2$ only involved the disappearance of those probabilities. Without matter or energy moving, there is no "action" being exerted on the particle by the wave.

We can now try to understand the nonseparability of two entangled particles in terms of single-particle nonlocality. The entangled particles share one volume of nonlocality, i.e., wherever the two-particle wave function has non-zero values of $|\Psi_{12}|^2$.

Quantum mechanics says that *either* particle has the same possibility (with calculable probability) of appearing at any particular location in this volume. Just as with the single-particle nonlocality, in standard quantum mechanics we cannot say where the two particles "are." Either one may be anywhere up to the moment of "collapse" of the two-particle wave function. But conservation principles require that whenever they finally do appear, it will be equidistant from the origin, in order to conserve linear momentum.

And more importantly, conservation principles and symmetry require that measurements of any particular property of the two particles find that they too are perfectly correlated, as we shall see in chapter 29.

Einstein's "objective reality" assumes that the particles simply have predictable paths from the start of the experiment to the final measurement(s), although the limits of quantum measurement may never allow us to "know" those paths.

It is the fundamental principle of conservation that governs the correlated outcome, not some hypothetical, faster than light, communication of information between the particles.

There are two cases, however, where the final outcomes are unknowable at the start. One is where a random interaction with the environment occurs. In this case said PAUL DIRAC, Nature makes a random choice. The other is WERNER HEISENBERG's "free choice" by the experimenter to intervene. This is the case for a measurement of entangled electrons (spin-1/2 particles) or photons (spin-1 particles), as we will see in the next few chapters.

Separability According to Quantum Theory

Quantum mechanics describes the probability amplitude wave function ψ_{12} of an entangled two-particle system as in a *superposition* of two-particle states.

$$\Psi_{12} = (1/\sqrt{2}) (|+-> - |-+>).$$

It is not separable into a product of single-particle states, and there is no information about individual particles traveling along observable paths.

The Copenhagen Interpretation, claims that quantum systems do not have properties until they are observed. And not merely measured by apparatus that records data. The result of the measurement must reach the *mind* of the experimenter, according to JOHN VON NEUMANN's "psycho-physical parallelism."

Einstein, however, frequently asked whether the particle has a position at the moment before it is measured? "Is the moon only there when we look at it," he quipped. And he famously told the philosopher Hilary Putnam, "Look, I don't believe that when I am not in my bedroom my bed spreads out all over the room, and whenever I open the door and come in it jumps into the corner."

Einstein took the Copenhageners as saying the two particles may actually be anywhere that Ψ_{12} is non-zero, then they jump to places that conserve the momentum only at the measurement.

The particles are thought to be in a superposition of all possible momentum or position eigenstates, as we see in the next chapter.

Now when entangled particles experience a random interaction with something in the environment (described as "decoherence"), or an experimental measurement by an observer, the two-particle wave function "collapses."

In the standard quantum physics view, all the possibilities/probabilities that are not actualized go to zero, just as with the single particle wave function. But now, two particles appear, *simultaneously* in a special frame in which their center of mass is not moving. In other moving frames, either particle may *appear to appear* before the other.

The two particles appear simultaneously, in a spacelike separation, now disentangled, and symmetrically located about the point of the interaction which entangled them.

If they did not appear as symmetrically as they had been at the beginning, both conservation laws and underlying principles of symmetry would be violated.

In Einstein's "objective reality" picture, no faster-than-light signaling is involved. There is no "action" going from one particle to the other. Their linear momenta, correlated at their moment of entanglement, always are correlated "locally" as they travel along at the particles' speed.

The fact that momenta, and most of their properties, are found synchronized, perfectly correlated, at later times, is because they are always correlated until a disturbance occurs, e.g., an interaction with the environment or a measurement by an observer.

It is only once a disentangling interaction occurs with either particle, that further interactions do nothing to the other, as Einstein requires for his *separability principle* (*Trennungsprinzip*).

> But on one supposition we should, in my opinion, absolutely hold fast: the real factual situation of the system S_2 is independent of what is done with the system S_1, which is spatially separated from the former. [1]

---

1   Einstein, 1949a, p.85

Chapter 27

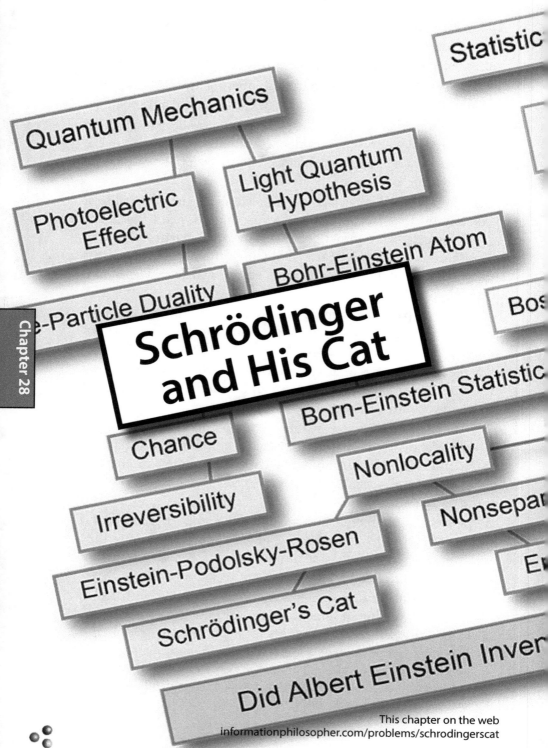

This chapter on the web
informationphilosopher.com/problems/schrodingerscat

Schrödinger and His Cat

A few weeks after the May 15, 1935 appearance of the EPR article in the *Physical Review* in the U.S., ERWIN SCHRÖDINGER wrote to Einstein to congratulate him on his "catching dogmatic quantum mechanics by its coat-tails."

In his EPR paper, Einstein cleverly introduced *two particles* instead of one. Schrödinger gave us a two-particle wave function that describes both particles. The particles are identical, indistinguishable, and with indeterminate positions, although EPR described them as widely separated, one "here" and measurable "now" and the other distant and to be measured "later."

Einstein now shows that the mysterious nonlocality that he first saw when the wave function for a single particle disappears everywhere at the instant the particle is found, can also be happening for two particles. But he maintained that "system S_2 is independent of what is done with the system S_1", as we saw in the last chapter.

Schrödinger, the creator of wave mechanics, surprised Einstein by challenging the idea that two systems that had previously interacted can at some point be treated as *separated*. And, he said, a two-particle wave function ψ_{12} cannot be factored into a product of separated wave functions for each system, ψ_1 and ψ_2.

Einstein called this a "separability principle" (*Trennungsprinzip*). But the particles cannot actually separate until another quantum interaction separates, decoheres, and disentangles them.

Schrödinger published a famous paper defining his idea of "entanglement" a few months later. It began:

> When two systems, of which we know the states by their respective representatives, enter into temporary physical interaction due to known forces between them, and when after a time of mutual influence the systems separate again, then they can no longer be described in the same way as before, viz. by endowing each of them with a representative of its own. I would not call that one but rather the characteristic trait of quantum mechanics, the one that enforces its entire departure from classical lines of thought. By the interaction the two

Chapter 28

representatives (or ψ-functions) have become entangled. They can also be disentangled, or decohered, by interaction with the environment (other particles). An experiment by a human observer is not necessary. To disentangle them we must gather further information by experiment, although we knew as much as anybody could possibly know about all that happened. Of either system, taken separately, all previous knowledge may be entirely lost, leaving us but one privilege: to restrict the experiments to one only of the two systems. After reestablishing one representative by observation, the other one can be inferred simultaneously. In what follows the whole of this procedure will be called the disentanglement...

Attention has recently [viz., EPR] been called to the obvious but very disconcerting fact that even though we restrict the disentangling measurements to one system, the representative obtained for the other system is by no means independent of the particular choice of observations which we select for that purpose and which by the way are entirely arbitrary. It is rather discomforting that the theory should allow a system to be steered or piloted into one or the other type of state at the experimenter's mercy in spite of his having no access to it. This paper does not aim at a solution of the paradox, it rather adds to it, if possible. [1]

Schrödinger says that the entangled system may become disentangled long before any measurements by a human observer. But if the particles continue on undisturbed, they may remain perfectly correlated for long times between measurements. Or they may decohere as a result of interactions with the environment, as proposed by decoherence theorists.

Schrödinger is perhaps the most complex figure in twentieth-century discussions of quantum mechanical uncertainty, ontological chance, indeterminism, and the statistical interpretation of quantum mechanics. His wave function and wave equation are the definitive tool for quantum mechanical calculations. They are of unparalleled accuracy. But Schrödinger's interpretations are extreme and in many ways out-of-step with standard quantum mechanics.

---

1   Schrödinger, 1935, p.555

Schrödinger denies quantum jumps and even the existence of objective particles, imagining them to be packets of his waves. He objects to Einstein's, and later Born's better known, interpretation of his waves as probability amplitudes. He denies uncertainty and is a determinist. His wave equation is deterministic.

Superposition

Schrödinger's wave equation is a *linear* equation. All its variables appear to the first power. This means that the sum of any two solutions to his equation is also a solution.

This property is what lies behind PAUL DIRAC's *principle of superposition* (chapter 19). Any wave function ψ can be a linear combination (or superposition) of multiple wave functions φ_n.

$$\psi = \Sigma_n c_n \varphi_n.$$

The φ_n are interpreted as possible eigenstates of a system, each with an eigenvalue E_n. The probability that the system is in eigenstate φ_n is c_n^2, provided their sum is normalized to unity,

$$\Sigma_n c_n^2 = 1.$$

If a system is in a superposition of two possible states, we can calculate the probabilities that in many experiments c_1^2 of them will be found in state φ_1 and c_2^2 of them will be found in state φ_2.

As Dirac explained, superposition is a mathematical tool that predicts the statistical outcomes of many identical experiments. But an individual system, for example a photon or material particle, is not actually in two states at the same time. Dirac said that's just a "manner of speaking."

> We have obtained a description of the photon throughout the experiment, which rests on a new rather vague idea of a photon being partly in one state and partly in another...
>
> The original state must be regarded as the result of a kind of superposition of the two or more new states, in a way that cannot be conceived on classical ideas...
>
> When we say that the photon is distributed over two or more given states the description is, of course, only qualitative...
>
> We must, however, get used to the new relationships between the states which are implied by this manner of speaking and must build up a consistent mathematical theory governing them.

The description which quantum mechanics allows us to give is merely a manner of speaking which is of value in helping us to deduce and to remember the results of experiments and which never leads to wrong conclusions. One should not try to give too much meaning to it. [2]

Nevertheless, around the time of EPR, Einstein began an attack on Dirac's principle of superposition, which was then amplified by ERWIN SCHRÖDINGER to become two of the greatest mysteries in today's quantum physics, Schrödinger's Cat, and Entanglement.

Before we discuss these, we will look at how Einstein and Schrödinger engaged in a major debate about the two particles in EPR. Can they act on one another "at a distance?" Do they ever separate as independent particles, when they interact with other particles, for example?

Schrödinger's Cat

Schrödinger's goal for his infamous cat-killing box was to discredit certain non-intuitive implications of quantum mechanics, of which his wave mechanics was the second formulation. Schrödinger's wave mechanics is more *continuous* and more deterministic than WERNER HEISENBERG's matrix mechanics.

Schrödinger never liked NIELS BOHR's idea of "quantum jumps" between Bohr's "stationary states" - the different "energy levels" in an atom. Bohr's second "quantum postulate" said that the jumps between discrete states emitted (or absorbed) energy in the amount $hv = E_m - E_n$.

Bohr did not accept ALBERT EINSTEIN's 1905 hypothesis that the emitted radiation is a *discrete* localized particle quantum of energy hv. Until well into the 1920's, Bohr (and MAX PLANCK, himself the inventor of the quantum hypothesis) believed radiation was a *continuous* wave. This was at the root of wave-particle duality, which Einstein saw as early as 1909.

It was Einstein who originated the mistaken suggestion that the *superposition* of Schrödinger's wave functions implies that two different physical states can exist at the same time. As we have seen, it was based on what PAUL DIRAC called a "manner of speaking" that a single system is "distributed" over multiple states. This was

---

2   Dirac, 1930, p.5

a serious interpretational error that plagues the foundation of quantum physics to this day.[3]

We never actually "see" or measure any system (whether a microscopic electron or a macroscopic cat) in two distinct states. Quantum mechanics simply predicts a significant probability of the system being found in these different states. And these probability predictions are borne out by the statistics of large numbers of identical experiments.

Einstein wrote to Schrödinger with the idea that the decay of a radioactive nucleus could be arranged to set off a large explosion. Since the moment of decay is unknown, Einstein argued that the superposition of decayed and undecayed nuclear states implies the superposition of an explosion and no explosion. It does not. In both the microscopic and macroscopic cases, quantum mechanics simply estimates the probability amplitudes for the two cases.

Schrödinger devised a variation of Einstein's provocative idea in which the random radioactive decay would kill a cat. Observers could not know what happened until the box is opened.

The details of the tasteless experiment include:

- a Geiger counter which produces an avalanche of electrons when an alpha particle passes through it
- a bit of radioactive material with a decay half-life likely to emit an alpha particle in the direction of the Geiger counter during a time T
- an electrical circuit energized by the electrons which drops a hammer
- a flask of a deadly hydrocyanic acid gas, smashed open by the hammer.

The gas will kill the cat, but the exact time of death is unpredictable and random because of the irreducible quantum indeterminacy in the time of decay (and the direction of the decay particle, which might miss the Geiger counter!).

---

3   See Dirac's "manner of speaking" in chapter 19.

This thought experiment is widely misunderstood. It was meant (by both Einstein and Schrödinger) to suggest that quantum mechanics describes the simultaneous (and obviously contradictory) existence of a live and dead cat. Here is the famous diagram with a cat both dead and alive.

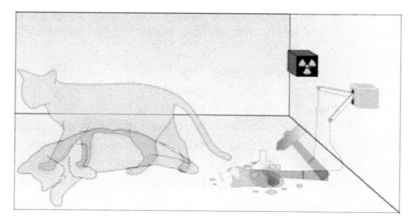

If we open the box at the time T when there is a 50% probability of an alpha particle emission, the most a physicist can know is that there is a 50% chance that the radioactive decay will have occurred and the cat will be observed as dead or dying.

If the box were opened earlier, say at T/2, there is only a 25% chance that the cat has died. Schrödinger's superposition of live and dead cats would look like this.

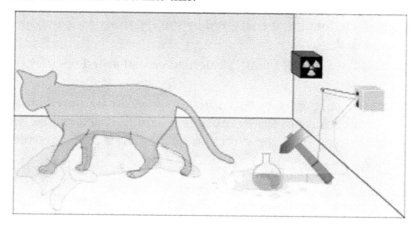

If the box were opened later, say at 2T, there is only a 25% chance that the cat is still alive. Quantum mechanics is giving us only statistical information - knowledge about probabilities.

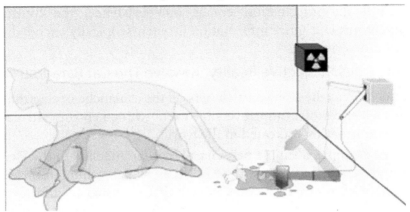

Schrödinger is simply wrong that the mixture of nuclear wave functions accurately describing decay can be magnified to world to describe a macroscopic mixture of live cat and dead cat wave functions and the simultaneous existence of live and dead cats.

Instead of a linear combination of pure quantum states, with quantum interference between the states, i.e.,

| Cat > = (1/√2) | Live > + (1/√2) | Dead >,

quantum mechanics tells us only that there is 50% chance of finding the cat in either the live or dead state, i.e.,

Cats = (1/2) Live + (1/2) Dead.

Just as in the quantum case, this probability prediction is confirmed by the statistics of repeated identical experiments, but no interference between these states is seen.

What do exist simultaneously in the macroscopic world are genuine alternative possibilities for future events. There is the real possibility of a live or dead cat in any particular experiment. Which one is found is irreducibly random, unpredictable, and a matter of pure chance.

Genuine alternative possibilities is what bothered physicists like Einstein, Schrödinger, and MAX PLANCK who wanted a return to deterministic physics. It also bothers determinist and compatibilist philosophers who have what WILLIAM JAMES calls an "antipathy to

Chapter 28

chance." Ironically, it was Einstein himself, in 1916, who discovered the existence of irreducible chance, in the elementary interactions of matter and radiation.

Until the information comes into existence, the future is indeterministic. Once information is macroscopically encoded, the past is determined.

How Does "Objective Reality" Resolve The Cat Paradox?

As soon as the alpha particle sets off the avalanche of electrons in the Geiger counter (an irreversible event with an entropy increase), new information is created in the world.

For example, a simple pen-chart recorder attached to the Geiger counter could record the time of decay, which a human observer could read at any later time. Notice that, as usual in information creation, energy expended by a recorder increases the entropy more than the increased information decreases it, thus satisfying the second law of thermodynamics.

Even without a mechanical recorder, the cat's death sets in motion biological processes that constitute an equivalent, if gruesome, recording. When a dead cat is the result, a sophisticated autopsy can provide an approximate time of death, because the cat's body is acting as an event recorder. There never is a superposition (in the sense of the simultaneous existence) of live and dead cats.

The cat paradox points clearly to the information physics solution to the problem of measurement. Human observers are not required to make measurements. In this case, information is in the cat's body, the cat is the observer.

In most physics measurements, any new information is captured by an apparatus well before any physicist has a chance to read any dials or pointers that indicate what happened. Indeed, in today's high-energy particle interaction experiments, the data may be captured but not fully analyzed until many days or even months of computer processing establishes what was observed. In this case, the experimental apparatus is the observer.

And, in general, the universe is its own observer, able to record (and sometimes preserve) the information created.

The basic assumption made in Schrödinger's cat thought experiments is that the deterministic Schrödinger equation describing a microscopic superposition of decayed and non-decayed radioactive nuclei evolves deterministically into a macroscopic superposition of live and dead cats.

But since the essence of a "measurement" is an interaction with another system (quantum or classical) that creates information to be seen (later) by an observer, the interaction between the nucleus and the cat is more than enough to collapse the wave function. Calculating the probabilities for that collapse allows us to estimate the probabilities of live and dead cats. These are probabilities, not probability amplitudes. They do not interfere with one another.

After the interaction, they are not in a superposition of states. We always have either a live cat or a dead cat, just as we always observe a complete photon after a polarization measurement and not a superposition of photon states, as Dirac explains so simply and clearly in his *Principles of Quantum Mechanics.*[4]

The original cat idea of Schrödinger, and Einstein, was to make fun of standard quantum mechanics. But the cat has taken on a life of its own, as we shall see in later chapters. Some interpretations of quantum mechanics, based entirely on a universal wave function, are puzzled by the absence of *macroscopic* superpositions. They say quantum mechanics involves *microscopic* superpositions like particles being in two places at the same time, going through both slits in the two-slit experiment for example. So why no macroscopic superpositions like Schrödinger's Cat?

The short answer is very simple. There are no microscopic superpositions either. As we saw in chapter 19, Dirac tells us that superpositions are just a "manner of speaking." Any real system is always in a single state. Treating it as in a superposition of some other basis states is a mathematical tool for making statistical predictions about large numbers of experiments.

The particular radioactive nucleus in Schrödinger's example is always either not yet decayed or already decayed!

---

4   Dirac, 1930, p.5

Statistic

Quantum Mechanics

Light Quantum
Hypothesis

Photoelectric
Effect

Bohr-Einstein Atom

e-Particle Duality

Bos

Entanglement
and Symmetry

Born-Einstein Statistic

Chance

Nonlocality

Irreversibility

Nonsepa

Einstein-Podolsky-Rosen

E

Schrödinger's Cat

Did Albert Einstein Inver

This chapter on the web
informationphilosopher.com/problems/entanglement

Entanglement and Symmetry

In his pioneering work on special and general relativity, Einstein's greatest work came from his use of fundamental "principles" to derive his new results. In special relativity, it was the principle that light has the same speed in all frames of reference. In general relativity, it was his equivalence principle, that an observer cannot distinguish between an accelerated frame and the force of gravity.

Each of these principles emerges from an underlying symmetry that produces an invariant quantity or a conservation law.

The speed of light is an invariant. The laws of physics are the same at different places in space-time. Otherwise we couldn't repeat experiments everywhere and discover the laws of nature.

Einstein discovered symmetries that helped him reformulate Maxwell's laws of electromagnetic fields. A few years later EMMY NOETHER (often described as the most important female mathematician) made a profound contribution to theoretical physics with her theorem on the fundamental relationship between symmetry and conservation principles.

For any property of a physical system that is symmetric, there is a corresponding conservation law.

For example, if a physical system is symmetric under rotations, its angular momentum is conserved. If symmetric in time, energy is conserved. If symmetric in space, momentum is conserved.

Noether's theorem allows physicists to gain powerful insights into any general theory in physics, by just analyzing the various transformations that would make the form of the laws involved invariant. No one understood the importance of these invariance principles better than Einstein. Nevertheless, Einstein introduced an odd asymmetry where none belongs in his EPR analysis of the behavior of two "entangled" particles.

Chapter 29

Einstein's Introduction of a False Asymmetry?

Almost every presentation of the EPR paradox and descriptions of entanglement begins with something like "Alice observes one particle..." and concludes with the question "How does the second particle get the information needed so that Bob's later measurements correlate perfectly with Alice's?"

There is a fundamental *asymmetry* in this framing of the EPR experiment. It is a surprise that Einstein, who was so good at seeing deep symmetries, did not consider how to remove the asymmetry.

Consider this reframing: Alice's measurement collapses the two-particle wave function Ψ_{12}. The two indistinguishable particles simultaneously appear at locations in a space-like separation. The frame of reference in which the source of the two entangled particles and the two experimenters are at rest is a *special frame* in the following sense. It is the frame in which their appearance is *simultaneous*. In this frame, the experiment is *symmetric*.

As Einstein knew very well, there are frames of reference moving with respect to the laboratory frame of the two observers in which the time order of the events can be reversed. In some moving frames Alice measures first, in others Bob measures first.

Einstein also knows well that two events in spacelike separation can have no causal influence on one another. They are not in one another's "light cone." No signals communicate between them.

If there is a *special* frame of reference (not a *preferred* frame in the relativistic sense), surely it is the one in which the origin of the two entangled particles is at rest.

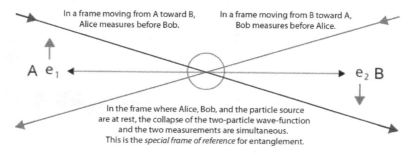

In a frame moving from A toward B, Alice measures before Bob.

In a frame moving from B toward A, Bob measures before Alice.

A e₁   e₂ B

In the frame where Alice, Bob, and the particle source are at rest, the collapse of the two-particle wave-function and the two measurements are simultaneous. This is the *special frame of reference* for entanglement.

Assuming that Alice and Bob are also at rest in this special frame and equidistant from the origin, we arrive at the simple picture in which any measurement that causes the two-particle wave function to collapse makes both particles appear simultaneously at determinate places with fully correlated properties (just the values that are needed to conserve energy, momentum, angular momentum, and spin).

Instead of the one particle making an appearance in Einstein's original case of nonlocality, in the two-particle case, when either particle is measured - or better, when the wave function is disturbed? - both particles appear.

The two-particle wave function splits into two single-particle wave functions.

$$\Psi_{12} => \Psi_1 \Psi_2$$

At this moment, the two-particle wave function decoheres (no longer shows interference properties), the particles are disentangled,

We know instantly those properties of the other particle that satisfy the conservation laws, including its location equidistant from, but on the opposite side of, the source, along with its other properties such as the spin, which must be equal and opposite to add up to the original spin = zero, for example.

When Alice detects the particle at t_0 (with say spin up), at that instant the other particle also becomes determinate (with spin down) at the same distance on the other side of the origin. The particles separate at t_0. Further measures of either particle will have no effect on the other!

Note that should Bob have measured before t_0, his would be the "first" measurement that causes the two-particle wave function to decohere and the particles to disentangle and finally separate.

We can also ask what happens if Bob is not at the same distance from the origin as Alice. This introduces a positional asymmetry. But there is still no time asymmetry from the point of view of the two-particle wave function collapse at to.

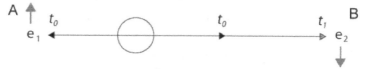

If observer A measures electron 1 with spin up at t_0, electron 2 *instantly* is spin down (Ψ collapses to $|+->$).

Electron 2 is then *determined* to be found with spin down if measured by observer B at a later time t_1.

But this was not *predetermined* before A's measurement at t_0.

What Did Einstein See? The Special Frame?

Remember Einstein's 1933 question to Leon Rosenfeld, "How can the final state of the second particle be influenced by a measurement performed on the first..."[1] Why did Einstein see something unusual in what we now call simply "knowledge-at-a-distance?"

The instantaneous nature of the "knowledge" is what Einstein saw as a potential violation of his principle of relativity. We argue that it picks out a *special frame* in which two events are "simultaneous."

Relativity denies simultaneity between separated events.

In 1927 at the Solvay conference the events were the detected particle on the screen and that mysterious second place on the screen.[2] In the 1935 EPR paper they were the "influence" of the first particle measurement on the second particle.

Between these two points is a space where Einstein thinks something is happening that violates his relativity principle. In the diagram above it's the line between Alice's observation at t_0 and the point t_0 on the line to Bob where the conserved momentum would locate the entangled particle on its way to Bob.

Events at those two points are "simultaneous" in the frame where the center of the experiment is at rest. There are very fast-moving frames coming from the right, where Bob's measurement at t_1 appears to happen *before* Alice's measurement at t_0.

Now these are the two points where electron spins (or photon polarizations) are measured in the tests of Bell's inequality (chapter 32), where Alice 's measurements "influence" Bob's.

---

1   See page 207.
2   See page 175

Einstein knew nothing about our puzzles in the "age of entanglement," yet his "spooky actions" are our concerns today!

His colleagues thought Einstein was too old to contribute anything new to quantum mechanics, but his contributions still zero in with a laser focus on today's most profound mysteries. How can his extraordinary mind have been so prophetic?

No Hidden Variables, but Hidden Constants!

We shall see in the next several chapters that many physicists hoped to confirm Einstein's criticisms of quantum mechanics by questioning the "foundations of quantum mechanics." They would offer either new "interpretations" of quantum mechanics, or new "formulations" that add or subtract elements to the theory.

In particular, they followed Einstein's argument that quantum mechanics is "incomplete," and might be completed by the discovery of additional variables.

There may be no "hidden variables," local or nonlocal. But there are "hidden constants." Hidden in plain sight, they are the "constants of the motion," *conserved* quantities like energy, momentum, angular momentum, and spin, both electron and photon. Created indeterministically when the particles are initially *entangled*, they then move locally with the now *apparently* separating particles.

In our extension of Einstein's "objective reality," we assume the particles have continuous paths from the start of the experiment to the final measurement(s), although the limits of quantum measurement never allow us to "know" those paths or any particular properties like the direction of spin components.

Conservation of momentum requires that positions where particles finally appear are equidistant from the origin, in order to conserve linear momentum. And every other conserved quantity also appears perfectly correlated at all *symmetric* positions. It is the fundamental principle of conservation that governs the correlated outcome, not some hypothetical, faster than light, communication of information between the particles at the time of measurement.

And in any case what would a particle as simple as an electron or a photon do with "information" from an identical particle? Indeed. how would the supposed "first" particle "communicate?"

Information is neither matter nor energy, though it needs matter to be embodied in an "information structure," and it needs energy to communicate information to other such structures.

Objective reality tells us that the two particles are (locally) carrying with them all the information that is needed for measurements to show perfect correlations. This is a major problem only because the Copenhagen Interpretation claims that the particles have no properties before their measurement, that each particle is in a superposition of states, so something is needed to bring their properties into agreement at the measurement.

Einstein's "objective reality" asks the simple question whatever could have caused the two particles to disagree? That is impossible without some physical interaction to change one or both of the particle properties. Such an interaction is of course the measurement by Alice (or Bob) that disentangles the particles.

Alice's "Free Choice" of Spin Direction

Following Einstein's false asymmetry that measurements of spacelike separated particles can be made "first" by one observer, it is widely but mistakenly said that Alice's outcome must be "influencing" Bob's.

What Alice does when she interacts with the two-particle wave function Ψ_{12} is to create new *information* that was not present when the particles were initially entangled. It cannot therefore be carried along locally with our "hidden constants" of the motion.

But the new information is created locally by Alice. The nonlocal two-particle wave function makes it available to both particles globally instantaneously, wherever they are.

The classic case of entangled electrons or photons is that they start in a state with total spin (or polarization) equal to zero (the so-called singlet state).

The singlet state is perfectly symmetric in all directions.

When Alice measures a polarization or spin direction, her measurement forces the two-particle system to have that overall preferred direction. This is what WOLFGANG PAULI called a "measurement of the second kind. PAUL DIRAC said the system is "projected" into this state. HENRY MARGENAU called it a "state preparation."

Quantum mechanically, the two-particle wave function is in a superposition of states in all directions and Alice's measurement projects it into Alice's freely chosen spin direction.

The two spins before her measurement were opposing one another but had no such preferred direction. Now they have opposite spins and in the direction chosen by Alice. This new information about polarization direction can not have been carried along locally with the hidden constants that conserve all physical properties, because that information did not exist until her measurement. .

Just because we cannot continuously measure positions, paths, and particle properties does not mean that they don't exist. And claiming they are not determined just before measurement asks the question of what forces exist to change them at the last moment?

The new preferred direction for the spins did not exist. They were the result of Alice's "free choice." But the Copenhagen Interpretation is simply wrong to extend the *non-existence* of Alice's new properties to those that travel "locally" with the particles

Our "hidden constants" traveling locally with the particles only require that the spins are always perfectly opposite. If Alice's measurement shows a spin component of +1/2 in her chosen z-direction, Bob will necessarily measure -1/2 in the z-direction.

Any other value would violate the conservation laws and break the symmetry.

Note that whether Alice measures +1/2 or -1/2 is random, the result of what Dirac calls "Nature's choice."

If Bob now "freely chooses" in any other angular direction, his correlations will be reduced by the cosine squared of the angular difference between him and Alice. This is the same physics that reduces the light coming through polarizers at different angles as we saw in chapter 19.

We shall see in chapter 32 that JOHN BELL strangely argued that "hidden variables" of the type imagined by Einstein or Bohm would produce correlations with a straight-line angular dependence, and not the familiar sinusoidal relationship .

Decades of Bell inequality tests claim to have shown that hidden variables must be nonlocal. "Hidden constants" like linear momentum and opposing spins are local! They are conserved properties that move along in the entangled particles at or below light speed.

The two-particle wave function is itself a global function encompassing the two particles (and beyond in the case of electrons).

When that two-particle wave function instantly acquires a preferred direction for its opposing spins it does so globally, giving the illusion of an effect or an "action" travelling from Alice to Bob.

But this is precisely the same "nonlocality" seen by Einstein in 1905 and reported by him first in 1927 at the fifth Solvay conference.

It is the mysterious and powerful global property of the wave function that Einstein called "ghostly" and a "guiding field." There is no "spooky action-at-a-distance" in the sense of one particle acting on the other, "influencing" it in some way.

It is the same "guiding" power of the wave function which in the two slit experiment statistically controls the locations of electrons or photons to show interference fringes, including null points where particles never appear.

This power of the wave function explains the mystery of entanglement, why Bob finds perfect correlations with Alice when she measures simultaneously or a moment before him so there is no time for knowledge of her freely chosen angle to travel to Bob.

There are two important moments to be understood, initial entangled formation and later disentangling measurement.

1) At formation, standard quantum mechanics usually describes the entangled two-particle wave function as in a superposition of up-down and down-up states,

$$\Psi_{12} = (1/\sqrt{2})\,(|+->-|-+>).$$

But PAUL DIRAC tells us an individual system is in just one of these states from the moment of formation.[3]

The singlet state, say $|+->$, is visualized as having no determinate spin direction as the particles travel apart. The spin state is isotropic, spherically symmetric.

2) The two-particle state collapses on Alice's measurement into a product of single-particle states, $|+>|->$.

---

3   See page 151.

When Alice measures her particle with her "free choice" of a definite spin direction, e.g., z+, it is the requirement to conserve total spin, not any communication, that projects Bob's particle, before his measurement, into z-. The particles are disentangled.

Just before Bob's measurement, his state has been *prepared* so that if he measures in Alice's direction, he will measure z- (say spin down) to her z+ (say spin up).

The two particles have been conserving zero total spin from the time of their singlet state preparation at the start of the experiment and, if undisturbed, they will be found in the same singlet state when they are measured. They have perfectly correlated opposing spins when(ever) they get jointly measured at the same angle.

The particular direction of spin is created by Alice.

One of Einstein's great principles was simplicity.[4] It is also known as the law of parsimony and Occam's Razor. The idea is that the simplest theory that fits all the known facts is the best theory. Einstein may have liked the idea that the most true theories would be beautiful in some sense, perhaps as the result of their symmetry.

Consider then the simplicity and parsimony of the idea that entangled particles, once "cross-linked" and sharing an antisymmetric two-particle wave function, are carrying with them at all times all the information needed for them to *appear to be* coordinating their actions - without communicating!

The information is "hidden" in the "constants of the motion." And where hidden variables are nonlocal, all hidden constants are local.

It is now fifty years since the first laboratory experiments were done to find whether quantum mechanics might be faulty, and hidden variables might be needed to explain entanglement.

There has been no evidence that anything is wrong with quantum mechanics. Isn't it time that we go back to Einstein's first principles and see whether the "objective reality" of continuous particle motions carrying with them all their conserved properties can give us a very simple, easy to explain, understanding of entanglement?

We can have entanglement without "action-at-a-distance."

_____

4   See chapter 35.

Information hidden in the constants of the motion is "locally real" at all times as the particles travel apart with no definite spin directions for either particle, but total spin always zero.

Can Conservation Laws Do It All?

But can conservation laws and symmetry explain the perfect correlation of every particle property to prove there is no instantaneous "action-at-a-distance" needed for entanglement?

All physicists know conservation works for linear momentum. Einstein used it in his 1933 letter to Leon Rosenfeld. But what about the properties tested in all modern experiments on entanglement, electron spin and photon polarization?

Can we show how these properties also are actually conserved as they are carried along with the particles, so there is no need for instantaneous communication between two widely separated entangled particles at the moment of their measurement, eliminating the conflict between quantum mechanics and special relativity?

The case of the photon is relatively straightforward, as we saw in Dirac's analysis (chapter 19). He said that an individual photon is not in a linear combination or superposition of states, as we assume when making predictions for a number of experiments.

We can simplify the two-particle state to either $| + - >$ or $| - + >$.

And since the two-particle, spin-zero, state has no preferred spin or polarization direction, we can say that they are in a superposition of possible spin or polarization components, and that the spin of one is in some average sense always opposite to that of the other.

"Objectively real" entanglement is in no sense a measurement of one particle "acting on" and causing a change in another distant particle. When Ψ_{12} decoheres, particles appear simultaneously in our *special* frame of reference. No properties are changing.

Einstein's "objective reality" requires that entangled particle properties are conserved from their initial state preparation to their ultimate measurements, giving the *appearance* of instantaneous communications, of Einstein's "spooky action-at-a-distance."

Pauli's Kinds of Measurement Again

When we describe the measurements of entangled particles that "collapse" the two-particle wave function, and which make the particles in a spacelike separation *appear* to interact instantaneously, infinitely greater than lightspeed, we must consider what kind of measurements are being made.

As we saw in chapter 19, WOLFGANG PAULI distinguished two kinds of measurements. The first is when we measure a system in a known state ψ. (It has been prepared in that state by a prior measurement.) If we again use a measurement apparatus with eigenvalues whose states include the known state, the result is that we again find the system in the known state ψ. No new information is created, since we knew what the state of the system was before the measurement. This Pauli called a *measurement of the first kind*.

Dirac noted that quantum mechanics is not always probabilistic. Measurements of the first kind are *certain*, like preparing a state and then measuring to see that it is still in that state. Today this is called a non-destructive measurement.

In Pauli's second case, the eigenstates of the system plus apparatus do not include the state ψ of the prepared system. Dirac's transformation theory says one should use a basis set of eigenstates appropriate to the new measurement apparatus, say the set φ_n.

In this case, the original wave function ψ can be expanded as a linear superposition of states φ_n with coefficients c_n,

$\psi = \sum_n c_n \varphi_n$,

where $c_n^2 = |<\psi | \varphi_n>|^2$ is the probability that the measurement will find the system in state φ_n.

Pauli calls this a *measurement of the second kind*. It corresponds to JOHN VON NEUMANN's Process 1, interpreted as a "collapse" or "reduction" of the wave function. Von Neumann said that new information is irreversibly recorded in the measuring apparatus.

In this measurement, all the unrealized possibilities are eliminated, and the one possibility that is actualized produces new information. We do not know which of the possible states becomes actual. That is a matter of ontological chance. If we did know in advance, there would be no new information.

Measurements of electron spin are done with Stern-Gerlach magnets. A stream of electrons with random spin directions passing through a magnet oriented in the z-direction separates into electrons deflected upward (z+) and those deflected downward (z-).

This is a measurement of the second kind, a state preparation. If we pass all those with z+ through a second magnet in the z-direction, they all are deflected upward again. This is a non-destructive measurement of the first kind. Information is preserved.

If those electrons in a known z+ state are passed through a magnet oriented in the x-direction, they are observed in a random distribution of x+ and x-. The z+ state information is lost.

At the initial entangled state preparation, neither electron has information about its spin components. Since there is no information, we can call this a measurement of the zeroth kind.

This describes the preparation of the entangled pair. We know nothing of the spin components of the electrons (or polarization of photons). But we do know that the spin of the left-going particle will be opposite to that of the right particle when they are measured.

Assume that Alice measures "first", which she does if she is closer to the center than Bob. This is a measurement of the second kind, because a preferred spin direction of the electron did not exist.

Alice makes a "free choice," as Heisenberg described it. The spin component value comes into existence. It did not necessarily have that value before her measurement. No matter which angle of orientation Alice measures, she will find spin randomly +1/2 or -1/2. Dirac called this "Nature's choice."

Between "Nature's choice" (quantum chance discovered by Einstein in 1916) and "free choice" (Einstein's "free creations of the human mind"), we untie the Gordian Knot of quantum mechanics! Neither we nor the universe are pre-determined.

If Bob measures the same angle as Alice (perhaps by prior agreement) and compares measurements later, he will find his data is perfectly correlated with Alice. Bob's measurement in the same direction as Alice is therefore a measurement of the first kind.

Alice prepares the state. Bob measures the same state.

If, however, Bob sets his apparatus to measure at a different angle, he finds a weaker correlation with Alice over several measurements.

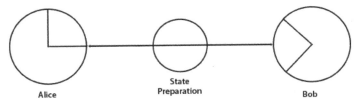

Alice   State Preparation   Bob

Bob also has a "free choice" as to what to measure. As he varies his angle away from Alice's, at first only a few measurements disagree, randomly but then disagreements increase, following the cosine dependence of light passing through rotating polarizers.[5]

John Bell made the very unphysical claim that the correlations would fall off linearly, in a straight line, and connected this "inequality" to Einstein's idea of additional ("hidden") variables.[6]

If Bob rotates his apparatus to 90°, spin in the x direction will be completely random. All correlations with Alice are now lost.

These measurements of the second kind project Bob's electron spin in a new direction . It prepares a new state. It does nothing to Alice's particle, since her measurement separated the electrons.

The reason Alice and Bob measure perfect entanglement when they measure in the same direction is because both spin directions were determined by Alice at the moment the two-particle wave function $| + - >$ collapsed and projected out the two values, $+1/2$ and $-1/2$, conserving the total spin as zero.

The total spin was zero before her measurement, but it had no definite spin component direction

This was not "spooky action-at-a-distance" traveling from Alice toward Bob. The collapse of Ψ_{12} is symmetric (or anti-symmetric) in all directions. It is this symmetry, and the conservation law for total electron spin, that completely explains entanglement.

The original state preparation of entangled particles created no new information about specific spin components. With some deep symmetry (photons) or anti-symmetry (electrons), it does not prepare the particles in definite states, as does Alice's measurement.

We could call this a measurement of the zeroth kind.

5   See Dirac's polarizers in chapter 19
6   See chapter 32.

Alice breaks the original symmetry, creating information about the new spin directions. If Bob measures at the same angle, it is a measurement of the first kind. If he measures at other angles, symmetry/anti-symmetry with Alice is broken and Bob's is a measurement of the second kind.

How Symmetry and Conservation Explain Entanglement

When a pair of electrons or photons is entangled, they are not prepared with spins that have definite components in specific spatial coordinate directions. But they must be such that if one is found to have spin +1/2 in any direction, the other will be -1/2. And these opposite directions will show up when Alice's measurement projects her electron and Bob's into definite directions.

The two electrons could be in a superposition of $| + - >$ and $| - + >$, as standard quantum mechanics likes to say. They may only acquire specific spin component directions when Alice's measurement projects the two-particle wave function into a definite direction.

Or it could be that Dirac is correct that they are in one or the other of these states from their entanglement. In this case, Einstein is right that they have all properties before they are measured. But they cannot yet have definite z spins. Einstein would understand this as the consequence of a new measurements.

Let's see how to visualize this in terms of Pauli's two kinds of measurements and a state creation that is *not a measurement* which leaves two entangled electrons in perfectly symmetric *directionless* spin states that together preserve total entropy zero.

First let's recall how measurements of spin in a Stern-Gerlach apparatus can distinguish electrons that are in a known state from those that are in a symmetric state with no definite direction.

The gray circle represents an unentangled electron with no specific spin direction. When that electron enters the magnet which is oriented in the z direction, it is either directed upward or downward. This a measurement of the second kind.

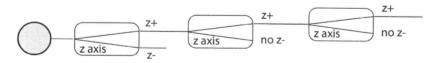

If it prepares a spin-up electron z+ and we pass it through a second magnet (or even a third) with the same z orientation, it does not change from z+. These are non-destructive measurements of the first kind. It never yields z- electrons

When we know a determined state goes in, the same comes out. Suppose we had a pair of entangled electrons with no determinate spin directions but with one carrying the positive spin and the other the negative. What happens as they pass through the magnets?

The positive spin electron, which has no determinate direction component, comes through the magnet projected into z+. Such a spatial directionless positive spin electron sent through an x-axis magnet produces only x+ electrons.

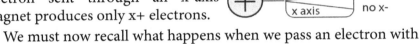

We must now recall what happens when we pass an electron with known spin z+ through a magnet oriented in the x direction.

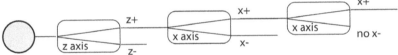

Both x directions are possible, and when a known x+ is produced, subsequent measurements of the first kind keep it the same x+. Now before we show how our entangled electron behaviors work to explain entanglement, we should show the loss of z+ spin when passed through a magnet oriented in the x direction, and the subsequent recovery of both z+ and z- components. An x+ electron contains the potential to produce both z+ and z- electrons.

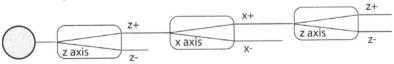

Finally, so we show all the amazing properties of electron spin, and add to understanding the idea of an electron with a spin value, but with no preferred spin direction, we can use a Stern-Gerlach magnet

to generate both z+ and z- and, providing we do not make a measurement, send them though in the opposite z direction to recreate the original.

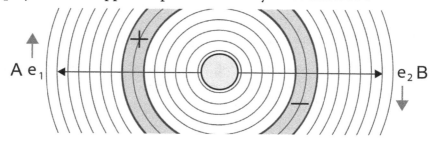

So let's see how our directionless spin states travel from their entanglement and then get projected into opposite spin directions by Alice and Bob .

At the start the two electrons are in the same small volume of phase space with their spins opposite, satisfying the Pauli exclusion principle, like the two electrons in the ground state of Helium.

A few moments later they travel apart in a $|+ ->$ state, with one electron having spin $+1/2$ and the other $-1/2$. But neither has a definite spatial spin component, in a given direction such as z+.

The *directionless* spin state is symmetric and isotropic, the same in all directions. It is rotationally invariant. The spin values of + and - are conserved quantities we can call *local* "hidden constants," traveling with the particles from their entanglement in the center.

Because they are entangled, the + spin in one electron is always perfectly opposite that of the - electron, though the spatial direction of the spins is entirely unknown.

These conserved spins provide the necessary information that hypothetical "hidden variables" could provide to the electrons at their moment of measurement. But no faster-than-light exchange of that information is involved, no "signaling" between the particles in a distant spacelike separation. Correlation information is carried along with the electrons at their speed. Their spins are always perfectly correlated, not suddenly correlated at the moment of measurement, as the Copenhagen Interpretation claims.

In her measurement, Alice *creates* new directional information that did not travel with the "hidden constants" of the motion. It was unknown beforehand. When Alice measures in the z direction, she "prepares" the state z+. But Einstein's "objective reality" view is correct that the system has most of its properties *before* her measurement.

In his original EPR, it was linear momentum that was conserved from the initial interaction. Conservation laws allowed him to know something about particle 2 simultaneous with his measurement of particle 1. This is not "action." This is just "knowledge-at-a-distance."

But there is one property the two particles could not have before Alice's measurement. It is something Einstein never thought about. That is the spatial direction of the polarization or electron spin imposed by Alice's "free choice" of which angle to measure.

If Bob also measures at Alice's angle, Bob's is a measurement of the first kind. The state that he measures was prepared by Alice. These are two perfectly correlated events that are *simultaneous* (in a "special frame") despite being in a spacelike separation.

When Einstein first saw this kind of nonlocal phenomenon in 1905 and described it in 1927, he thought it violated his special theory of relativity, and his idea of the *impossibility of simultaneity.*

Nevertheless, this is one more amazing insight into nature that Einstein was the first person to see, even if it bothered him.

These simultaneous spatially separated events are a consequence of the two-particle wave function Ψ_{12} collapsing into the product of two single-particle wave functions Ψ_1 and Ψ_2

The Ψ_{12} wave function has decohered, the particles are disentangled, they acquire their opposite spin component directions, + spin goes to z+, - spin to z-.

In all entanglement experiments, these simultaneous values of opposing spins or polarizations that appear now have definite spatial directions, which is new information. The z+ and z- values are "nonlocal." The +1/2 and -1/2 spins came with the particles, as Einstein hoped to show. They are "local," like the particle momenta.

We turn to "interpretations" of quantum mechanics and their failed attempts to explain the "mystery" underlying nonlocality.

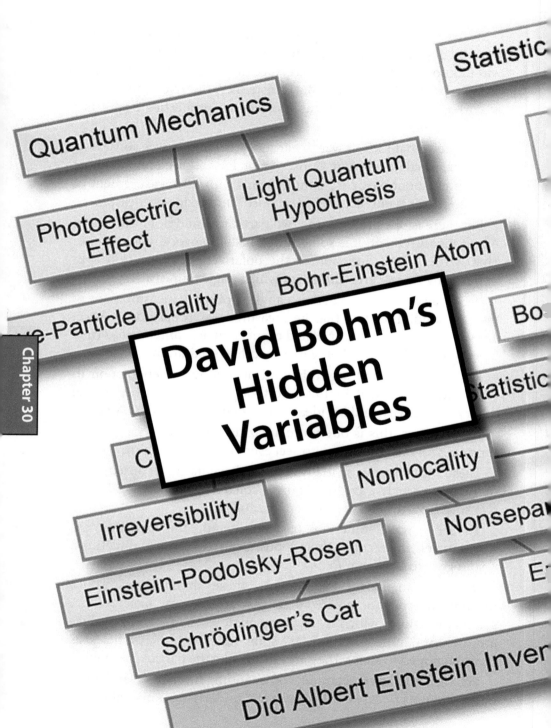

Statistic

Quantum Mechanics

Light Quantum
Hypothesis

Photoelectric
Effect

Bohr-Einstein Atom

e-Particle Duality

Bo

David Bohm's
Hidden
Variables

Statistic

C

Nonlocality

Irreversibility

Nonsepa

Einstein-Podolsky-Rosen

E-

Schrödinger's Cat

Did Albert Einstein Inver

David Bohm's Hidden Variables

DAVID BOHM is perhaps best known for new experimental methods to test Einstein's suggestion of "additional variables" that would explain the EPR paradox by providing the information needed at the distant "entangled" particle, so it can coordinate its properties perfectly with the "local" particle. Bohm proposed the information would be transmitted by a new vector or "quantum" potential that travels faster than the speed of light.

Bohm wrote in 1952,

> The usual interpretation of the quantum theory is based on an assumption having very far-reaching implications, viz., that the physical state of an individual system is completely specified by a wave function that determines only the probabilities of actual results that can be obtained in a statistical ensemble of similar experiments. This assumption has been the object of severe criticisms, notably on the part of Einstein, who has always believed that, even at the quantum level, there must exist precisely definable elements or dynamical variables determining (as in classical physics) the actual behavior of each individual system, and not merely its probable behavior. Since these elements or variables are not now included in the quantum theory and have not yet been detected experimentally, Einstein has always regarded the present form of the quantum theory as incomplete, although he admits its internal consistency. [1]

Bohm's new supraluminal signaling would communicate extra variables he called "hidden" that would "complete" quantum mechanics, restoring the determinism of classical physics that Bohm mistakenly thought Einstein was looking for.

Five years later, Bohm and his Israeli student Yakir Aharonov reformulated the original EPR argument in terms of electron spin. They said experimental tests with continuous variables are much more difficult than tests with discrete quantities, such as the spin of electrons or polarization of photons. They wrote:

---

1   Bohm 1952, p.166

We consider a molecule of total spin zero consisting of two atoms, each of spin one-half. The wave function of the system is therefore

$$\psi = (1/\sqrt{2}) [\psi+ (1) \psi- (2) - \psi- (1) \psi+ (2)]$$

where $\psi+ (1)$ refers to the wave function of the atomic state in which one particle (A) has spin $+\hbar/2$, etc. The two atoms are then separated by a method that does not influence the total spin. After they have separated enough so that they cease to interact, any desired component of the spin of the first particle (A) is measured. Then, because the total spin is still zero, it can immediately be concluded that the same component of the spin of the other particle (B) is opposite to that of A. [2]

Einstein may have encouraged his Princeton colleague Bohm to develop hidden variables to "complete" quantum mechanics and possibly restore determinism. Einstein had heartily approved of Bohm's textbook and was initially supportive of Bohm's new mechanics. Einstein thought Bohm was young enough and smart enough to produce the mathematical arguments that the older generation of "determinist" physicists like ERWIN SCHRÖDINGER, MAX PLANCK, and others had not been able to accomplish.

But when Bohm finished the work, based on LOUIS DE BROGLIE's 1923 "pilot-wave" idea (which Einstein had supported), Einstein rejected it, as he always had rejected nonlocality in the form of instantaneous "action-at-a-distance." Bohm's work was simply inconsistent with Einstein 's theory of relativity. It still involved the "impossible" simultaneity of events in a spacelike separation.

No "Hidden Variables," but Hidden Constants?

There may be no hidden variables, local or nonlocal. But as we saw in the previous chapter, there are "hidden constants." Hidden in plain sight, they are the "constants of the motion," *conserved* quantities like energy, momentum, angular momentum, and spin, both electron and photon. Created indeterministically when the particles are *entangled*, they then move along with the *apparently* separating particles, conserving total spin zero.

In our application of Einstein's "objective reality," we assume the particles have continuous paths from the start of the experiment to the final measurement(s), although the limits of quantum measurement never allow us to "know" those paths.

---

2   Bohm and Aharonov, 1957, p. 1070

Conservation of momentum requires that positions where they finally do appear are equidistant from the origin, in order to conserve linear momentum. And every other conserved quantity, like angular momentum, electron or photon spin, as well as energy, also appear perfectly correlated at all *symmetric* positions.

But the particles appear to not have definite values of electron or photon spin *before* their first measurement by Alice or Bob. This state preparation created no new information about definite spin directions. It was not a "measurement" that leaves the particles in a definite state, as will Alice's measurement.

We call it a measurement of the zeroth kind.

Once particles are in a definite state of | + - > or | - + > it is the fundamental principle of conservation that governs the correlated outcome, not some hypothetical, faster than light, communication of information between the particles at the time of measurement.

Einstein's "objective reality" means that conservation laws hold at every position along the path, from the first measurement by Alice or Bob to their second measurement. Just because we cannot measure positions and paths does not mean that they don't exist.

The hidden constants of the motion include electron spins, which were suggested by Bohm as the best test for the hidden variables needed to support nonseparability and entanglement. The two particles conserve the same opposing spins up to the time of their measurement by Alice or Bob.

Unfortunately, hidden constants are not able to explain the "simultaneous" assignments of the spin components. Although Einstein never considered two opposing spins that conserve total spin zero, his thinking applies perfectly. And Alice's measurement direction corresponding exactly to Bob's is one more case of what Einstein saw first in 1905- his "impossible" simultaneity.

Bohmian Mechanics

Bohm is also well known for his "Bohmian Mechanics," a formulation of non-relativistic quantum mechanics that emphasizes the motion of particles and promises to restore causality to physics. It is a deterministic theory, one of several "interpretations" that are today's most popular alternatives to the Copenhagen Interpretation.

By emphasizing the motion of particles, Bohmian mechanics de-emphasizes the wave function Ψ, limiting its role to *guiding* the motion of the particles, in comparison to competing interpretations that deny the existence of particles altogether.

Bohmian mechanics includes a mechanism whereby physical effects can move faster than light, providing an explanation for Einstein's *nonlocality*. But as we saw in the last chapter, Einstein's "objective reality " provides a simpler solution that removes any conflict between relativity and quantum mechanics.

It's a surprise Einstein did not agree with Bohm, because Bohmian mechanics describes particles as moving along continuous paths, just as we visualize for Einstein's "objective reality." In the famous two-slit experiment, Bohm's particles always move through just one slit, even as the guiding wave function moves through both slits when both are open.

> we must use the same wave function as is used in the usual interpretation... We do not in practice, however, control the initial location of the particle, so that although it goes through a definite slit, we cannot predict which slit this will be. [3]

The Bohmian mechanics solution involves three simple steps:
First, close slit 1 and open slit 2. The particle goes through slit 2.
It arrives at x on the plate with probability $|\psi_2(x)|^2$,
where ψ_2 is the wave function which passed through slit 2.
Second, close slit 2 and open 1. The particle goes through slit 1.
It arrives at x on the plate with probability $|\psi_1(x)|^2$,
where ψ_1 is the wave function which passed through slit 1.
Third, open both slits. The particle goes through slit 1 or slit 2.
It arrives at x with probability $|\psi_1(x)+\psi_2(x)|^2$.
Now observe that in general, $|\psi_1(x)+\psi_2(x)|^2 =$
$|\psi_1(x)+\psi_2(x)|^2 = |\psi_1(x)|^2 + |\psi_2(x)|^2 + 2R\psi^*_1(x)\,\psi_2(x)$.
The last term comes from the *interference* of the wave packets ψ_1 and ψ_2 which passed through slit 1 and slit 2.
The probabilities of finding particles when both slits are open are different from the sum of slit 1 open and slit 2 open separately.
The wave function determines the probabilities of finding particles, just as Einstein first proposed. [4]

Chapter 30

3   Bohm 1952, p.174
4   Dürr and Teufel, 2009, p.9

This reduces RICHARD FEYNMAN's "one" mystery. We need not worry as he did about how a particle can go through both slits. But there remains the deeper mystery, how an abstract probabilities function (mere information) can influence the motions of the particles to produce the interference patterns. A wave in one place influencing the particle in another is "impossible" simultaneity.

Bohm's explanation of the two-slit experiment is completely compatible with Einstein's "objective reality." It does not solve the "deep mystery" of how the wave function "guides" the particles.

Irreversibility

In his excellent 1951 textbook, *Quantum Theory*, Bohm described the necessity for irreversibility in any measurement. Bohm followed JOHN VON NEUMANN's measurement theory in which recorded data is *irreversible*. A measurement has only been made when new information has come into the world and adequate entropy has been carried away to ensure the stability of the new information, long enough for it to be observed by a "conscious" observer.

> From the previous work it follows that a measurement
> process is irreversible in the sense that, after it has occurred,
> re-establishment of definite phase relations between the
> eigenfunctions of the measured variable is overwhelmingly
> unlikely. This irreversibility greatly resembles that which appears
> in thermodynamic processes, where a decrease of entropy is also
> an overwhelmingly unlikely possibility...
>
> Because the irreversible behavior of the measuring apparatus
> is essential for the destruction of definite phase relations and
> because, in turn, the destruction of definite phase relations is
> essential for the consistency of the quantum theory as a whole,
> it follows that thermodynamic irreversibility enters into the
> quantum theory in an integral way. [5]

But Bohmians today have a different view on irreversibility. As Dürr and Teuful describe it in their book, *Bohmian Mechanics*,

> The second law of thermodynamics captures irreversibility, and
> at the same time points towards the problem of irreversibility,
> which is to justify the special atypical initial conditions on
> which, according to Boltzmann, the second law is based... What
> is the physics behind the selection? We do not know. That
> ignorance of ours deserves to be called an open problem: the
> problem of irreversibility. [6]

Chapter 30

5   Bohm, 1951, p.168
6   Dürr and Teufel, 2009, p.90. See our chapter 12.

Statistic

Quantum Mechanics

Light Quantum
Hypothesis

Photoelectric
Effect

Bohr-Einstein Atom

ave-Particle Duality

Bos

Hugh Everett III's Many Worlds

Bohr-Einstein Statistic

istic

Nonlocality

Irreversibility

Nonsepa

Einstein-Podolsky-Rosen

E

Schrödinger's Cat

Did Albert Einstein Inven

Hugh Everett III's Many Worlds

HUGH EVERETT III was one of JOHN WHEELER's most famous graduate students. Others included RICHARD FEYNMAN. Wheeler supervised more Ph.D. theses than any Princeton physics professor.

Everett took mathematical physics classes with EUGENE WIGNER, who argued that human consciousness (and perhaps some form of cosmic consciousness) was essential to the "collapse" of the wave function.

Everett was the inventor of the "universal wave function" and the "relative state" formulation of quantum mechanics, later known as the "many-worlds interpretation."

The first draft of Everett's thesis was called "Wave Mechanics Without Probability." Like the younger ALBERT EINSTEIN and later ERWIN SCHRÖDINGER, Everett was appalled at the idea of indeterministic events. For him, it was much more logical that the world was entirely deterministic.

Everett began his thesis by describing JOHN VON NEUMANN's "two processes."

> Process 1: The discontinuous change brought about by the observation of a quantity with eigenstates φ_1, φ_2,..., in which the state ψ will be changed to the state φ_j with probability $|\psi, \varphi_j|^2$

> Process 2: The continuous, deterministic change of state of the (isolated) system with time according to a wave equation $\delta\psi/\delta t = U\psi$, where U is a linear operator. [1]

Everett then presents the internal contradictions of observer-dependent collapses of wave functions with examples of "Wigner's Friend," an observer who observes another observer. For whom does the wave function collapse?

Everett considers several alternative explanations for Wigner's paradox, the fourth of which is the standard statistical interpretation of quantum mechanics, which was criticized (correctly) by Einstein as not being a *complete* description.

Alternative 4: To abandon the position that the state function

---

[1]   DeWitt and Graham, 1973, p.3

Chapter 31

is a complete description of a system. The state function is to be regarded not as a description of a single system, but of an ensemble of systems, so that the probabilistic assertions arise naturally from the incompleteness of the description.

It is assumed that the correct complete description, which would presumably involve further (hidden) parameters beyond the state function alone, would lead to a deterministic theory, from which the probabilistic aspects arise as a result of our ignorance of these extra parameters in the same manner as in classical statistical mechanics. [2]

For the most part, Everett seems to represent Einstein's "ensemble" or statistical interpretation, but he also is following DAVID BOHM. In order to be "complete," "hidden variables" would be necessary.

Everett's "theory of the universal wave function" is the last alternative, in which he rejects process 1, wave function collapse:

Alternative 5: To assume the universal validity of the quantum description, by the complete abandonment of Process 1.
The general validity of pure wave mechanics, without any statistical assertions, is assumed for all physical systems, including observers and measuring apparata. Observation processes are to be described completely by the state function of the composite system which includes the observer and his object-system, and which at all times obeys the wave equation (Process 2). [3]

Everett says this alternative has many advantages.

It has logical simplicity and it is complete in the sense that it is applicable to the entire universe. All processes are considered equally (there are no "measurement processes" which play any preferred role), and the principle of psycho-physical parallelism is fully maintained. Since the universal validity of the state function description is asserted, one can regard the state functions themselves as the fundamental entities, and one can even consider the state function of the whole universe. In this sense this theory can be called the theory of the "universal wave function," since all of physics is presumed to follow from this function. [4]

Chapter 31

---

2   DeWitt and Graham, 1973, p.8
3   *ibid.*
4   *ibid.*

Information and Entropy

In a lengthy chapter, Everett develops the concept of information - despite the fact that his deterministic view of physics allows no alternative possibilities. For CLAUDE SHANNON, the developer of the theory of communication of information, there can be no information created ad transmitted without possibilities. Everett correctly observes that in classical mechanics information is a conserved property, a constant of the motion. No new information can be created in such a deterministic universe.

> As a second illustrative example we consider briefly the classical mechanics of a group of particles. The system at any instant is represented by a point...in the phase space of all position and momentum coordinates. The natural motion of the system then carries each point into another, defining a continuous transformation of the phase space into itself. According to Liouville's theorem the measure of a set of points of the phase space is invariant under this transformation. This invariance of measure implies that if we begin with a probability distribution over the phase space, rather than a single point, the total information,... which is the information of the joint distribution for all positions and momenta, remains constant in time. [5]

Everett correctly notes that if total information is constant, the total entropy is also constant.

> if one were to define the total entropy to be the negative of the total information, one could replace the usual second law of thermodynamics by a law of conservation of total entropy, where the increase in the standard (marginal) entropy is exactly compensated by a (negative) correlation entropy. The usual second law then results simply from our renunciation of all correlation knowledge (*stosszahlansatz*), and not from any intrinsic behavior of classical systems. The situation for classical mechanics is thus in sharp contrast to that of stochastic processes, which are intrinsically irreversible.

---

5   *ibid.*, p.31

The *Appearance* of Irreversibility in a Measurement

There is another way of looking at this apparent irreversibility within our theory which recognizes only Process 2. When an observer performs an observation the result is a superposition, each element of which describes an observer who has perceived a particular value. From this time forward there is no interaction between the separate elements of the superposition (which describe the observer as having perceived different results), since each element separately continues to obey the wave equation. Each observer described by a particular element of the superposition behaves in the future completely independently of any events in the remaining elements, and he can no longer obtain any information whatsoever concerning these other elements (they are completely unobservable to him).

The irreversibility of the measuring process is therefore, within our framework, simply a subjective manifestation reflecting the fact that in observation processes the state of the observer is transformed into a superposition of observer states, each element of which describes an observer who is irrevocably cut off from the remaining elements. While it is conceivable that some outside agency could reverse the total wave function, such a change cannot be brought about by any observer which is represented by a single element of a superposition, since he is entirely powerless to have any influence on any other elements.

There are, therefore, fundamental restrictions to the knowledge that an observer can obtain about the state of the universe. It is impossible for any observer to discover the total state function of any physical system, since the process of observation itself leaves no independent state for the system or the observer, but only a composite system state in which the object-system states are inextricably bound up with the observer states. [6]

This is Everett's radical thesis that the observation "splits" the single observer into a "superposition" of multiple observers, each one of which has knowledge only of the new object-system state or "relative state" (interpreted later by Bryce DeWitt as different "universes") As soon as the observation is performed, the composite state is split into a superposition for which each element describes

----

a different object-system state and an observer with (different) knowledge of it. Only the totality of these observer states, with their diverse knowledge, contains complete information about the original object-system state - but there is no possible communication between the observers described by these separate states. Any single observer can therefore possess knowledge only of the relative state function (relative to his state) of any systems, which is in any case all that is of any importance to him.

In the final chapter of his thesis, Everett reviews five possible "interpretations, the "popular", the "Copenhagen", the "hidden variables", the "stochastic process", and the "wave" interpretations.

a. The "popular" interpretation. This is the scheme alluded to in the introduction, where ψ is regarded as objectively characterizing the single system, obeying a deterministic wave equation when the system is isolated but changing probabilistically and discontinuously under observation.[7]

b. The Copenhagen interpretation. This is the interpretation developed by Bohr. The ψ function is not regarded as an objective description of a physical system (i.e., it is in no sense a conceptual model), but is regarded as merely a mathematical artifice which enables one to make statistical predictions, albeit the best predictions which it is possible to make. This interpretation in fact denies the very possibility of a single conceptual model applicable to the quantum realm, and asserts that the totality of phenomena can only be understood by the use of different, mutually exclusive (i.e., "complementary") models in different situations. All statements about microscopic phenomena are regarded as meaningless unless accompanied by a complete description (classical) of an experimental arrangement.[8]

c. The "hidden variables" interpretation. This is the position (Alternative 4 of the Introduction) that ψ is not a complete description of a single system. It is assumed that the correct complete description, which would involve further (hidden) parameters, would lead to a deterministic theory, from which the probabilistic aspects arise as a result of our ignorance of these extra parameters in the same manner as in classical statistical mechanics.[9]

7   *ibid.*, p.110
8   *ibid.*
9   *ibid.*, p.111.

Everett says that here the ψ-function is regarded as a description of an ensemble of systems rather than a single system. Proponents of this interpretation include Einstein and Bohm.

> d. The stochastic process interpretation. This is the point of view which holds that the fundamental processes of nature are stochastic (i.e., probabilistic) processes. According to this picture physical systems are supposed to exist at all times in definite states, but the states are continually undergoing probabilistic changes. The discontinuous probabilistic "quantum-jumps" are not associated with acts of observation, but are fundamental to the systems themselves. [10]

This is very close to our information interpretation of quantum mechanics, which claims that collapses of the wave function result from interactions between quantum systems, independent of any observers or measurement processes.

> e. The wave interpretation. This is the position proposed in the present thesis, in which the wave function itself is held to be the fundamental entity, obeying at all times a deterministic wave equation. [11]

Everett says that his thesis follows most closely the view held by Erwin Schrödinger, who denied the existence of "quantum jumps" and collapses of the wave function. See Schrödinger's *Are There Quantum Jumps?*, Part I and Part II (and, years after Everett, John Bell (1987) and H. Dieter Zeh (1993) who wrote articles with similar themes.

On the "Conscious Observer"

Everett proposed that the complicated problem of "conscious observers" can be greatly simplified by noting that the most important element in an observation is the recorded information about the measurement outcome in the memory of the observer. He proposed that human observers could be replaced by automatic measurement equipment that would achieve the same result. A measurement would occur when information is recorded by the measuring instrument.

---

10   *ibid.*, p.114
11   *ibid.*, p.115.

It will suffice for our purposes to consider the observers to possess memories (i.e., parts of a relatively permanent nature whose states are in correspondence with past experience of the observers). In order to make deductions about the past experience of an observer it is sufficient to deduce the present contents of the memory as it appears within the mathematical model.

As models for observers we can, if we wish, consider automatically functioning machines, possessing sensory apparatus and coupled to recording devices capable of registering past sensory data and machine configurations. [12]

Everett's observer model is a classic example of artificial intelligence.

We can further suppose that the machine is so constructed that its present actions shall be determined not only by its present sensory data, but by the contents of its memory as well. Such a machine will then be capable of performing a sequence of observations (measurements), and furthermore of deciding upon its future experiments on the basis of past results. If we consider that current sensory data, as well as machine configuration, is immediately recorded in the memory, then the actions of the machine at a given instant can be regarded as a function of the memory contents only, and all relevant experience of the machine is contained in the memory. [13]

Everett's observer model has what might be called artificial consciousness.

For such machines we are justified in using such phrases as "the machine has perceived A" or "the machine is aware of A" if the occurrence of A is represented in the memory, since the future behavior of the machine will be based upon the occurrence of A. In fact, all of the customary language of subjective experience is quite applicable to such machines, and forms the most natural and useful mode of expression when dealing with their behavior, as is well known to individuals who work with complex automata. [14]

---

12   ibid., p.64.
13   ibid.
14   ibid.

Everett's model of machine memory completely solves the problem of "Wigner's Friend." As in the information interpretation of quantum mechanics, it is the recording of information in a "measurement" that makes a subsequent "observation" by a human observer possible.

Bryce De Witt

Everett stepped away from theoretical physics almost entirely even before his thesis was finally accepted under JOHN WHEELER and published in the July 1957 issue of *Reviews of Modern Physics*. along with an accompanying article by Wheeler.

Without the strong interest in the many-worlds interpretation of quantum mechanics by Bryce DeWitt, it might have much less interest and influence today.

In 1970, DeWitt wrote an article on Everett's "relative-state" theory for *Physics Today*. A few years later he compiled a collection of Everett's work, including the 1957 paper and the much longer "The Theory of the Universal Wave Function,"along with interpretive articles, by DeWitt, Wheeler, and others.

Summary of Everett's Ideas

Everett's idea for the "universal validity of the quantum description" can be read as saying that quantum mechanics applies to all physical systems, not merely microscopic systems. This is correct. Then the transition to "classical" mechanics emerges in the limit of the Planck quantum of action $h \rightarrow 0$, or more importantly, $h/m \rightarrow 0$ (since h never changes), so that classical physics appears in large massive objects (like human beings) because the indeterminacy is too small to measure.

Like Einstein, Everett says that the ψ-function is a description of an ensemble of systems rather than a single system. It is true that the phenomenon of wave interference is only inferred from the results of many single particle experiments. We never "see" interference in single particles directly. Probabilistic assertions arise naturally from the incompleteness of the description.

Everett correctly observes that in classical mechanics information is a conserved property, a constant of the motion. No new information can be created in a classical universe. But the observed universe has clearly been gaining new information structures since the origin. Indeed, both information and entropy have been increasing and continue to increase today. This cannot be explained by Everett.

Everett's automatic measuring equipment that stores information about measurements in its "memory" nicely solves von Neumann's problem of "psycho-physical parallelism" in "conscious-observer"-dependent quantum mechanics, like the Bohr-Heisenberg "Copenhagen Interpretation."

The Everett theory preserves the "appearance" of possibilities as well as all the results of standard quantum mechanics. It is an "interpretation" after all. So even wave functions "appear" to collapse. Note that if there are many possibilities, whenever one becomes actual, the others disappear instantly. In Everett's theory, they become other possible worlds

Unfortunately, as DeWitt and most modern followers of Everett see it, alternative possibilities are in different, inaccessible universes. In each deterministic universe, there is only one possible future.

Many of Everett's original ideas become central in later deterministic interpretations of quantum mechanics, such as the decoherence program of H.DIETER ZEH and WOJCIECH ZUREK.

Some of Everett's important new ideas show up also in the work of JOHN BELL, to which we now turn.

Chapter 31

Statistic

Quantum Mechanics

Light Quantum
Hypothesis

Photoelectric
Effect

Bohr-Einstein Atom

ave-Particle Duality

Bo

John Bell's
Inequality

Born-Einstein Statistic

Chance

Nonlocality

Irreversibility

Nonsepa

Einstein-Podolsky-Rosen

E

Schrödinger's Cat

Did Albert Einstein Inver

John Bell's Inequality

In 1964 John Bell showed how the 1935 "thought experiments" of Einstein, Podolsky, and Rosen (EPR) could be made into real experiments. He put limits on DAVID BOHM's "hidden variables" in the form of what Bell called an "inequality," a *violation* of which would confirm standard quantum mechanics. Bell appears to have hoped that Einstein's dislike of quantum mechanics could be validated by hidden variables, returning to physical determinism.

But Bell lamented late in life...

> It just is a fact that quantum mechanical predictions and experiments, in so far as they have been done, do not agree with [my] inequality. And that's just a brutal fact of nature... that's just the fact of the situation; the Einstein program fails, that's too bad for Einstein, but should we worry about that?

> I cannot say that action at a distance is required in physics. But I can say that you cannot get away with no action at a distance. You cannot separate off what happens in one place and what happens in another. Somehow they have to be described and explained jointly. [1]

Bell himself came to the conclusion that *local* "hidden variables" will never be found that give the same results as quantum mechanics. This has come to be known as Bell's Theorem.

Bell concluded that all theories that reproduce the predictions of quantum mechanics will be "nonlocal." But as we saw in chapter 23, Einstein's nonlocality defined as an "action" by one particle on another in a spacelike separation ("at a distance") at speeds faster than light, simply *does not exist*. What does exist is Einstein's "impossible simultaneity" of events in a spacelike separation.

We have seen that the ideas of nonlocality and *nonseparability* were invented by Einstein, who disliked them, just as he disliked his discovery of chance. ERWIN SCHRÖDINGER also disliked chance, but his wave mechanics can explain the perfect correlations of the properties of entangled particles. See chapter 29.

We explained entanglement as the consequence of "hidden constants" that are "local" in the sense that they are carried along with the moving particles, conserving all the particles' properties so they remain perfectly correlated whenever they are measured.

Chapter 32

---

1   Transcript of CERN talk. http://www.youtube.com/watch?v=V8CCfOD1iu8

These pre-existing local constants can not explain the perfect correlation of Alice and Bob's measurements in a specific spatial direction. This we attribute to the projection of the directionless and symmetric two-particle wave function into a specific spin direction by Alice's measurement.

Experiments to test Bell's inequality have done more to prove the existence of entangled particles than any other work. As a result, many people credit Bell with the very idea of entanglement. Our efforts to restore credit to Einstein for this and most other exotic effects in quantum mechanics is therefore not an easy task.

This is particularly difficult because Einstein did not like much of what he was first person to see - single-particle nonlocality, two-particle nonseparability, and other fundamental elements of quantum mechanics, notably its statistical nature, indeterminism, and ontological chance.

We saw in chapter 30 that DAVID BOHM developed a version of quantum theory that would restore determinism to quantum mechanics as well as explaining nonlocality. This was the beginning of a trend among young physicists to question the *foundations* of quantum mechanics. No one was more supportive of this trend than Bell, though he warned all his younger colleagues that questioning the "orthodox" Copenhagen Interpretation could compromise their academic advancement.

We have chosen Bohm, Hugh Everett, Bell, and the decoherence theorists as the leading members of the effort to challenge "standard" quantum mechanics, although there are several others. Ironically, they all base their work on trying to support Einstein's criticisms of quantum mechanics, especially his early hopes for restoring determinism, whereas Einstein in his later life had moved on to his worries about nonlocality violating relativity.

From his earliest work, Bell followed Bohm's deterministic and nonlocal alternative to standard quantum mechanics. He also followed Schrödinger's denial of quantum jumps and even the existence of particles. Decoherence theorists agree on this denial of Dirac's *projection postulate*. Like Schrödinger, they use a misinterpretation of Dirac's *principle of superposition, viz.,* that particles can be in multiple states at the same time.

Bell's Theorem

In his classic 1964 paper "On the Einstein-Podolsky-Rosen Paradox," Bell made the case for *nonlocality*.

> The paradox of Einstein, Podolsky and Rosen was advanced as an argument that quantum mechanics could not be a complete theory but should be supplemented by additional variables. These additional variables were to restore to the theory causality and locality. In this note that idea will be formulated mathematically and shown to be incompatible with the statistical predictions of quantum mechanics. It is the requirement of locality, or more precisely that the result of a measurement on one system be unaffected by operations on a distant system with which it has interacted in the past, that creates the essential difficulty. There have been attempts [by von Neumann] to show that even without such a separability or locality requirement no 'hidden variable' interpretation of quantum mechanics is possible. These attempts have been examined [by Bell] elsewhere and found wanting. Moreover, a hidden variable interpretation of elementary quantum theory has been explicitly constructed [by Bohm]. That particular interpretation has indeed a gross non-local structure. This is characteristic, according to the result to be proved here, of any such theory which reproduces exactly the quantum mechanical predictions.
>
> With the example advocated by Bohm and Aharonov, the EPR argument is the following. Consider a pair of spin one-half particles formed somehow in the singlet spin state and moving freely in opposite directions. Measurements can be made, say by Stern-Gerlach magnets, on selected components of the spins σ_1 and σ_2. If measurement of the component $\sigma_1 \cdot a$, where a is some unit vector, yields the value + 1 then, according to quantum mechanics, measurement of $\sigma_2 \cdot a$ must yield the value — 1 and vice versa. Now we make the hypothesis, and it seems one at least worth considering, that if the two measurements are made at places remote from one another the orientation of one magnet does not influence the result obtained with the other.
>
> Since we can predict in advance the result of measuring any chosen component of σ_2, by previously measuring the same component of σ_1, it follows that the result of any such measurement must actually be predetermined. Since the initial

Chapter 32

quantum mechanical wave function does not determine the result of an individual measurement, this predetermination implies the possibility of a more complete specification of the state. [2]

"pre-determination" is too strong a term. The "previous" measurement just "determines" the later measurement.

As we showed in chapter 29, there are in fact many properties that are determined at the initial entanglement and are conserved from that moment to the measurement of σ1 • a. We call them "hidden constants." They are *local* quantities that travel with the particles.

Experimental Tests of Bell's Inequality

Bell experiments are usually described as the distant measurements of electron spins or photon polarizations by Alice and Bob, when their polarization or spin detectors are set at different angles.

Electrons in an entangled "singlet" spin state have spins in opposite directions. As Bell said above, when measured at the same angle (0°), spins are anti-correlated. The correlation is -1. If measured in opposite directions (180°) , the correlation is +1.

Measurements at 90° are completely uncorrelated. With photons, a vertically polarized photon will be completely absorbed by a horizontal polarizer.

Measurements will be decorrelated randomly at a small angle from 0°, say 1°. Since Bell assumes (with no physical reason) that measurements at 1° more (now 2°) are statistically independent of those in the first 1° angle, they should be no more than twice the decorrelation of the first 1° angle. Bell therefore predicts that the correlations at other angles will yield a straight-line relationship.

But it is well known that when polarizers are rotated, the correlations fall off as the cosine (amplitude) or cosine² (intensity). Measuring the components of spins or polarization at intermediate angles shows a "violation" of what Bell called his inequality. Instead of his

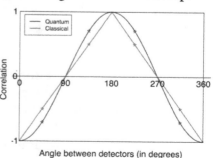

Angle between detectors (in degrees)

---

2   Bell, 1964, p.195

physically unrealistic straight-line correlation for hidden variables, we see the quantum results tracing out a sinusoid.

The most important intermediate angle, where the deviation from Bell's straight line is the greatest, is 22.5°.

At that angle, one-quarter of the way to 90° where the correlation will be 0, Bell's hidden variables prediction is a correlation of only 75%. The quantum physics correlation is $\cos^2(22.5°) = 85\%$.

We can display the above curves inside a unit square of possible correlations, with an inside square of Bell's local hidden variables, and then the circular region of quantum mechanics correlations, which are the same as Bell's at the corners, but move out to the circle at intermediate angles.

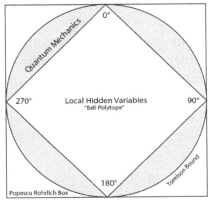

In 1976, Bell knew very well that the behavior of his local hidden variables at the corners has a physically unrealistic sharp "kink."[3] He said unlike the quantum correlation, which is a smooth curve stationary in θ at $\theta = 0$, the hidden variable correlation must have a kink there. He illustrated the unrealistic "kink."

What is the origin of this kink? It is buried in Bell's assumptions about his "hidden variables," that they are random, hidden in pre-existing conditions at the start of the experiment, and they can predict all the outcomes. Bell assumed that the variables can be specified

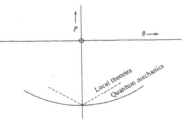

completely by means of parameters λ, where λ has a "uniform probability distribution"[4] over angles, It is this uniform distribution that leads to his unrealistic straight line prediction.

Bell's inequality for hidden variables is not based on physics as much as his assumed distribution of probabilities. By contrast, there are good physical reasons to think that we can *visualize* the

Chapter 32

angular dependence of correlations by recalling PAUL DIRAC's work with polarizers crossed at various angles (chapter 19). When Bob measures at the same angle as Alice, or even at angles 180° apart, the polarized light will pass straight through (a non-destructive measurement of the first kind). As we turn one polarizer away from the parallel or anti-parallel angles, some of the light is absorbed in the polarizer, but not very much at first, then falling off more quickly as we approach 90° where all the light is absorbed, There is no "kink" at 0° or 180°.

The earliest measurements were done in the hope of finding hidden variables and showing quantum physics to be "incomplete." As early as 1969 John Clauser, Michael Horne, Abner Shimony, and Richard Holt had shown Bell's hidden variable prediction had been violated and quantum physics was validated.

Here is the apparatus for the classic CHSH experiment. [5]

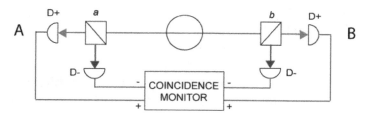

The coincidence monitor accumulates N++, N+-, N-+, and N--. As B's polarizer turns away from parallel, where perfect correlation is say, $| + - >$ or $| - + >$, we start to get randomness that produces results like $| + + >$ or $| - - >$. At 22.5°, Bell's straight-line hidden variables predicts 75% of measurements will be correlated + - or - +, the other 25% a random mixture of + +, - -, + -, - +.

Here are some experimental results using protons in a singlet state that confirm the 85% correlation predicted for quantum mechanics. [6]

In particular, note the confirmation of the curved sinusoidal (or cosine) shape and not Bell's physically

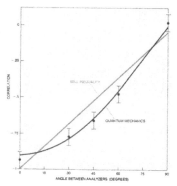

---

5   Clauser et al. 1969

6   d'Espagnat, 1979, p.174

unrealistic set of straight lines with sharp kinks at the corners that Bell's inequality predicts.

With quantum mechanics confirmed, why didn't Bell and his many supporters simply give up the search for hidden variables that he claimed could validate Einstein? How can Bell inequality tests still be considered important after so many years of success? It is probably the continued dissatisfaction with quantum mechanics

As early as 1970, EUGENE WIGNER, who became a lifelong supporter of attempts to provide new foundations for quantum mechanics, had clearly explained what the results would be of a Bell inequality test, well before the CHSH results were published.

> Bell does introduce, however, the postulate that the hidden variables determine the spin component of the first particle in any of the ω directions and that this component is independent of the direction in which the spin component of the second particle is measured. Conversely, the values of the hidden variables also determine the spin component of the second particle in any of the three directions ω_1, ω_2, ω_3, and this component is independent of the direction in which the component of the spin of the first particle is measured. These assumptions are very natural since the two particles may be well separated spatially so that the apparatus measuring the spin of one of them will not influence the measurement carried out on the other. Bell calls, therefore, the assumption just introduced the locality assumption...

> Wigner says that the angular dependence of correlations can be derived also by observing that the singlet state is spherically symmetric so that the total probability of the first particle's spin being in the direction ω_i (rather than the opposite direction) is ½|. If the measurement of the first particle's ω_i component gives a positive result, the measurement of this component of the second particle necessarily gives a negative result. Hence, the measurement of the spin of this particle in the ω_2 direction gives a positive result with the probability $\cos^2 \tfrac{1}{2}\theta$, where θ is the angle between the $-\omega_i$ and the ω_2 direction.[7]

John Bell surely knew enough physics to recognize that his straight line "inequality" would never be found and that the sinusoidal correlations of quantum mechanics would be confirmed. Yet he encouraged young experimenters to try, in the vain hopes that they would overturn quantum mechanics and become world famous.

---

7   Wigner, 1970, p.1007

As it turned out. they (and so Bell) did become world famous, not for disproving quantum mechanics, but for discovering the kind of nonlocality and nonseparability that Einstein had seen and feared.

Experimenters noted the low quality of the results and significant sources of errors in older laboratory technology, which might contain "loopholes" that would allow "Einstein's" hidden variables and return to determinism. Their search continued for decades, attracting vast amounts of publicity for the "age of entanglement."

Most all the loopholes have now been closed, but there is one loophole that can never be closed because of its metaphysical/philosophical nature. That is the "(pre-)determinism loophole." Bell called it "superdeterminism.

If every event occurs for reasons that were established at the beginning of the universe, then the experimenters lack any free will or "free choice" and all their experimental results are meaningless.

Bell's Superdeterminism

During a mid-1980's interview by BBC Radio 3 organized by P. C. W. Davies and J. R. Brown, Bell proposed the fanciful idea of "superdeterminism" that could explain the correlation of results in two-particle experiments without the need for faster-than-light signaling. The two measurements by Alice and Bob need only have been pre-determined by causes reaching both experiments from an earlier time.

> Davies: I was going to ask whether it is still possible to maintain, in the light of experimental experience, the idea of a deterministic universe?

> Bell: You know, one of the ways of understanding this business is to say that the world is super-deterministic. That not only is inanimate nature deterministic, but we, the experimenters who imagine we can choose to do one experiment rather than another, are also determined. If so, the difficulty which this experimental result creates disappears.

> Davies: Free will is an illusion - that gets us out of the crisis, does it?

> Bell: That's correct. In the analysis it is assumed that free will is genuine, and as a result of that one finds that the intervention

of the experimenter at one point has to have consequences at a remote point, in a way that influences restricted by the finite velocity of light would not permit. If the experimenter is not free to make this intervention, if that also is determined in advance, the difficulty disappears. [8]

Bell's superdeterminism would deny the important "free choice" of the experimenter (originally suggested by NIELS BOHR and WERNER HEISENBERG) and later explored by JOHN CONWAY and SIMON KOCHEN. Conway and Kochen claim that the experimenters' free choice requires that electrons themselves must have free will, something they call their "Free Will Theorem."

Following Bell's ideas, NICHOLAS GISIN and ANTOINE SUAREZ argue that something might be coming from "outside space and time" to correlate results in their own experimental tests of Bell's Theorem. ROGER PENROSE and STUART HAMEROFF have proposed causes coming "backward in time" to achieve the perfect EPR correlations, as has philosopher HUW PRICE.

In his 1997 book, *Time's Arrow and Archimedes' Point*, Price proposes an Archimedean point "outside space and time" as a solution to the problem of nonlocality in the Bell experiments in the form of an "advanced action." [9]

Rather than a "superdeterministic" common cause coming from "outside space and time" (as proposed by Bell, Gisin, Suarez, and others), Price argues that there might be a cause coming backwards in time from some interaction in the future. Penrose and Hameroff have also promoted this idea of "backward causation," sending information backward in time in BENJAMIN LIBET's experiments and in the EPR experiments.

JOHN CRAMER's Transactional Interpretation of quantum mechanics and other Time-Symmetric Interpretations like that of Yakir Aharonov and K. B Wharton also search for Archimedean points "outside space and time."

All these wild ideas designed to return physical determinism are in many ways as extravagant as Hugh Everett's "many worlds."

Chapter 32

Bell's Preferred Frame

A little later in the same BBC interview, Bell suggested that a *preferred* frame of reference might explain nonseparability and entanglement. And there is something valuabe in this picture.

> [Davies] Bell's inequality is, as I understand it, rooted in two assumptions: the first is what we might call objective reality - the reality of the external world, independent of our observations; the second is locality, or non-separability, or no faster-than-light signalling. Now, Aspect's experiment appears to indicate that one of these two has to go. Which of the two would you like to hang on to?
>
> [Bell] Well, you see, I don't really know. For me it's not something where I have a solution to sell! For me it's a dilemma. I think it's a deep dilemma, and the resolution of it will not be trivial; it will require a substantial change in the way we look at things. But I would say that the cheapest resolution is something like going back to relativity as it was before Einstein, when people like Lorentz and Poincare thought that there was an aether - a preferred frame of reference - but that our measuring instruments were distorted by motion in such a way that we could not detect motion through the aether. Now, in that way you can imagine that there is a preferred frame of reference, and in this preferred frame of reference things do go faster than light. But then in other frames of reference when they seem to go not only faster than light but backwards in time, that is an optical illusion. [10]

The standard explanation of entangled particles usually begins with an observer A, often called Alice, and a distant observer B, known as Bob. Between them is a source of two entangled particles. The two-particle wave function describing the indistinguishable particles cannot be separated into a product of two single-particle wave functions, at least until the wave function is measured..

The problem of faster-than-light signaling arises when Alice is said to measure particle A and then puzzle over how Bob's (later) measurements of particle B can be perfectly correlated, when there is not enough time for any "influence" to travel from A to B.

Now as John Bell knew very well, there are frames of reference moving with respect to the laboratory frame of the two observers in

10   Davies and Brown, 1993, p.48-9

which the time order of the events can be reversed. In some moving frames Alice measures first, but in others Bob measures first.

Back in the 1960's, C. W. RIETDIJK and HILARY PUTNAM considered observers A and B in a "spacelike" separation and moving at high speed with respect to one another. ROGER PENROSE developed a similar argument in his book *The Emperor's New Mind*. He called it the Andromeda Paradox. [11]

If there is a preferred or "special" frame of reference, surely it is the one in which the origin of the two entangled particles is at rest. Assuming that Alice and Bob are also at rest in this special frame and equidistant from the origin, we arrived in chapter 29 at the simple picture in which any measurement that causes the two-particle wave function Ψ_{12} to collapse makes both particles appear simultaneously at determinate places (just what is needed to conserve energy, momentum, angular momentum, and spin).

Bell became world-famous as the major proponent of quantum entanglement, understood as the instantaneous transmission of a signal between quantum systems, however far apart.

> In a theory in which parameters are added to quantum
> mechanics to determine the results of individual measurements,
> without changing the statistical predictions, there must be a
> mechanism whereby the setting of one measuring device can
> influence the reading of another instrument, however remote.
> Moreover, the signal involved must propagate instantaneously, so
> that such a theory could not be Lorentz invariant. [12]

Einstein would surely have rejected this argument, as he had rudely dismissed that of David Bohm, because it violates relativity with an "impossible simultaneity." Bell's continued defense of hidden variables was motivated in part by his objections to JOHN VON NEUMANN's "proof" that hidden variables are "impossible." He was also a critic of von Neumann's theory of measurement, especially the "collapse" in von Neumann's "process 1" and the need for a "conscious observer."

---

11   Penrose, 1989, p.303
12   Bell, 1964, p.199

As we saw in chapter 25, von Neumann developed WERNER HEISENBERG's idea that the collapse of the wave function requires a "cut" (*Schnitt* in German) between the microscopic quantum system and the observer. Von Neumann said it did not matter where this cut was placed along the "psycho-physical" path between the experiment, the observer's eye, and the observer's mind, because the mathematics would produce the same experimental results. Bell called this a "shifty split."

Bell's "Shifty Split"

We can identify Bell's "shifty split" with the "moment" at which the boundary between the quantum and classical worlds occurs. It is the moment that *irreversible* observable *information* enters the universe.

In Bell's drawing of possible locations for his "shifty split" we can identify the correct moment - when irreversible new information appears, independent of an observer's mind.

In our information solution to the problem of measurement, the timing and location of Bell's "shifty split" (the "cut" or "Schnitt" of Heisenberg and von Neumann) are identified with the interaction between quantum system and classical apparatus that leaves the apparatus in an *irreversible* stable state providing information to the observer.

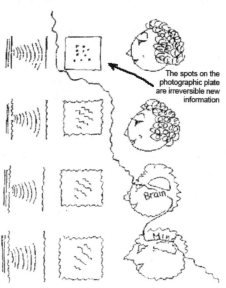

The spots on the photographic plate are irreversible new information

As Bell should have seen, it is therefore not a "measurement" by a conscious observer that is needed to "collapse" wave functions. It is the irreversible interaction of the quantum system with another system, whether quantum or approximately classical. The interaction must be one that changes the information about the system. And that means a local entropy decrease and overall entropy increase to make the information stable enough to be observed by an experimenter and therefore be a measurement.

We can identify the "cut" as the moment information is recorded in the universe, and so available to an observer. In Bell's diagram, it is the appearance of spots on the photogra[phic plate or CCD.

Are There Quantum Jumps?

In 1987, Bell contributed an article to a centenary volume for Erwin Schrödinger entitled "Are There Quantum Jumps?" Schrödinger had always denied such jumps or any collapses of the wave function. Bell's title was inspired by two articles with the same title by Schrödinger in 1952 (Part I, Part II). [13]

Just a year before Bell's death in 1990, physicists assembled for a conference on "62 Years of Uncertainty" (referring to WERNER HEISENBERG's 1927 principle of indeterminacy).

John Bell's contribution to the conference was an article called "Against Measurement." In it he attacked the statistical interpretation of quantum mechanics.

> In the beginning, Schrödinger tried to interpret his wavefunction as giving somehow the density of the stuff of which the world is made. He tried to think of an electron as represented by a wavepacket — a wave-function appreciably different from zero only over a small region in space. The extension of that region he thought of as the actual size of the electron — his electron was a bit fuzzy. At first he thought that small wavepackets, evolving according to the Schrödinger equation, would remain small. But that was wrong. Wavepackets diffuse, and with the passage of time become indefinitely extended, according to the Schrödinger equation. But however far the wavefunction has extended, the reaction of a detector to an electron remains spotty. So Schrödinger's 'realistic' interpretation of his wavefunction did not survive. [14]

> Then came the Born interpretation. The wavefunction gives not the density of stuff, but gives rather (on squaring its modulus) the density of probability. Probability of what exactly? Not of the electron being there, but of the electron being found there, if its position is 'measured.'

> Why this aversion to 'being' and insistence on 'finding'? The founding fathers were unable to form a clear picture of things

13   Schrödinger, 1952
14   Miller, 2012, p.29. We saw this in chapter18.

on the remote atomic scale. They became very aware of the intervening apparatus, and of the need for a 'classical' base from which to intervene on the quantum system.

As we saw in chapter 20, It was Einstein who first interpreted the light wave as the probability of finding particles and as "guiding" the motion of particles. Once the Schrödinger wave function was invented, MAX BORN said that $|\psi|^2$ gives us precisely the probability of finding particles. Why did Bell dislike this powerful idea?

> In the picture of de Broglie and Bohm, every particle is attributed a position x(t). Then instrument pointers — assemblies of particles have positions, and experiments have results. The dynamics is given by the world Schrödinger equation plus precise 'guiding' equations prescribing how the x(t)s move under the influence of Ψ.

In the Bohmian mechanics picture, particles are traveling along distinct paths. Einstein's "objective reality" is a similar view. If the particles are conserving "constants of the motion," they correlate properties in Bell experiments without nonlocal "hidden variables."

We have seen how the "guiding" wave function produces perfectly correlated spin directions for Alice and Bob measurements,.in chapter 29. How it can guide individual particles to produce the statistical interference patterns in the two-slit experiment we will explain in the next chapter.

On the 22nd of January 1990, Bell gave a talk at CERN in Geneva summarizing the situation with his inequalities. He gives three reasons for not worrying.

- Nonlocality is unavoidable, even if it looks like "action at a distance." [It also looks like an "impossible simultaneity"]
- Because the events are in a spacelike separation, either one can occur before the other in some relativistic frame, so no "causal" connection can exist between them.
- No faster-than-light signals can be sent using entanglement and nonlocality.

Bell concluded:

> So as a solution of this situation, I think we cannot just say 'Oh oh, nature is not like that.' I think you must find a picture

in which perfect correlations are natural, without implying determinism, because that leads you back to nonlocality. And also in this independence as far as our individual experiences goes, our independence of the rest of the world is also natural. So the connections have to be very subtle, and I have told you all that I know about them. Thank you.

John Bell Today

Bell is revered as a founder of the "second revolution" in quantum mechanics. He is also a major figure in the call for new "foundations of quantum mechanics." Bell's Theorem has been described as the founding result of quantum information theory.

His fame rests on the idea that there is something wrong with quantum mechanics and that Einstein's call for additional variables to "complete" quantum mechanics is part of the solution.

Einstein was bothered by the claim of the Copenhagen Interpretation that nothing can be known about an "objective reality" independent of human observers. Even more extreme was the anthropo-centered idea that human observers are creating reality, that nothing exists until we measure it.

We have seen that the "free choice" of the experimenter does indeed create aspects of physical reality, in Bell's case it is the preferred angles of Alice and Bob that are the core idea of entangled particles in a spacelike separation that acqure values instantaneously, simultaneously, appearing to violate Einstein's principle of relativity..

Einstein worried about this nonlocality from his *annus mirabilis* in 1905 to the end of his life. But Bell's "inequality,"a physically unrealistic straight-line and linear dependence of correlations between Alice and Bob as they rotate their polarizers, is nothing Einstein would ever have accepted. For Bell to call it "Einstein's program," and pronounce it a failure, is a great disservice to Einstein.

Nevertheless, it is poetic justice that Bell returns Einstein to the center of attention in "quantum physics 2.0," the second revolution.

Two entangled particles are now known as "EPR pairs," in four possible "Bell states." These pairs are also called "qubits," the fundamental unit of quantum computing and communication.

Chapter 32

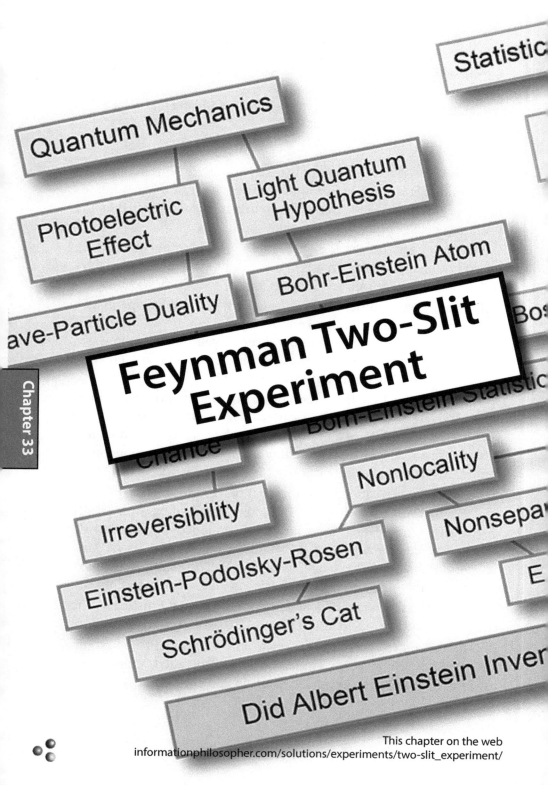

Chapter 33

Feynman Two-Slit Experiment

RICHARD FEYNMAN said that the two-slit experiment contains "all of the mystery" of quantum mechanics.

> I will take just this one experiment, which has been designed to contain all of the mystery of quantum mechanics, to put you up against the paradoxes and mysteries and peculiarities of nature one hundred per cent. Any other situation in quantum mechanics, it turns out, can always be explained by saying, 'You remember the case of the experiment with the two holes? It's the same thing' I am going to tell you about the experiment with the two holes. It does contain the general mystery; I am avoiding nothing; I am baring nature in her most elegant and difficult form. [1]

We will show that the two-slit experiment does contain the key mystery of quantum mechanics, but it's not exactly what Feynman described in 1964. It is connected to the new mystery of "entanglement." Feynman's mystery was simply how a particle can interfere with itself if it goes through only one slit. Our view is that the particle goes through one slit. We show that it is the probability amplitude of the wave function that is interfering with itself.

We are making use of Einstein's vision of an "objective reality." We say the motion of an individual particle of matter or energy obeys fundamental conservation principles - conservation of all a particle's properties. This means the particle path exists and it is smooth and continuous in space and time, even if it impossible to measure the path, to determine its position without disturbing it.

This claim is very controversial, because WERNER HEISENBERG's description of the Copenhagen Interpretation insists that "the path only comes into existence when we measure it."

Einstein said that claiming a particle has no position just before we measure it is like saying the moon only exists when we are looking at it! That it is impossible *to know* the path of a particle without measuring it does not mean that a path does not exist.

---

1   Feynman, 1967, chapter 6

Chapter 33

We are left with the mystery as to how mere "probabilities" can influence (statistically control) the positions of material particles - how *immaterial* information can affect the *material* world. This remains the deep *metaphysical* mystery in quantum mechanics.

There is something similar in quantum entanglement, where measurement of one particle appears to transmit something to the other "entangled" particle. In the two-slit experiment it is the value of the wave function at one place "influencing" the location where the particle appears. In entanglement, the collapse of the two-particle wave function leaves the spin components ot the two particles correlated perfectly.

Like Einstein's 1927 description of nonlocality, both of these involve the "impossible" simultaneity of events in a spacelike separation.

In the two-slit experiment, just as in the Dirac Three Polarizers experiment,[2] the critical case to consider is just one photon or electron at a time in the experiment.

With one particle at a time (whether photon or electron), the quantum object is mistakenly described as interfering with itself, when interference is never seen in a single event. It only shows up in the statistics of large numbers of experiments. Indeed, interference fringes are visible even in the one-slit case, although this is rarely described in the context of the quantum mysteries.

It is the fundamental relation between a particle and the associated wave that controls its probable locations that raises the "local reality" question first seen in 1905 and described in 1909 by Einstein. Thirty years later, the EPR paper and Erwin Schrödinger's insights into the wave function of two entangled particles, first convinced a few physicists that there was a deep problem .

It was not for another seventeen years that David Bohm suggested an experimental test of EPR and thirty years before John Stewart Bell in 1964 imagined an "inequality" that could confirm or deny quantum mechanics. Ironically, the goal of Bell's "theorem" was to invalidate the non-intuitive aspects of quantum mechanics and restore Einstein's hope for a more deterministic picture of an "objective reality" at, or perhaps even underlying below, the microscopic level of quantum physics.

---

2   See chapter 19.

At about the same time, in his famous *Lectures on Physics* at Cal Tech and the Messenger Lectures at Cornell, Feynman described the two-slit experiment as demonstrating what has since been described as the "only mystery" of quantum mechanics.

How, Feynman asked, can the particle go through both slits? We will see that if anything goes through both slits it is only *immaterial information* - the probability amplitude wave function. The particle itself always goes through just one slit. A particle cannot be divided and in two places at the same time. It is the probability amplitude wave function that interferes with itself.

A highly localized particle can not be identified as the wave widely distributed in space. We will show that the wave function is determined by the boundary conditions of the measuring apparatus. It has nothing to do with whether or not a particle is in the apparatus, though it depends on the wavelength of the particle.

The immaterial wave function exerts a causal influence over the particles, one that we can justifiably call "mysterious." It results in the statistics of many experiments agreeing with the quantum mechanical predictions with increasing accuracy as we increase the number of identical experiments.

It is this "influence," no ordinary "force," that is at the heart of Feynman's "mystery" in quantum mechanics.

We will show that the probability of finding particles at different places in the two-slit experiment is determined by solving the Schrödinger equation for its eigenvalues and eigenfunctions (wave functions and probability amplitudes), given the boundary conditions of the experiment.

The wave function and its probabilities depend on the boundary conditions, such as *whether one slit is open or two*. They do not depend on whether a particle is actually present, though the calculations depend on the wavelength of a particle.

The two-slit experiment shows better than any other experiment that a quantum wave function is a probability amplitude that *interferes with itself*, producing some places where the probability (the square of the absolute value of the complex probability amplitude) of finding a quantum particle is actually zero.

Perhaps the most non-intuitive aspect of the two-slit experiment is this. When we see the pattern of light on the screen with just one slit open, then open the second slit - admitting more light into the experiment - we observe that some places on the screen where there was visible light, have now gone dark! And this happens even when we are admitting only one particle of light at a time.

Let's remind ourselves about how the crests and troughs of water waves interfere, and then how Feynman presented the two-slit experiment to students in his famous *Lectures on Physics*.

Let's look first at the one-slit case. We prepare a slit that is about the same size as the wavelength of the light in order to see the interference of waves most clearly. Parallel waves from a distant source fall on the slit from below. The diagram shows how the wave from the left edge of the slit interferes with the one from the right edge. If the slit width is d and the photon wavelength is λ, at an angle $\alpha \approx \lambda/2d$ there will be destructive interference.

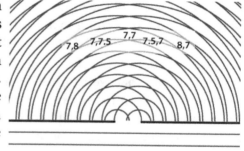

At an angle $\alpha \approx \lambda/d$, there is constructive interference (which shows up as the fanning out of light areas in the interfering waves in the illustration). The diagram indicates constructive interference between the 7th and 8th waves from the left and right sides of the slit.

Feynman began with a description of bullets fired at a screen with two holes, arguing that bullets do not interfere, he showed that the pattern with two holes open is simply the sum of the results from one hole or the other hole open.

$$P_{12} = P_1 + P_2$$

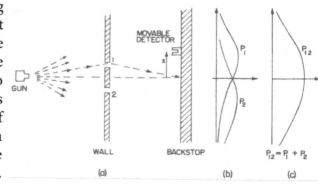

He then described the results for water waves.

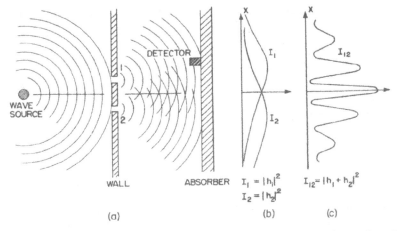

$I_1 = |h_1|^2$
$I_2 = |h_2|^2$

$I_{12} = |h_1 + h_2|^2$

(a)   (b)   (c)

Here the individual results I_1 and I_2 for one or the other hole open do not simply add up. The individual wave intensities are the squares of the amplitudes - $I_1 = |h_1|^2$, $I_2 = |h_2|^2$. Instead they show the cancellation of crests and troughs that produce constructive and destructive interference. The formula is $I_{12} = |h_1 + h2|^2$. This has the same pattern of bright and dark areas that are found in the "fringes" of light at the sharp edges of an object.

Feynman next shows how a two-slit experiment using electrons does not behave like bullets, but instead looks just like water waves, or light waves. He then shows that the mathematics is the same as for water waves. But he says "It is all quite mysterious. And the more

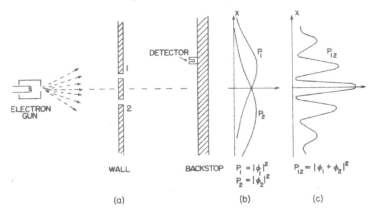

$P_1 = |\phi_1|^2$
$P_2 = |\phi_2|^2$

$P_{12} = |\phi_1 + \phi_2|^2$

(a)   (b)   (c) you look at it the more mysterious it seems." "How can such an interference come about?", he asks. "Perhaps...it is not true that the lumps go either through hole 1 or hole 2." He says

> We conclude the following: The electrons arrive in lumps, like particles, and the probability of arrival of these lumps is distributed like the distribution of intensity of a wave. It is in this sense that an electron behaves "sometimes like a particle and sometimes like a wave"...
>
> The only answer that can be given is that we have found from experiment that there is a certain special way that we have to think in order that we do not get into inconsistencies. What we must say (to avoid making wrong predictions) is the following.
>
> If one looks at the holes or, more accurately, if one has a piece of apparatus which is capable of determining whether the electrons go through hole 1 or hole 2, then one can say that it goes either through hole 1 or hole 2. But, when one does not try to tell which way the electron goes, when there is nothing in the experiment to disturb the electrons, then one may not say that an electron goes either through hole 1 or hole 2. If one does say that, and starts to make any deductions from the statement, he will make errors in the analysis. This is the logical tightrope on which we must walk if we wish to describe nature successfully.

Einstein was deeply bothered by this Copenhagen thinking that claims that we cannot know the particle path, that a path does not even exist until we make a measurement, that the particle may be in more than one place at the same time, maybe dividing and going through both slits, etc.

So let's combine *conservation principles* with Einstein's view that it is the wave function that determines the probability and the statistics of particle positions for a large number of experiments (he called it an "ensemble").

We can then argue, corresponding to Einstein's idea of an "objective reality," that the particle of matter or energy always *goes through just one slit* in a continuous, though *unknown* path.

But whichever slit the particle enters, the probability of finding it at a specific location inside the apparatus is determined by the square of the absolute value $|\Psi|^2$ of the complex probability amplitude at that location.

The probability amplitude is the solution to the Schrödinger equation given the boundary conditions. And the boundary conditions depend on whether one or two slits are open!

We can thus overcome Feynman's difficulties, his inconsistencies, his "special way to think," and his "logical tightrope." Mostly, Einstein's reality view denies an electron behaves "sometimes like a particle and sometimes like a wave." The particle is real. The wave is an accurate theory about the particle's behavior.

We may never be able to measure the specific location of an electron in an atomic orbit. But the wave function gives us all the information we need about atomic orbitals to do the quantum mechanics of atoms and possible molecules, with their nodal surfaces, just like the nodes in the two-slit interference pattern.

Let's compare the wave functions inside the two-slit apparatus when one slit or two slits are open.

With one slit open we see the classic Fraunhofer pattern with their light zones of constructive interference and dark zones where the waves are one-half wavelength different, so the crest of one wave cancels the trough of the other. Many texts mistakenly say that interference is only possible with two slits open.

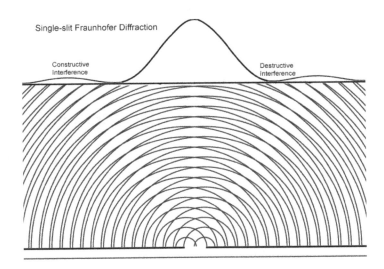

Single-slit Fraunhofer Diffraction

Constructive
Interference

Destructive
Interference

With two slits open we can still see the overall shape of the single-slit Fraunhofer pattern with its broad central maximum, but now

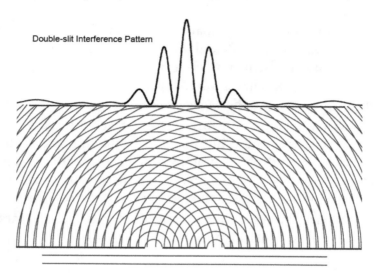

Double-slit Interference Pattern

multiple interference fringes appear.

We claim that this interference pattern *does not depend on which slit the particle enters*, but only on the probability amplitude of the wave function that solves the Schrödinger equation inside the experimental apparatus, given the boundary conditions, viz., which slits are open. [3]

While this picture eliminates the question of which slit the particle enters, it does not eliminate the deeper *metaphysical* mystery of how the *immaterial* information in the wave function can influence the particle paths and positions, one particle at a time, to produce the distribution of particles observed in the statistics of large numbers of particles.

But Einstein always said quantum mechanics is a statistical theory. And he was first to say very clearly that the waves, later the wave functions, are *guiding* the particles. He said the waves are a *guiding field* - a *Führungsfeld*.

It is this mystery, how abstract information can control concrete objects, not Feynman's worry about how a single particle can go through both slits, that is the deepest mystery in quantum mechanics.

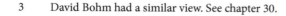

3   David Bohm had a similar view. See chapter 30.

Feynman's Path-Integral Formulation of Quantum Mechanics

In 1948 Feynman developed his "sum over paths" approach to quantum mechanics. It was based on a 1933 article by P. A. M. Dirac to formulate quantum mechanics using a Lagrangian function rather than the standard Hamiltonian, and to use a variational method to solve for the least action. It involves calculations over all space.

The idea of a single path for a quantum system (for example, the path of an electron or photon in the two-slit experiment) is replaced with a sum over an infinity of quantum-mechanically possible paths to compute the probability amplitude. The path-integral method is equivalent to the other formalisms of quantum mechanics but its visualization shows how it can sense when both slits are open.

Feynman's calculation of the probability amplitude for a particle entering say the left slit, and arriving at a specific point on the detector screen, is the result of adding together contributions from all possible paths in configuration space, however strange the paths.

Each path contributes a function of the time integral of the Lagrangian along the path. In Feynman's approach and in the transaction interpretations of quantum mechanics by John Cramer and Ruth Kastner, some paths explore the open slits.

The resulting probability amplitude is different at the back screen when one or both slits are open, just as we see in Einstein's "objective reality" way of analyzing the problem.

In order for the state of the slits to "influence" the motion of each individual particle to produce the statistical interference pattern that shows up for many particles, the wave function has to "know" its value at every point inside the two-slit experiment.

Chapter 33

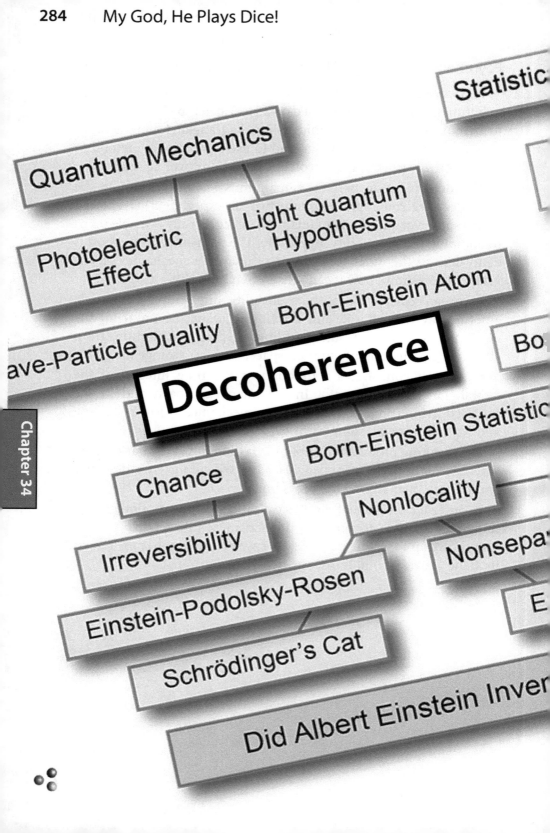

Decoherence

Decoherence is the study of interactions between a quantum system (generally a very small number of microscopic particles like electrons, photons, atoms, molecules, etc. - often just a single particle) and the larger macroscopic environment, which is normally treated "classically," that is, by ignoring quantum effects, but which decoherence theorists study quantum mechanically.

Decoherence theorists attribute the absence of *macroscopic* quantum effects like interference (which is a coherent process) to interactions between a quantum system and the larger macroscopic environment. They maintain that no system can be completely isolated from the environment. The decoherence (which accounts for the disappearance) of macroscopic quantum effects is shown experimentally to be correlated with the loss of isolation.

Niels Bohr maintained that a macroscopic apparatus used to "measure" quantum systems must be treated classically. John von Neumann, on the other hand, assumed that everything is made of quantum particles, even the mind of the observer. This led him and Werner Heisenberg to say that a "cut" must be located somewhere between the quantum system and the mind, which would operate in a sort of "psycho-physical parallelism." John Bell drew a diagram with locations for what he called the "shifty split" between the experiment and the mind of the observer.[1]

A main characteristic of quantum systems is the appearance of wavelike interference effects. These only show up in large numbers of repeated identical experiments that make measurements on single particles at a time. Interference is never directly "observed" in a single experiment. When interference is present in a system, the system is called "coherent." Decoherence then is the loss or suppression of that interference.

Interference experiments require that the system of interest is extremely well isolated from the environment, except for the "measurement apparatus." This apparatus must be capable of

---

1   See chapter 32.

Chapter 34

recording the information about what has been measured. It can be a photographic plate or an electron counter, anything capable of registering a quantum event, usually by releasing a cascade of metastable processes that amplify the quantum-level event to the macroscopic world, where an "observer" can see the result.

This does not mean that specific quantum level events are determined by that observer (as noted by several of the great quantum physicists - MAX BORN, PASCUAL JORDAN, ERWIN SCHRÖDINGER, PAUL DIRAC, and textbook authors Landau and Lifshitz, Albert Messiah, and Kurt Gottfried, among others). Quantum processes are happening all the time. Most quantum events are never observed, though they can be inferred from macroscopic phenomenological observations.

To be sure, those quantum events that are "measured" in a physics experiment which is set up to measure a certain quantity are dependent on the experimenter and the design of the experiment. To measure the electron spin in a Stern-Gerlach experiment, the experimenter is "free to choose" to measure, for example, the z-component of the spin, rather than the x- or y-component. This will influence quantum level events in the following ways:

The experimental outcome will produce a definite value for the z-component of the spin (either +1/2 or -1/2). We do not create the particular value for the z-component of spin. This is a random choice made by Nature, as Dirac put it.

The x-component after the measurement will be indeterminate, described as in a superposition of +1/2 or -1/2 states

$$| \psi > = (1/\sqrt{2}) | +1/2 > + (1/\sqrt{2}) | -1/2 >$$

It is in this sense that Bohr and Heisenberg describe properties of the quantum world as not existing until we make a measurement. We are "free to choose" the experiment to perform. If we measure position for example, the precise position value may not exist in some sense immediately before the measurement, according to the Copenhagen Interpretation. ALBERT EINSTEIN challenged this idea. His "objective reality" imagined a world in which particles and their continuous paths really exist.

The Decoherence Program

The "decoherence program" of H. DIETER ZEH, Erich Joos, WOJCIECH ZUREK, JOHN WHEELER, MAX TEGMARK, and others has multiple aims -

- to show how classical physics emerges from quantum physics. They call this the "quantum to classical transition."
- to explain the lack of macroscopic superpositions of quantum states (e.g., Schrödinger's Cat as a superposition of live and dead cats).
- in particular, to identify the mechanism that suppresses ("decoheres") interference between states as something involving the "environment" beyond the system and measuring apparatus.
- to explain the *appearance* of particles following paths (They say there are no "particles," and maybe no paths).
- to explain the *appearance* of discontinuous transitions between quantum states (Decoherentists say there are no "quantum jumps" either).
- to champion a "universal wave function" (as a superposition of states) that evolves in a "unitary" fashion (i.e., deterministically) according to the Schrödinger equation.
- to clarify and perhaps solve the measurement problem, which they define as the lack of macroscopic superpositions.
- to explain the "arrow of time."
- to revise the foundations of quantum mechanics by changing some of its assumptions, notably challenging the "collapse" of the wave function or "projection postulate."

Decoherence theorists say that they add no new elements to quantum mechanics (such as "hidden variables") but they do deny one of the three basic assumptions - namely Dirac's projection postulate. This is the method used to calculate the probabilities of various outcomes, which probabilities are confirmed to several significant figures by the statistics of large numbers of identically prepared experiments.

Decoherentists accept (even overemphasize) Dirac's principle of superposition. Some also accept the axiom of measurement, although some question the link between eigenstates and eigenvalues.

The decoherence program hopes to offer insights into several other important phenomena:

- What Zurek calls the "einselection" (environment-induced superselection) of preferred states (the so-called "pointer states") in a measurement apparatus.
- The role of the observer in quantum measurements.
- Nonlocality and quantum entanglement (which is used to "derive" decoherence).
- The origin of irreversibility (by "continuous monitoring").
- The approach to thermal equilibrium.
- The decoherence program finds unacceptable the following aspects of the standard quantum theory:
- Quantum "jumps" between energy eigenstates.
- The "apparent" collapse of the wave function.
- In particular, explanation of the collapse as a "mere" increase of information.
- The "appearance" of "particles."
- The "inconsistent" Copenhagen Interpretation, i.e., quantum "system," classical "apparatus."
- The "insufficient" Ehrenfest Theorems.

Decoherence theorists admit that some problems remain to be addressed, especially the "problem of outcomes." Without the collapse postulate, it is not clear how definite outcomes are explained. In a universe with a single wave function, nothing ever happens.

As Tegmark and Wheeler put it:

> The main motivation for introducing the notion of wave-function collapse had been to explain why experiments produced specific outcomes and not strange superpositions of outcomes...it is embarrassing that nobody has provided a testable deterministic equation specifying precisely when the mysterious collapse is supposed to occur. [2]

---

2   *Scientific American*, February 2001, p.75.

Some of the controversial positions in decoherence theory, including the denial of collapses and particles, come straight from the work of ERWIN SCHRÖDINGER, for example in his 1952 essays "Are There Quantum Jumps?" (Part I and Part II), where he denies the existence of "particles," claiming that everything can be understood as waves. John Bell wrote an article with the same title.

Other sources include: HUGH EVERETT III and his "relative state" or "many world" interpretations of quantum mechanics; EUGENE WIGNER's article on the problem of measurement; and Bell's reprise of Schrödinger's arguments on quantum jumps.

Decoherence theorists therefore look to other attempts to formulate quantum mechanics. Also called "interpretations," these are more often reformulations, with different basic assumptions about the foundations of quantum mechanics. Most begin from the "universal" applicability of the *unitary* time evolution that results from the Schrödinger wave equation.

They include these formulations:

- DeBroglie-Bohm "pilot-wave" or "hidden variables".
- Everett-DeWitt "relative-state" or "many worlds".
- Ghirardi-Rimini-Weber "spontaneous collapse".

Note that these "interpretations" are often in serious conflict with one another. Where Schrödinger thinks that waves alone can explain everything (there are no particles in his theory), David Bohm thinks that particles not only exist but that every particle has a definite position carrying a "hidden parameter" of his theory.

H. Dieter Zeh, the founder of decoherence, sees

one of two possibilities: a modification of the Schrödinger equation that explicitly describes a collapse (also called "spontaneous localization") or an Everett type interpretation, in which all measurement outcomes are assumed to exist in one formal superposition, but to be perceived separately as a consequence of their dynamical autonomy resulting from decoherence. While this latter suggestion has been called "extravagant" [by John Bell] (as it requires myriads of co-existing quasi-classical "worlds"), it is similar in principle to the conventional (though nontrivial) assumption, made tacitly in all classical descriptions of observation, that consciousness is localized in certain semi-stable and sufficiently complex sub-systems (such as human brains or parts thereof) of a much larger external world. Occam's razor, often applied to the "other worlds", is a dangerous instrument: philosophers of the past used it to deny the existence of the interior of stars or of the back side of the moon, for example. So it appears worth mentioning at this point that environmental decoherence, derived by tracing out unobserved variables from a universal wave function, readily describes precisely the apparently observed "quantum jumps" or "collapse events." [3]

We briefly review the standard theory of quantum mechanics and compare it to the "decoherence program," with a focus on the details of the measurement process. We divide measurement into several distinct steps, in order to clarify the supposed "measurement problem" (for decoherentists it is mostly the lack of macroscopic state superpositions) and perhaps "solve" it.

The most famous example of probability-amplitude-wave interference is the two-slit experiment. Interference is between the probability amplitudes whose absolute value squared gives us the probability of finding the particle at various locations behind the screen with the two slits in it.

Finding the particle at a specific location is said to be a "measurement."

In standard quantum theory, a measurement is made when the quantum system is "projected" or "collapsed" or "reduced" into a single one of the system's allowed states. If the system was "prepared" in one of these "eigenstates," then the measurement will find it in that state with probability one (that is, with certainty).

However, if the system is prepared in an arbitrary state ψ_a, it can be represented as being in a linear combination of the measuring system's basic energy states φ_n.

$$\psi_a = \Sigma c_n \mid n >.$$
where
$$c_n = < \psi_a \mid \varphi_n >.$$

3   Joos et al. 2013, p.22

It is said to be in "superposition" of those basic states. The probability P_n of its being found in state φ_n is

$$P_n = < \psi_a \mid \varphi_n >^2 = c_n^2 .$$

As Dirac forcefully told us,[4] this does not mean an individual system is in more than one of those states. That is just a "manner of speaking." It means that measurements of many similar systems will be found distributed among the states with the probabilities P_n.

Between measurements, the time evolution of a quantum system in such a superposition of states is described by a unitary transformation $U(t, t_0)$ that preserves the same superposition of states as long as the system does not interact with another system, such as a measuring apparatus. As long as the quantum system is isolated from any external influences, it evolves continuously and deterministically in an exactly predictable (causal) manner.

This we take to be a central fact of Einstein's "objective reality." A system prepared in a state with certain properties (such as spin) conserves all those properties as it evolves without decohering.

Whenever the quantum system does interact however, with another particle or an external field, its behavior ceases to be causal and it evolves discontinuously and indeterministically. This acausal behavior is uniquely quantum mechanical. It is the *origin of irreversibility*. Nothing like it is possible in classical mechanics. Attempts to "reinterpret" or "reformulate" quantum mechanics are attempts to eliminate this discontinuous acausal behavior and replace it with a deterministic process.

We must clarify what we mean by "the quantum system" and "it evolves" in the previous two paragraphs. This brings us to the mysterious notion of "wave-particle duality." In the wave picture, the "quantum system" refers to the deterministic time evolution of the complex probability amplitude or quantum state vector ψ_a, according to the "equation of motion" for the probability amplitude wave ψ_a, which is the Schrödinger equation,

$$ih \, \delta\psi_a/\delta t = H \, \psi^a.$$

The probability amplitude looks like a wave and the Schrödinger equation is a wave equation. But the wave is an abstract complex

Chapter 34

---

4   See chapter 19.

quantity whose absolute square is the probability of finding a quantum particle somewhere. It is distinctly not the particle, whose exact position is unknowable while the quantum system is evolving deterministically. It is the probability amplitude wave that interferes with itself, going through both slits, for example. Particles, as such, never interfere (although they may collide).

Note that we never "see" a superposition of particles (or fragments of a particle) in distinct states. Particles are *not in two places at the same time* just because there is a probability of finding it in those two places! And note that a particle may be following a property-conserving path, although we cannot know that path.

When the particle interacts, with the measurement apparatus for example, we always find the whole particle. It suddenly appears. For example, an electron "jumps" from one orbit to another, absorbing or emitting a discrete amount of energy (a photon). When a photon or electron is fired at the two slits, its appearance at the photographic plate is sudden and discontinuous. The probability wave instantaneously becomes concentrated at the new location.

There is now unit probability (certainty) that the particle is located where we find it to be. This is described as the "collapse" of the wave function. Where the probability amplitude might have evolved under the unitary transformation of the Schrödinger equation to have significant non-zero values in a very large volume of phase space, all that probability suddenly "collapses" (faster than the speed of light, which deeply bothered Einstein as nonlocal behavior) to the newly found location of the particle.

Einstein worried that some mysterious "spooky action-at-a-distance" must act to prevent the appearance of a second particle at a distant point where a finite probability of appearing had existed just an instant earlier. (See chapter 23.)

But the distributed probability at all other places is not something physical and substantial that must "move" to the newly found location. It is just abstract information.

Decoherence and the Measurement Problem

For decoherence theorists, the unitary transformation of the Schrödinger equation cannot alter a superposition of microscopic states. Why then, when microscopic states are time evolved into macroscopic ones, don't macroscopic superpositions emerge?

According to H. D. Zeh:

> Because of the dynamical superposition principle, an initial superposition $\Sigma\, c_n \mid n >$ does not lead to definite pointer positions (with their empirically observed frequencies). If decoherence is neglected, one obtains their entangled superposition $\Sigma\, c_n \mid n > \mid \Phi_n >$, that is, a state that is different from all potential measurement outcomes.[5]

And according to Erich Joos, another founder of decoherence:

> It remains unexplained why macro-objects come only in narrow wave packets, even though the superposition principle allows far more "nonclassical" states (while micro-objects are usually found in energy eigenstates). Measurement-like processes would necessarily produce nonclassical macroscopic states as a consequence of the unitary Schrödinger dynamics. An example is the infamous Schrödinger cat, steered into a superposition of "alive" and "dead".[6]

The fact that we don't see superpositions of macroscopic objects *is* the "measurement problem," according to Zeh and Joos.

An additional problem is that decoherence is a completely unitary process (Schrödinger dynamics) which implies time reversibility. What then do decoherence theorists see as the origin of irreversibility? Can we time reverse the decoherence process and see the quantum-to-classical transition reverse itself and recover the original coherent quantum world?

To "relocalize" the superposition of the original system, we need only have complete control over the environmental interaction. This is of course not practical, just as LUDWIG BOLTZMANN found in the case of JOSEF LOSCHMIDT's reversibility objection.

Does *irreversibility* in decoherence have the same rationale - "not possible for all practical purposes" - as in classical statistical mechanics?

According to more conventional thinkers, the measurement problem is the failure of the standard quantum mechanical formalism (Schrödinger equation) to completely describe the nonunitary "collapse" process. Since the collapse is irreducibly indeterministic, the time of the collapse is completely unpredictable and unknowable.

---

5   *Decoherence and the Appearance of a Classical World in Quantum Theory*, p.20
6   *ibid.*, p.2.

Indeterministic quantum jumps are one of the defining characteristics of quantum mechanics, both the "old" quantum theory, where Bohr wanted continuous radiation to be emitted and absorbed discontinuously when his atom jumped between staionary states, and the modern standard theory with the Born-Jordan-Heisenberg-Dirac "projection postulate."

To add new terms to the Schrödinger equation in order to control the time of collapse is to misunderstand the irreducible chance at the heart of quantum mechanics, as first seen clearly, in 1917, by Einstein. When he derived his A and B coefficients for the emission and absorption of radiation, he found that an outgoing light particle must impart momentum hv/c to the atom or molecule, but the direction of the momentum can not be predicted! Nor can the theory predict the *time* when a light quantum will be emitted.

Such a random time was not unknown to physics. When ERNEST RUTHERFORD derived the law for radioactive decay of unstable atomic nuclei in 1900, he could only give the probability of decay time. Einstein saw the connection with radiation emission:

> "It speaks in favor of the theory that the statistical law assumed for [spontaneous] emission is nothing but the Rutherford law of radioactive decay.[7]

But the inability to predict both the time and direction of light particle emissions, said Einstein in 1917, is "a weakness in the theory..., that it leaves time and direction of elementary processes to chance (Zufall, ibid.)." It is only a weakness for Einstein, of course, because his God does not play dice. Decoherence theorists too appear to have what WILLIAM JAMES called an "antipathy to chance."

What Decoherence Gets Right

Allowing the environment to interact with a quantum system, for example by the scattering of low-energy thermal photons or high-energy cosmic rays, or by collisions with air molecules, surely will suppress quantum interference in an otherwise isolated experiment. But this is because large numbers of uncorrelated (incoherent) quantum events will "average out" and mask the

7   Pais, 1982, p.411

quantum phenomena. It does not mean that wave functions are not collapsing. They are, at every particle interaction.

Decoherence advocates describe the environmental interaction as "monitoring" of the system by continuous "measurements."

Decoherence theorists are correct that every collision between particles entangles their wave functions, at least for the short time before decoherence suppresses any coherent interference effects of that entanglement.

But in what sense is a collision a "measurement." At best, it is a "pre-measurement." It changes the information present in the wave functions from information before the collision. But the new information may not be recorded anywhere (other than being implicit in the state of the system).

All interactions change the state of a system of interest, but not all leave the "pointer state" of some measuring apparatus with new information about the state of the system.

So environmental monitoring, in the form of continuous collisions by other particles, is changing the specific information content of both the system, the environment, and a measuring apparatus (if there is one). But if there is no recording of new information (negative entropy created locally), the system and the environment may be in thermodynamic equilibrium.

Equilibrium does not mean that decoherence monitoring of every particle is not continuing.

It is. There is no such thing as a "closed system." Environmental interaction is always present.

If a gas of particles is not already in equilibrium, they may be approaching thermal equilibrium. This happens when any non-equilibrium initial conditions (Zeh calls these a "conspiracy") are being "forgotten" by erasure of path information during collisions.

Without that erasure, information about initial conditions woould remain in the paths of all the particles, as LUDWIG BOLTZMANN feared. This means that, in principle, the paths could be reversed to return to the initial, lower entropy, conditions (Loschmidt paradox).

Statistic

Quantum Mechanics

Light Quantum
Hypothesis

Photoelectric
Effect

Bohr-Einstein Atom

ave-Particle Duality

Bo

Einstein's Principles

Tra

Born-Einstein Statistic

Chance

Nonlocality

Irreversibility

Nonsepa

Einstein-Podolsky-Rosen

E

Schrödinger's Cat

Did Albert Einstein Inver

Chapter 35

Einstein's Principles

While the young ALBERT EINSTEIN learned a great deal from ERNST MACH's notion that theories are "economic summaries of experience," in his later years he attacked theories that were simply designed to fit the available facts. Einstein challenged the idea that induction from a number of examples can lead to fundamental theories.

Positivists and empiricists declared that any theory not built from sense data about our experiences was mere *metaphysics*.

Einstein disagreed. The best theories should be based on "principles," he argued, perhaps biased by the astonishing success of his 1905 principle of relativity and 1916 equivalence principle?

Special relativity dazzled the world with its predictions that measured lengths of an object depend on the observer's speed relative to the object, and that events separated in space can have their time order reversed depending on the speed of the observer.

When all Einstein's amazing predictions were confirmed by experiment, many rushed to the subjectivist conclusion that everything is relative to one's point of view. But Einstein saw a deeper and *absolute* version of his principle, namely that the speed of light is an *invariant*, independent of the speed of the observer.

His theory of general relativity was based on his equivalence principle, that no experiment can distinguish between gravity and an accelerating force.

Einstein in no way denied the critical importance of experience, especially the experiments that test the validity of any theory and the principles it is based upon.

But here Einstein parted ways with physicists who believe that their theories, having been grounded in worldly experience, must actually *exist* in the real world. He startled many philosophers of science by declaring theories to be fictions, *inventions* by thinkers and not *discoveries* about the material contents of the universe.

Chapter 35

Inspired by the great nineteenth-century mathematician RICHARD DEDEKIND, Einstein often described theories and their underlying principles as "free creations of the human mind."

A contemporary of Dedekind, LEOPOLD KRONECKER, had made the powerful claim that "God made the integers, all else is the work of man." Einstein may have felt that even the integers were created by human beings.

Einstein described his ideas about theories based on principles in 1919, shortly after his great success with general relativity, and long before the work of the so-called "founders" of quantum mechanics.

> There are several kinds of theory in physics. Most of them are constructive. These attempt to build a picture of complex phenomena out of some relatively simple proposition. The kinetic theory of gases, for instance, attempts to refer to molecular movement the mechanical thermal, and diffusional properties of gases. When we say that we understand a group of natural phenomena, we mean that we have found a constructive theory which embraces them.

> But in addition to this most weighty group of theories, there is another group consisting of what I call theories of principle. These employ the analytic, not the synthetic method. Their starting-point and foundation are not hypothetical constituents, but empirically observed general properties of phenomena, principles from which mathematical formula are deduced of such a kind that they apply to every case which presents itself. Thermodynamics, for instance, starting from the fact that perpetual motion never occurs in ordinary experience, attempts to deduce from this, by analytic processes, a theory which will apply in every case. The merit of constructive theories is their comprehensiveness, adaptability, and clarity, that of the theories of principle, their logical perfection, and the security of their foundation...

> Since the time of the ancient Greeks it has been well known that in describing the motion of a body we must refer to another body. The motion of a railway train is described with reference to the ground, of a planet with reference to the total assemblage of visible fixed stars. In physics the bodies to which motions are spatially referred are termed systems of coordinates. The laws of mechanics of Galileo and Newton can be formulated only by using a system of coordinates. [1]

---

1 *Science*, 51 (No. 1305); January 2, 1920; originally published in *The Times* (London), 28 November 1919, pp. 13–14.

What were Einstein's Principles?

Some of his principles were held by many earlier thinkers, such as the law of parsimony or simplicity, also known as Occam's Razor, that the simplest theory that fits all the known facts is the best theory. He may have liked the idea that the most true theories would be beautiful in some sense, for example their symmetry.

Others of Einstein's principles were the accepted laws of classical physics and chemistry. They were postulated relations between physical quantities that proved correct in experimental tests.

They include Newton's three laws of motion, his law of universal gravitation, Maxwell's and Faraday's laws of electromagnetism, and the four laws of thermodynamics. Einstein would have accepted Kirchhoff's Law that the spectrum of blackbody radiation does not depend on the material that is radiating. He himself proved the Stefan-Boltzmann law that radiated energy is proportional to the fourth power of the temperature T.

Now the first law of thermodynamics is also a *conservation* principle, specifically the conservation of energy. It was not fully understood until motion energy was seen to be converted into heat by frictional forces in the early nineteenth century. The conservation of other quantities like linear and angular momentum had been understood from motions of the planets, which show no obvious frictional forces. Einstein mentioned the lack of perpetual motion machines, which embodies the conservation of energy.

As we mentioned in the introduction, the great mathematician EMMY NOETHER stated a theorem that each of these conservation principles is the result of a symmetry property of a physical system.

Laws of physics are thought be independent of time and place. That they are independent of the time results in the conservation of energy. Independence of place leads to the conservation of momentum. Independence of angle or direction produces the conservation of angular momentum.

These great symmetries and conservation laws are sometimes described as *cosmological* principles. At the grandest universe scale, there is no preferred direction in space. The ultimate reference "to which motions are spatially referred" is most often the center of mass of nearby material objects, or as Mach expected, the entire matter in the universe, not an *immaterial* "system of coordinates."

The average density of galaxies appears the same in all directions, and the remote cosmic microwave background of radiation shows no asymmetries. There was thought to be no preferred time until the twentieth-century discovery of the Big Bang.

We shall see that Einstein did not fully apply these conservation principles in his work on the *nonlocal* behaviors shown by entangled particles. And despite being quite familiar with Noether's work, we have seen that he abandoned fundamental symmetry principles in his 1935 analysis of the Einstein-Podolsky-Rosen Paradox.[2]

One great principle that every physicist accepted in the early twentieth century was *causality*, the simple idea that every effect has a cause. Causality in turn implies that identical causes will produce identical effects, leading to the physical and philosophical idea of determinism.

Determinism is the idea that there is but one possible future, because all the events at any moment are the complete causes of the immediately following events and those events the immediate causes of the next events. The only possibilities are those that actually occur. Until he became convinced of the statistical nature of quantum mechanics in the late 1920's, Einstein was a determinist.

Some work that Einstein saw as lacking principles were attempts to fit equations to observed data, like Wien's distribution and displacement laws, and Planck's radiation law.

Einstein may have elevated the *continuum* to a principle, though 1) he was instrumental in disproving the hypothesis of an ether as the medium for electromagnetism, and 2) his work on Brownian motion established the atomic hypothesis which disproved the idea of continuous matter, just as his light quantum hypothesis disproved continuous energy.

In any case Einstein knew that all principles, and the laws of physics based on them, began as ideas, free creations of the human mind, and they only acquired their status as laws when confirmed by repeated experiments.

---

2   See chapter 26.

The Absolute Principles of Physics

Some of the *absolute* principles in physics are the conservation laws for mass/energy, momentum, angular momentum, and electron spin. The constant velocity of light is another.

Emmy Noether's theorem says these conservation principles are the consequence of deep *symmetry* principles of nature. She said for any property of a physical system that is symmetric, there is a corresponding conservation law.

Noether's theorem allows physicists to gain powerful insights into any general theory in physics, by just analyzing the various transformations that would make the form of the laws involved *invariant.*

For example, if a physical system is symmetric under rotations, its angular momentum is conserved. If it is symmetric in space, its momentum is conserved. If it is symmetric in time, its energy is conserved. Now locally there is time symmetry, but cosmically the expansion of the universe gives us an *arrow of time* connected to the increase of entropy and the second law of thermodynamics.

The conservation of energy was the *first law* of thermodynamics.

The famous *second law* says entropy rises to a maximum at thermal equilibrium. It was thought by many scientists, especially MAX PLANCK, to be an *absolute* law. But as we saw in chapter 3, JAMES CLERK MAXWELL and LUDWIG BOLTZMANN considered it a *statistical* law.

Einstein called Boltzmann's expression for the entropy "Boltzmann's Principle." $S = k \log W$. At the 1911 Solvay Conference, Einstein wrote,'

> the question arises, on the validity of which general principles
> of physics we may hope to rely in the field of concern to us. In
> the first place we are all agreed that we should retain the energy
> principle.
> A second prnciple to the validity of which, in my opinion, we
> absolutely have to adhere is Boltzmann's definition of entropy by
> means of probability. [3]

---

3   Stachel, 2002, p.375

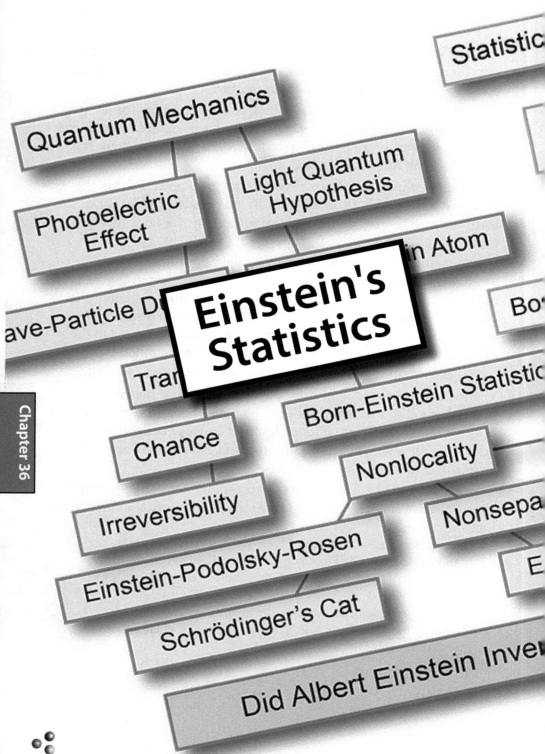

Einstein's Statistics

We saw in chapter 5 that Einstein rederived all of classical statistical mechanics between 1902 and 1904, going beyond the kinetic theory of gases developed by LUDWIG BOLTZMANN in the nineteenth century. Twenty years later, Einstein discovered quantum statistics (chapter 15). All of this *before* the "founders" of quantum mechanics discovered the equations that allow us to *calculate* quantum properties to extraordinary levels of accuracy.

Einstein did not care much for the details of calculation, except to prove a fundamental theory. He oversaw the transition from classical statistics to quantum statistics. Just two years later, after WERNER HEISENBERG had developed matrix mechanics and ERWIN SCHRÖDINGER created wave mechanics, Einstein generously allowed his friend MAX BORN to take full credit for the "statistical interpretation" of quantum mechanics, which Einstein had seen qualitatively at least a decade earlier (chapter 20).

To be sure, Born identified Einstein's *qualitative* probability with the calculated squared modulus of Schrödinger's wave function $|\psi|^2$. This made the statistical interpretation *quantitative*.

As we have seen so well, Einstein was very unhappy about the ontological implications of the statistics he discovered. He said many times to Born over the next few decades, "God does not play dice," But over those decades Born never noticed that Einstein had embraced *indeterminism* in quantum mechanics . Einstein's criticisms were mostly directed to *nonlocality* (chapter 23).

In his early work on statistical mechanics, Einstein showed that small *fluctuations* in the motions of gas particles are constantly leading to departures from equilibrium. It is like the departures from the smooth analytic bell curve for any finite number of events. The entropy does not rise smoothly to a maximum and then stay there indefinitely. It fluctuates randomly. The second law is not absolute.

Chapter 36

The second law of thermodynamics is unique among the laws of physics because of its *irreversible* behavior. Heat flows from hot into cold places until they come to the same equilibrium temperature. The one-directional nature of *macroscopic* thermodynamics (with its gross "phenomenological" variables temperature, energy, entropy) is in fundamental conflict with the assumption that *microscopic* collisions between molecules, whether fast-moving or slow, are governed by dynamical, deterministic laws that are time-reversible. But is this not correct.

At the atomic and molecular level, if collisions were time reversible, there would be no arrow of time, but we see that Einstein's 1916 work on transition probabilities for emission and absorption of radiation shows that particle collisions are not reversible when the interaction with radiation is taken into account (chapter 12).

Statistical "laws" grow out of examples in which there are very large numbers of entities. Large numbers make it impractical to know much about the individuals, but we can say a lot about *averages* and the probable distribution of values around the averages. This is the "quantum-to-classical transition."

Boltzmann's Principle

Einstein's work in statistical mechanics was grounded in Boltzmann's relationship between entropy and probability.

$S = k \ln W$

The entropy S is the logarithm of the number of ways W the particles can be arranged in the available phase-space cells, multiplied by a constant k that MAX PLANCK called Boltzmann's constant. Boltzmann knew this relationship, but wrote the constant as the universal gas constant R divided by the number N of particles in one molecular weight of a gas.

If there is only one way that a given macroscopic system can be arranged in phase space, then $W = 1$, $\ln W = 0$, and entropy is the absolute minimum, $S = 0$.

Quantum Mechanics a Statistical Theory

Einstein and the "founders" of quantum mechanics engaged in fruitless debates for many years about the "completeness" of quantum mechanics. Quantum mechanics is incomplete because it cannot determine the exact properties of individual systems. For Einstein, that limits quantum mechanics to statistics.

> The statistical character of the present theory would then have to be a necessary consequence of the incompleteness of the description of the systems in quantum mechanics. [1]

Einstein's "objective reality" depends on the applicability of *absolute* conservation laws to individual quantum systems, even though properties of individual systems can only be studied *statistically.*

The great second law of thermodynamics is only a *statistical* law. The approach of entropy to a maximum at thermal equilibrium is always subject to statistical fluctuations

Quantum Statistics

Perhaps Einstein's most profound work in statistics was his 1924 discovery of *quantum statistics.* Prompted by a new derivation of Planck's radiation distribution law by SATYENDRA NATH BOSE, Einstein showed that the distribution of photons differs from Boltzmann's molecular distribution by the addition of a -1 in the denominator.

Shortly after Einstein's paper, PAUL DIRAC showed that fermions (spin 1/2 particles) also depart from the Boltzmann distribution, by the addition of a +1 in the denominator.

No (bosons) $\approx (1 / (e^{hv/kT} - 1).$

No (atoms/molecules) $\approx (1 / (e^{hv/kT}).$

No (fermions) $\approx (1 / (e^{hv/kT} + 1).$

Einstein in 1905 proved material particles (atoms) exist and hypothesized that light particles exist. In 1924 he discovered bosons. His quantum statistics gave us the first examples of the two fundamental kinds of particle in the standard model of particle physics - fermions and bosons. See chapter 15.

---

1   Schilpp, 1949, p.87

Chapter 37

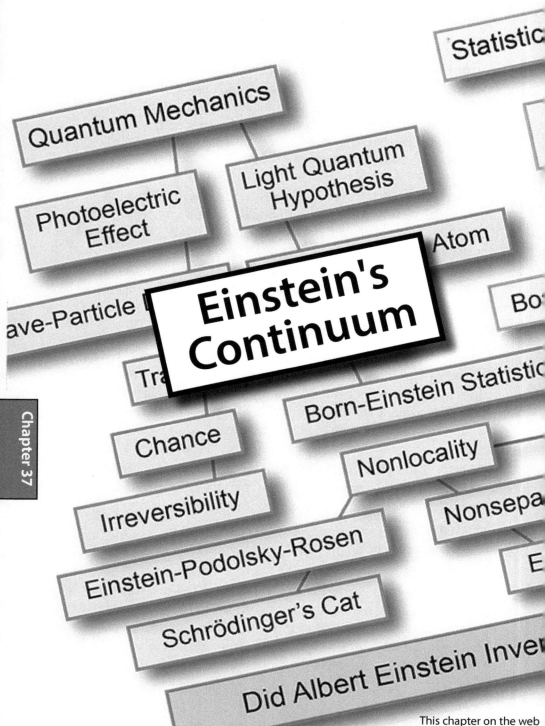

This chapter on the web
informationphilosopher.com/

Einstein's Continuum

Is the Nature of Reality Continuous or Discrete?

Is it possible that the physical world is made up of nothing but discrete discontinuous *particles*? Are continuous *fields* with well-defined, arbitrarily accurate, values for the field at all places and times simply theoretical constructs, confirmed only by averages over large numbers of particles?

Space and time themselves have well-defined values everywhere, but are these just the abstract *information* of the *ideal* coordinate system that allows us to keep track of the positions and motions of particles? Space and time are physical, but they are not *material*.

We use material things, rulers and clocks, to measure space and time. We use the abstract mathematics of real numbers and assume there are an *infinite number* of real points on any line segment and an infinite number of moments in any time interval. But are these continuous functions of space and time nothing but *immaterial* ideas with no material substance?

The two great physical theories at the end of the nineteenth century, ISAAC NEWTON's classical mechanics and JAMES CLERK MAXWELL's electrodynamics, are *continuous field theories*.

Solutions of their field equations determine precisely the exact forces on any material particle, providing complete information about their past and future motions and positions. Field theories are generally regarded as *deterministic* and *certain*.

Although the dynamical laws are "free inventions of the human mind," as Einstein always said,[1] and although they ultimately depend on experimental evidence, which is always *statistical*, the field theories have been considered superior to merely statistical laws. Dynamical laws are thought to be *absolute*, based on *principles*.

---

1   "Geometry and Experience," in *Ideas and Opinions*, p.234

Chapter 37

We will find that the continuous, deterministic, and analytical laws of classical dynamics and electromagnetism, expressible as differential equations, are idealizations that "go beyond experience."

A continuum is approached in the limit of large numbers of particles, when the random *fluctuations* of individual events can be averaged over. But this is like the limit theorems of the differential calculus, when large numbers are allowed to go to infinity, and infinitesimals are assumed to reach zero.

All field theories use *continuous* functions that introduce mathematical infinities and infinitesimals. Einstein suspected these infinities may only "exist" in human minds. He learned this from the great mathematicians LEOPOLD KRONECKER and RICHARD DEDEKIND.

Einstein discovered his favorite phrase "free creations of the human mind" in the work of Dedekind (*freie Schöpfungen des menschlichen Geistes*) so Einstein also knew very well Dedekind's argument that all the axioms of Euclid's geometry can be proven with no reference to a continuum between geometric points. A discrete algebraic theory would be equally good, said Dedekind.

> If anyone should say that we cannot conceive of space as anything else than continuous, I should venture to doubt it and to call attention to the fact that a far advanced, refined scientific training is demanded in order to perceive clearly the essence of continuity and to comprehend that besides rational quantitative relations, also irrational, and besides algebraic, also transcendental quantitative relations are conceivable. [2]

God Created the Integers

Einstein was assuredly also familiar with Kronecker's famous quote "God has made the integers, all else is the work of man." (*Die ganzen Zahlen hat der liebe Gott gemacht, alles andere ist Menschenwerk*). These ideas must have given Einstein a healthy skepticism about his work on *continuous* field theories. In his later

---

2   Dedekind, 1901, p.38

years, Einstein gave thought to algebraic or *discrete* difference equations, not continuous differential equations.

Einstein may have even doubted the "existence" of the integers. He and Leopold Infeld wrote in the book, *The Evolution of Physics,*

> Science is not just a collection of laws, a catalogue of unrelated facts. It is a creation of the human mind, with its freely invented ideas and concepts... "Three trees" is something different from "two trees." Again "two trees" is different from "two stones." The concepts of the pure numbers 2, 3, 4..., freed from the objects from which they arose, are creations of the thinking mind which describe the reality of our world. [3]

Experiments that support physical laws are always finite in number. The number of particles in the observable universe is finite. Experimental evidence is always *statistical*. It always contains *errors* distributed randomly around the most probable result, like the fluctuations of entropy around its maximum.

The smooth Gaussian bell curve approached when a very large number of independent random events is plotted is clearly an idealization. That Bell curve is clearly an "idea," a "free creation of the human mind. "

Einstein was gravely concerned that there is nothing in his "objective reality" corresponding to this continuum.

> From the quantum phenomena it appears to follow with certainty that a finite system of finite energy can be completely described by a finite set of numbers (quantum numbers). This does not seem to be in accordance with a continuum theory, and must lead to an attempt to find a purely algebraic theory for the description of reality. [4]

To Leopold Infeld, Einstein wrote in 1941,

> "I tend more and more to the opinion that one cannot come further with a continuum theory." [5]

3   Einstein and Infeld, 1961, p.294
4   Einstein, 1956, p.165
5   Pais, 1982, p.467

Quantum Mechanics

Statistic

Photoelectric
Effect

Light Quantum
Hypothesis

ave-Particle Duality

Bohr-Einstein Atom

Einstein's
Field Theory

Bos

Tr

Born-Einstein Statistic

Chance

Nonlocality

Irreversibility

Nonsepa

Einstein-Podolsky-Rosen

E

Schrödinger's Cat

Did Albert Einstein Inven

Einstein's Field Theory

In the last thirty years of his life Einstein's main mission was to create a *unified field theory* that would combine the gravitational field of Newton (or Einstein), the electromagnetic field of Maxwell, and perhaps the probability field of quantum mechanics.

But he also worried much of his life that continuous fields are only theories, purely abstract information, whereas discrete particles have a more substantial reality, arranging themselves in material *information structures*.

But the ideal and pure information of continuous field theories clearly has causal powers over the "discrete" material world, as we saw in the two-slit experiment (chapter 33).

Einstein in his later years grew quite pessimistic about the possibilities for deterministic continuous field theories, by comparison with indeterministic and *statistical* discontinuous particle theories like those of quantum mechanics.

Einstein deeply believed that any physical theory must be based on a continuous field. For Einstein, physical objects must be described by continuous functions of field variables in four-dimensional space-time coordinates. In quantum field theory (QFT), particles are functions of (singularities in) these fields. In quantum electrodynamics (QED), fields are merely properties of aggregated particles. Which then are the more fundamental?

It appears to be particles, especially today when the last fundamental particle predicted by the standard theory (the Higgs boson) has been found. Einstein suspected that his dream of a unified field theory may not be possible.

In his 1949 autobiography for his volume in Paul Schilpp's *Library of Living Philosophers*, Einstein asked about the theoretical foundation of physics in the future, "Will it be a field theory [or] will it be a statistical [particles] theory?"

> "Before I enter upon the question of the completion of the general theory of relativity, I must take a stand with reference to the most successful physical theory of our period, viz.,

Chapter 38

the statistical quantum theory which, about twenty-five years ago, took on a consistent logical form (Schrödinger, Heisenberg, Dirac, Born). This is the only theory at present which permits a unitary grasp of experiences concerning the quantum character of micro-mechanical events. This theory, on the one hand, and the theory of relativity on the other, are both considered correct in a certain sense, although their combination has resisted all efforts up to now. This is probably the reason why among contemporary theoretical physicists there exist entirely differing opinions concerning the question as to how the theoretical foundation of the physics of the future will appear. Will it be a field theory; will it be in essence a statistical theory? I shall briefly indicate my own thoughts on this point. [1]

Castle In The Air

In 1954 Einstein wrote his friend Michele Besso to express his lost hopes for a continuous field theory like that of electromagnetism or gravitation,

"I consider it quite possible that physics cannot be based on the field concept, i.e:, on continuous structures. In that case, nothing remains of my entire castle in the air, gravitation theory included, [and of] the rest of modern physics." [2]

In the same year, he wrote to David Bohm,

I must confess that I was not able to find a way to explain the atomistic character of nature. My opinion is that if the objective description through the field as an elementary concept is not possible, then one has to find a possibility to avoid the continuum (together with space and time) altogether. But I have not the slightest idea what kind of elementary concepts could be used in such a theory. [3] (Einstein to David Bohm, 28 October 1954).

Einstein sees a conflict between relativity and quantum mechanics

Again in the same year, he wrote to H.S.Joachim,

it seems that the state of any finite spatially limited system may be fully characterized by a finite number of numbers. This speaks against the continuum with its infinitely many

1   Schilpp, 1949, p.81
2   Pais, 1982, p.467
3   Stachel, 1986, p.380

degrees of freedom. The objection is not decisive only because one doesn't know, in the contemporary state of mathematics, in what way the demand for freedom from singularity (in the continuum theory) limits the manifold of solutions. [4]

The fifth edition of Einstein's *The Meaning of Relativity* included a new appendix on his field theory of gravitation. In the final paragraphs of this work, his last, published posthumously in 1956, Einstein wrote,

"Is it conceivable that a field theory permits one to understand the atomistic and quantum structure of reality ? Almost everybody will answer this question with "no"...

"One can give good reasons why reality cannot at all be represented by a continuous field. From the quantum phenomena it appears to follow with certainty that a finite system of finite energy can be completely described by a finite set of numbers [quantum numbers]. This does not seem to be in accordance with a continuum theory, and must lead to an attempt to find a purely algebraic theory for the description of reality. But nobody knows how to obtain the basis of such a theory." [5]

No one has described Einstein's doubts about continuous field theories better that JOHN STACHEL, one of the early editors of the Collected Papers of Albert Einstein. Stachel speculated about "another Einstein" with doubts about a continuum and field.

Stachel points to Einstein's 1923 article "Does Field Theory Offer Possibilities for the Solution of the Quantum Problem?," in which Einstein points out that the great successes of quantum theory over the last quarter of a century should not be allowed to conceal the lack of any logical foundation for the theory.

He quotes Einstein...

The essential element of the previous theoretical development, which is characterized by the headings mechanics, Maxwell-Lorentz electrodynamics, theory of relativity, lies in the circumstance that they work with differential equations that uniquely determine events [das Geschehen] in a four-

Chapter 38

---

4   *ibid.581*
5   Einstein, 1956, pp.165-66

dimensional spatio-temporal continuum if they are known for a spatial cross-section...In view of the existing difficulties, one has despaired of the possibility of describing the actual processes by means of differential equations. [6]

The linear Schrödinger differential equation for waves cannot give us the details of individual particles, only the statistics of ensembles of particles. Stachel provides several powerful statements from 1935 to Einstein's posthumous writings pointing toward discrete "algebraic" theories of particles replacing continuum field theories.

In modern terms, the arrangement of particles would be described by integers, the quantum numbers as "bits" of information in a "digital" theory, not the continuum of an "analog" theory.

> In any case one does not have the right today to maintain that the foundation must consist in a field theory in the sense of Maxwell. The other possibility, however, leads in my opinion to a renunciation of the time-space continuum and to a purely algebraic physics. Logically this is quite possible (the system is described by a number of integers; "time" is only a possible viewpoint [Gesichtspunkt], from which the other "observables" can be considered—an observable logically coordinated to all the others. Such a theory doesn't have to be based upon the probability concept. For the present, however, instinct rebels against such a theory (Einstein to Paul Langevin, 3 October 1935). [7]

> It has been suggested that, in view of the molecular structure of all events in the small, the introduction of a space-time continuum may be considered as contrary to nature. Perhaps the success of Heisenberg's method points to a purely algebraical method of description of nature, to the elimination of continuous functions from physics. Then, however, we must also give up, on principle, the utilization of the space-time continuum. It is not inconceivable that human ingenuity will some day find methods that will make it possible to proceed along this path. Meanwhile, however, this project resembles the attempt to breathe in an airless space ("Physics and Reality,"

6   Stachel, 2002, p.149
7   *ibid.*, p.140

[1936], cited from Einstein Ideas and Opinions 1954, 319). [8]

In present-day physics there is manifested a kind of battle between the particle-concept and the field-concept for leadership, which will probably not be decided for a long time. It is even doubtful if one of the two rivals finally will be able to maintain itself as a fundamental concept (Einstein to Herbert Kondo, 11 August 1952). [9]

Objective reality does not lead to Einstein's "Unified Field Theory," but it does leave us with three distinct fields, the electromagnetic, the gravitational, and the quantum mechanical probability field, all generating *abstract* information that makes quantitative predictions about the behavior of *real* particles.

Einstein's "castle in the air," "breathing in empty space," should not lead us to despair about quantum field theories, but only to see them more clearly as Einstein first described a wave, as a "ghost field" and a "guiding field."

Perhaps we should say that where the particles are "real," the fields are imaginary - "free creations of the human mind."

Particles are actual. They are involved in actions and interactions.

Fields are possibilities. Wave functions allow us to calculate the probabilities for each possibiity, making predictions to degrees of accuracy unheard of in the other sciences.

In short, fields are theories, mere ideas, abstract information about *continuous* functions across infinite space and time.

Particles are facts, derived from *discrete* concrete experiments done in the here and now.

Chapter 38

8 *ibid.*, p.150
9 *ibid.*, p.150

Statistic

Quantum Mechanics

Light Quantum
Hypothesis

Photoelectric
Effect

Bohr-Einstein Atom

ave-Particle Duality

Bos

Trans

Einstein's
Objective
Reality

Statistic

Chance

Nonlocality

Irreversibility

Nonsepa

Einstein-Podolsky-Rosen

E

Schrödinger's Cat

Did Albert Einstein Inven

Einstein's Objective Reality

In his search for an "objective reality," Einstein asked whether a particle has a determinate position just before it is measured. The Copenhagen view is that a particle's position, path, and other properties only come into existence when they are measured.

Let's assume that material particles have definite paths as they travel from collision to collision, as LUDWIG BOLTZMANN'S statistical mechanics assumed. They are not brought into existence by the actions of a physicist, as WERNER HEISENBERG claimed, although some values, like spin components, may be created by the "free choice" of the experimenter as to what to measure.

In an objective reality, particle paths and their instantaneous positions are always *determinate* in principle, though not *determinable* in practice without experimental measurements, which might alter the particle's properties irreversibly.

Let's identify Einstein's "objective reality" with his "local reality," in which all "actions" or "interactions" are "local." These include classical "actions-at-a-distance" in Newtonian mechanics and Maxwell electromagnetism that are mediated by electromagnetic or gravitational fields, understood as the interchange of particles at speeds less than or equal to the speed of light.

As we saw in chapter 23, "nonlocality" usually means what Einstein discovered as early as 1905 and much later called "spooky action-at-a-distance," because it *appears* to require a particle or its associated wave at one point in space to act on another point far away in a spacelike separation.

"Nonlocality" defined this way as *actions* by one particle on another at a distance simply does not exist.

But "entangled" particles in a spacelike separation *appearing* to be changing their properties "simultaneously" in at least one frame of reference certainly does exist. A measurement by Alice or Bob to determine the electron spin components in a specific spatial direction is a measurement of the second kind.

This is nonlocality in the original sense of Einstein in 1905 and 1927. It appears to violate his "impossibility of simultaneity."

Chapter 39

Entanglement and Objective Reality

In our application of Einstein's "objective reality" to such entanglement (chapters 26 to 29), we have shown that such purportedly "nonlocal actions" do not involve any interchanges, nothing material or energetic is moving, no information can be sent between the particles, etc.

The *appearance* of instantaneous interactions between objects in a spacelike separation arises because "orthodox" quantum physics claims that objects do not have properties until they are measured. It assumes that perfectly correlated properties in two separated particles are newly created when they are measured, instead of being already present in the particles as they "objectively" and "locally" travel from their initial entanglement.

In chapter 29 we showed that most properties of each particle have traveled with them from the moment of their entanglement.

To be sure, some new property values may be created in a measurement, because the observer has a "free choice" as to what to measure. The paradigm example is a measurement of electron spin or photon polarization in a definite spatial direction.

We can still use Einstein's demands for conservation of spin and symmetry to explain why the two measurements by Alice and Bob always conserve the total spin as zero. But it is not obvious how two events in a spacelike separation that appear simultaneously (in the special frame in which the measurement apparatus is at rest) can correlate arbitrary spin component *directions* perfectly.

They violate Einstein's "impossibility of simultaneity."

Our best explanation is to credit perfect correlation to the deeply mysterious power of the wave function ψ to "influence" events at great spacelike separations.

This was Schrödinger's immediate reaction to Einstein's EPR paper in 1935. The coherent two-particle wave function is not separable into the product of two single-particle wave functions, but when it does decohere, the property of the chosen spin directions is conserved for each electron.

The Two-Slit Experiment and Objective Reality

Einstein's "objective reality" visualizes particles as having continuous paths. In particlular, the path of a particle in the two-slit experiment always goes through just one of the slits. [1]

The quantum wave function, by comparison, goes through both slits when they are open, producing an interference pattern quite different from those with only one of the slits open.

This view explains the two-slit experiment completely, without worrying, as RICHARD FEYNMAN did on his "logical tightrope," how a particle might go through both splits, for example, by being in two places at the same time. (See chapter 33.)

But Feynman is nevertheless right that the two-slit experiment contains "one" deep mystery in quantum mechanics.

How does the quantum wave function "influence" the motion of particles so that they reproduce (statistically) the interference patterns seen in the two-slit experiment?

The squared modulus of the wave function $|\psi|^2$ is a probability field. Gravitational and electromagnetic fields allow us to calculate the forces on a test particle, then solve for the particle motion. But a probability field exerts no known force. And if it were a force, it would need to act statistically, where gravitational and electromagnetic forces are deterministic.

Irreversibility and Objective Reality

Einstein's "objective reality" allows us to visualize colliding particles as having determinate but not determinable paths. LUDWIG BOLTZMANN and his colleagues saw that those paths might conserve the path information. That would, if we could reverse the paths, lead to a decrease in entropy in violation of the second law of thermodynamics.

To this "local reality" of paths conserving information we can add Einstein's 1917 discovery of ontological chance when light interacts with matter, absorbing or emitting radiation. Photon emission and absorption during molecular collisions deflect the molecules randomly from their paths.

Chapter 39

---

1   Bohmian mechanics agrees with this. See chapter 30.

This destroys the path information and molecular correlations, justifying Boltzmann's assumption of "molecular chaos" (*molekular ungeordnete*) as well as Maxwell's earlier assumption that molecular velocities may not actually be correlated as determinism suggests.

Of the dozen or so mysteries and paradoxes in quantum mechanics described in our preface, Einstein's "objective reality" analysis contributes to solutions for some of the most important - nonlocality, nonseparability, entanglement, the two-slit experiment, and microscopic irreversibility. It also sheds light on others, but we need now to see how Einstein's excellent understanding of quantum physics can resolve a few more..

The wave functions of quantum mechanics produce only *predictions* of the probability of finding the particles themselves at different positions in space, as Einstein himself was first to see. Those probabilities depend on the boundary conditions, like a box confining the standing waves of a harmonic oscillator, the slits in the two-slit experiment, or the nodes in atomic and molecular orbitals confined by the nuclear attraction.

But there is nothing substantial at those points unless a discrete particle is there. And Einstein suspected that reality might consist only of discrete particles. Even space and time might be nothing (i.e., not things). In his 1949 autobiography, he wrote

> Physics is an attempt conceptually to grasp reality as it is thought independently of its being observed. In this sense one speaks of "physical reality." In pre-quantum physics there was no doubt as to how this was to be understood. In Newton's theory reality was determined by a material point in space and time; in Maxwell's theory, by the field in space and time. In quantum mechanics it is not so easily seen. [2]

Einstein knows that waves, now wave functions, exert an "influence" over material particles. To Einstein the influence looked like simultaneous events in a spacelike separation, which his theory of relativity thought impossible.

---

2   Schilpp, 1949, p.81

Whether it is the wave function in the two-slit experiment influencing the locations on the screen, or the collapse of the two-particle wave function into two single-particle wave functions, each with the perfectly correlated spin components needed to conserve total spin, Einstein's "objective reality" lets us see "hidden constants" that act to conserve all those properties and maintain existing symmetries.

> If one asks: does a ψ-function of the quantum theory represent a real factual situation in the same sense in which this is the case of a material system of points or of an electromagnetic field, one hesitates to reply with a simple "yes" or "no"...Does the individual system not have this q-value before the measurement, but only after a measurement when it randomly jumps into this position from somewhere else? But what about the single measured value of q? Did the respective individual system have this q-value even before the measurement? To this question there is no definite answer within the framework of the [quantum] theory, since the measurement is a process which implies a finite disturbance of the system from the outside; it would therefore be thinkable that the system obtains a definite numerical value for q (or p), i.e., the measured numerical value, only through the measurement itself. [3]

But as Werner Heisenberg thought, there are definitely times when an experimenter creates specific values, using her "free choice" of which property to measure. When Alice chooses the angle for her measurement, she disentangles the two-particle wave function. We now have simultaneous events in a spacelike separation. Einstein's symmetry and conservation principles are at work to ensure that Bob's measurement at the same angle conserves the total spin.

Einstein's insight into his EPR paradox never involved this subtle complexity of spinning electrons, although he was the discoverer of quantum statistics that Paul Dirac used to explain electron spins, but his objectively real picture can explain much of what is going on.

The puzzle of the wave function's influence over matter is the remaining "deep metaphysical mystery" of quantum mechanics.

---

3   Schilpp, 1949, p.81

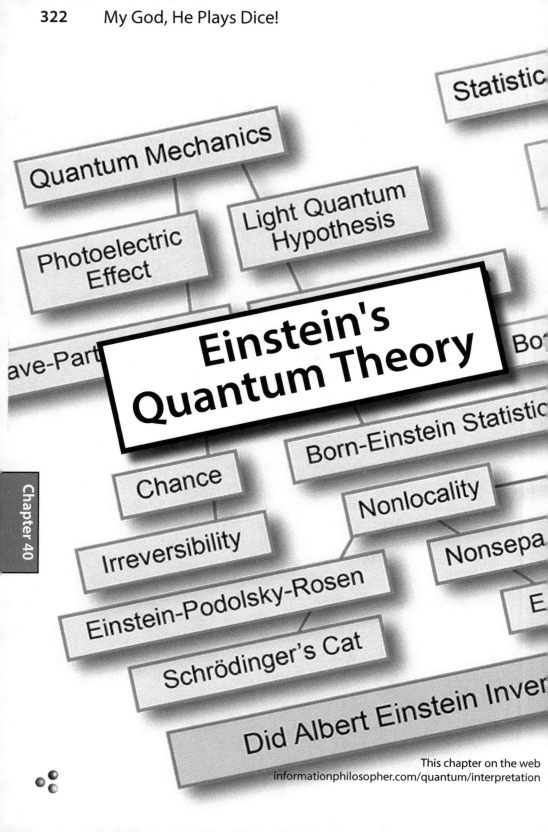

Statistic

Quantum Mechanics

Light Quantum
Hypothesis

Photoelectric
Effect

ave-Part

Einstein's
Quantum Theory

Bo

Born-Einstein Statistic

Chance

Nonlocality

Irreversibility

Nonsepa

Einstein-Podolsky-Rosen

E

Schrödinger's Cat

Did Albert Einstein Inver

This chapter on the web
informationphilosopher.com/quantum/interpretation

Einstein's Quantum Theory

We have noted Einstein's view that *principles* are the best basis for scientific theories (chapter 35?) and that quantum mechanics is fundamentally a *statistical* theory - despite Einstein's doubts about the reality of chance (chapter 36?),. We have also elaborated Einstein's suspicions about the continuum (chapter 37?) and his concerns about continuous field theories (chapter 38?).

We now propose to examine a quantum theory that would embrace Einstein's hope for an "objective reality" underlying quantum mechanics. The only "real "elements will be the particles.

We also suggest that the quantum wave function might be considered a continuous "field" that can be put alongside the gravitational and electromagnetic fields, not in a single "unified field theory" as Einstein hoped, but as a field nevertheless and with mind-boggling power over the particles.

Continuous fields of gravitation and of electromagnetism allow us to calculate precisely the forces on a test particle at a geometric point, should a particle be there. The quantum wave function is also a continuous field. It describes the probability of finding a particle at a given point in continuous space and time. All these continuous fields are determined for all space and time by the distribution of particulate matter and energy in space, the so-called boundary conditions and initial conditions.

Just as general relativity can be seen as curving space, so quantum theory can be seen to add a property to space that "influences" the discrete particles. In RICHARD FEYNMAN's path-integral formulation of quantum mechanics, the principle of least action explores all space to establish the quantum probabilities everywhere.

But infinities arise when we represent space and time with a continuum We imagine an infinite number of infinitesimal points between any two points on a line. Long before Einstein, LUDWIG BOLTZMANN had his doubts about the continuum and its infinities.

Chapter40

Should space and time be merely "free creations of the human mind," should they not "exist" in the same sense that matter and energy particles exist, and should they be only ideal and not "real", then the implications for quantum theory are profound.

If the "objectively real" (chapter 39) includes only material and energy particles obeying the most established laws of physics such as the conservation laws and the principles of symmetry behind them, we must reassess quantum theory, and we must follow Einstein's extraordinary insights wherever they lead, despite his well-known doubts about violations of his relativity.

Einstein's main objection to the Copenhagen Interpretation of quantum mechanics was its claim that a particle has no position, or indeed any other observable property, until the particle is measured. This is mostly anthropomorphic nonsense

His second objection was taking the superposition of states to describe "objectively real" superpositions, so that particles can be in two places at the same time.

Einstein's idea is that there is an "objective reality" in nature where particles have definite positions and paths, definite energies, momenta, and spins, even if quantum mechanics limits our ability to know them with the perfect precision of classical mechanics.

Despite his reputation as the major critic of quantum mechanics, Einstein came to accept its indeterminism and statistical nature. As we have seen, he had himself discovered these aspects of quantum mechanics (chapters 6, 11, and 12).

If the theory were merely *constructed* on data derived from experience, he said, quantum mechanics can only be approximate.

He wanted a better theory based on principles.

Einstein always hoped to discover - or better invent - a more fundamental theory, preferably a field theory like the work of Newton and Maxwell and his own relativity theories. He dreamed of a single theory that would unite the gravitational field, the electromagnetic field, the "spinor field," and even what he called the "ghost field" or "guiding field" of quantum mechanics.

Such a theory would use partial differential equations to predict field values continuously for all space and time. That theory would be a "free creation of the human mind." Pure thought, he said, mere ideas, could comprehend the real, as the ancients dreamed. [1]

Einstein wanted a field theory based on absolute *principles* such as the constant velocity of light, the conservation laws for energy and momentum, symmetry principles, and Boltzmann's principle that the entropy of a system depends on the possible distributions of its components among the available phase-space cells.

We can now see the limits of Einstein's interpretation, because fields are not substantial, like particles. A field is abstract immaterial *information* that simply predicts the behavior of a particle at a given point in space and time, should one be there!

Fields are *information*. Particles are *information structures*.

A gravitational field describes paths in curved space that moving particles follow. An electromagnetic field describes the forces felt by an electric charge at each point. The wave function Ψ of quantum mechanics - we can think of it as a possibilities field - provides probabilities that a particle will be found at a given point.

In all three cases *continuous immaterial* information accurately describes causal influences over *discrete material* objects.

In chapter 39, we showed that Einstein's insights about an "objective reality" can explain

1) nonlocality, which appears to violate his principle of relativity,

2) the two-slit experiment, which RICHARD FEYNMAN described as the "one mystery" of quantum physics,

3) entanglement, which ERWIN SCHRÖDINGER thought was "the characteristic trait" of quantum mechanics,

and 4) Ludwig Boltzmann's "molecular disorder," the origin of macroscopic irreversibility in thermodynamics..

Einstein's work also illuminates a few other quantum puzzles such as wave-particle duality, the metaphysical question of ontological chance, the "collapse" of the wave function, the "problem of measurement," the role of a "conscious observer," the conflict between relativity and quantum mechanics, and even the puzzle of Schrödinger's Cat.

Let's see how Einstein can help us understand these quantum puzzles and mysteries.

---

1   On The Method of Theoretical Physics, p.167

Einstein's "Objectively Real" Quantum Mechanics

Note that the local values of any field depends on the distribution of matter in the rest of space, the so-called "boundary conditions." Curvature of space depends on the distribution of masses. Electric and magnetic fields depend on the distribution of charges. And a quantum probability field depends on whether there are one or two slits open in the mysterious two-slit experiment. No particle has to travel through both slits in order for interference fringes to appear.

The quantum probability field $|\Psi|^2$, calculated from the deterministic Schrödinger equation, is a property of space. Like all fields, it has a value at each point whether or not there is a particle present there. Like all fields, it is determined by the distribution of nearby matter in space. These are the boundary conditions for the field. It has continuous values at every point, whether or not any particle is present at a given point.

1. Individual particles have the usual classical properties, like position and momentum, plus uniquely quantum properties, like spin, but all these properties can only be established *statistically*. The quantum theory gives us only statistical information about an individual particle's position and momentum, consistent with WERNER HEISENBERG's uncertainty principle, and only probable values for all possible properties.

But "objectively," a particle like an electron is a compact information structure with a definite, albeit unknown, position and momentum, both of which cannot be measured together with arbitrary accuracy. And it has other definite properties, such as the spatial components of electron spin, or of photon polarization, which also can not be measured together.

Just because we cannot measure an individual particle path with accuracy does not mean the particle does not follow a continuous path, let alone be in two places at the same time. And along this path, Einstein's "objective reality" requires that all the particle's properties are conserved, as long as there is no interaction with the external environment.

What is at two (or more) places at one time is the quantum wave function ψ, whose squared modulus $|\psi|^2$ gives us the non-zero probability of finding the particle at many places. But the matter/energy particle is not identical to the *immaterial* wave function!

Einstein and Schrödinger were strongly critical of the Copenhagen Interpretation's implication that superpositions represent real things. Tongue in cheek, Einstein suggested a superposition of explosives that would both explode and not explode. Schrödinger turned Einstein's criticism into a cat that is in a superposition of dead and alive.

It is testimony to the weirdness in modern quantum theory that Schrödinger's Cat is today one of the most popular ideas in quantum mechanics, rarely seen as a trenchant criticism of the theory.

2. The quantum wave functions are *fields*. Einstein called them *ghost fields* or *guiding fields*. The fields are *not* the particles. Fields have values in many places at the same time, indeed an *infinite* number of places. But particles are at one place at a time. Quantum field values are complex numbers which allow interference effects, causing some places to have no particles. Fields are *continuous* variables and not localized. Einstein showed that a particle of matter or energy is always *discrete* and localized. Light quanta are emitted and absorbed only as whole units, for example when one light quantum ejects an electron in the photoelectric effect.

Einstein was the first physicist to see wave-particle duality. And he was first to interpret the wave as the probability of finding a particle. MAX BORN's identification of the probability as the squared modulus $|\psi|^2$ of the wave function only made Einstein's qualitative identification quantitative and calculable.

The Copenhagen notion of *complementarity*, that a quantum object is both a particle and a wave, or sometimes one and sometimes the other, depending on the measurements performed, is confusing and simply wrong. A particle is always a particle and the wave behavior of its probability field is simply one of the particle's properties, like its mass, charge, spin, etc. Just as the gravitational field gives us the gravitational force on the particle, $|\Psi|^2$ gives us the probability of finding the particle at every point.

For Einstein, attempts to describe quantum objects as *nothing but* waves was absurd.

3. Because quantum physics does not give us precise information about a particle's location, Einstein was right to call it *incomplete, especially* when compared to classical physics. Quantum mechanics is a *statistical* theory and contains only probable information about an individual particle. Einstein's example of incompleteness was very simple. If we have one particle in two possible boxes, an *incomplete* theory gives us the probabilities of being found in each box. A *complete* theory would say for example, "the particle is in the first box."

4. While the probability wave field is abstract and *immaterial* information (Einstein's "ghost field") it *causally* influences the particle (Einstein's "guiding field"), just as the particle's spin dramatically alters its quantum statistics, another Einstein discovery. In particular. ψ somehow controls a particle's allowed positions though not by exerting any known forces. These non-intuitive behaviors are simply impossible in classical physics, and the empirical evidence for them is only seen (statistically) in large numbers of experiments, never in a single experiment.

In Einstein's quantum theory, there is no evidence that a single particle ever violates conservation principles by changing its position or any other property discontinuously. Changes in a particle's properties are always the results of interacting with other particles.

5. Although NIELS BOHR deserves credit for arranging atoms in the periodic table, the deep reasons for *two* particles in the first shell and *eight* in the second only became clear after Einstein discovered spin statistics in 1924, following a suggestion by S. N. BOSE, and after PAUL DIRAC and ENRICO FERMI extended the work to electrons. .

6. In the two-slit experiment, Einstein's localized particle *always goes through one slit or the other*, but when the two slits are open the probability wave function, which influences where the particle can be, is different from the wave function when one slit is open. The possibilities field (a wave) is determined by the boundary conditions of the experiment, which are different when only one slit is open. The particle does not go through both slits. It does not "interfere with itself." It is *never in two places at the same time*.

This agrees with Bohmian mechanics, which says that the wave function goes through both slits, even as the particle "objectively"always goes through only one slit.

7. The experiment with *two* entangled particles was introduced by Einstein in the 1935 EPR paradox paper. The Copenhagen assumption that each particle is in a random unknown combination of spin up and spin down, independent of the other particle, simply because we have not yet measured either particle, is wrong and the source of the EPR "paradox." Just as a particle has an unknown but definite position, entangled particles have definite spins, conserved since their initial preparation, even if the spins are unknown individually, they are interdependent jointly to conserve total spin.

When the particles travel away from the central source, with total spin zero, the two spins are opposite at all times. Or at a minimum, the spin is undefined for each particle because it is rotationally invariant and isotropic the same in all directions. When Alice chooses an angle to measure the spin, she adds new information that was not present at the original entanglement.

One operative principle for Einstein's "objective reality" is conservation. To assume that their spins are independent is to consider the absurd outcome that spins could be found both up (or both down), a violation of a conservation principle that is more egregious than the amazing fact spins are always perfectly correlated in any measurements.

8. ERWIN SCHRÖDINGER explained to Einstein in 1936 that two entangled particles share a single wave function that can not be separated into the product of two single-particle wave functions, at least not until there is an interaction with another system which *decoheres* their perfect correlation. This is intuitively understandable because conservation laws preserve their perfect correlation unless one particle is disturbed, for example by environmental decoherence, by some interaction with the environment.

9. Einstein ultimately accepted the indeterminism in quantum mechanics and the uncertainty in pairs of conjugate variables, despite the clumsy attempt by his colleagues Podolsky and Rosen to challenge uncertainty and restore determinism in the EPR paper.

Chapter40

10. In 1931 Einstein called Dirac's transformation theory "the most perfect exposition, logically, of this [quantum] theory" even though it lacks "enough information to enable one to decide" a particle's exact properties.[2] In 1933 Dirac reformulated quantum physics using a Lagrangian rather than the standard Hamiltonian representation.   The time integral of the Lagrangian has the dimensions of action, the same as Planck's quantum of action h. And the *principle of least action* visualizes the solution of dynamical equations like Hamilton's as exploring all paths to find that path with minimum action.

Dirac's work led RICHARD FEYNMAN to invent the path-integral formulation of quantum mechanics. The transactional interpretations of JOHN CRAMER and RUTH KASTNER have a similar view. The basic idea of exploring all paths is in many ways equivalent to saying that the probabilities of various paths are determined by a solution of the wave equation using the boundary conditions of the experiment. As we saw above, such solutions involve whether one or two slits are open, leading directly to the predicted interference patterns, given only the wavelength of the particle.

11. In the end, of course, Einstein held out for a continuous field theory, one that could not be established on the basis of any number of empirical facts about measuring particles, but must be based on the discovery of principles, logically simple mathematical conditions which determine the field with differential equations. His dream was a "unified field theory," one that at least combined the gravitational field and electromagnetic field, and one that might provide an underpinning for quantum mechanics someday.

Einstein was clear that even if his unified field theory was to be deterministic and causal, the statistical indeterminism of quantum mechanics itself would have to be preserved.

This seemingly impossible requirement is easily met in Einstein's "objectively real" quantum theory if we confine determinism to Einstein's *continuous* fields, which are pure abstract immaterial information. Einstein's 1917 discovery of indeterminism and the

___

statistical nature of physics need apply only to particles, which are *discrete* information structures.

It is therefore most significant to note that the mathematics of Schrödinger's wave equation and his wave function is entirely deterministic.

Quantum systems are often pictured as evolving in two ways, thought to be logically inconsistent by many physicists and philosophers:

- The first is the continuous wave function deterministically exploring all the possibilities for interaction (cf. von Neumann process 2).

- The second is the particle randomly choosing one of those possibilities to become actual (cf. von Neumann process 1).

No knowledge can be gained by a "conscious observer" unless new information has previously been irreversibly recorded in the universe. Such new information can be created and recorded in three places:

- In the target quantum system,

- In the combined target system and measuring apparatus,

- It can then, and only then, become knowledge recorded in the observer's mind. See John Bell's "shifty split" in chapter 32.

The measuring apparatus is material and quantum mechanical, not deterministic or "classical." It need only be statistically determined and capable of recording the *irreversible* information about an interaction. The apparatus is on the "classical" side of the "quantum to classical transition." The human mind is similarly only statistically determined.

- There is only one world.

- It is a quantum world.

Ontologically, the quantum world is indeterministic, but in our everyday common experience it appears to be causal and deterministic, the so-called "classical" world. The "quantum-to-classical transition" occurs for any large macroscopic object that contains a large number of atoms. For large enough systems, independent quantum events are "averaged over." The uncertainty in position x and velocity v of the object becomes less than the observational uncertainty.

$\Delta v \, \Delta x \geq h / m$ becomes immeasurably small as m increases and h / m goes to zero.

It is an error to compare h going to zero in quantum mechanics with v being small compared to c in relativity theory. Velocity v can go to zero. Planck's quantum of action h is constant so it cannot.

The classical laws of motion, with their apparently strict causality, emerge when objects are large enough so that microscopic events can be ignored, but this determinism is fundamentally statistical and physical causes are only probabilistic, however near to certainty.

Information philosophy interprets the wave function ψ as a "possibilities" field. With this simple change in terminology, the mysterious process of a wave function "collapsing" becomes a much more intuitive discussion of ψ providing all the possibilities (with mathematically calculable probabilities), followed by a single actuality, at which time the probabilities for all non-actualized possibilities go to zero (they "collapse") instantaneously. But no matter, no energy, and in particular, no information is transferred anywhere!

Einstein's "objectively real" quantum theory is standard quantum physics, though freed of some absurd Copenhagen Interpretations. It accepts the Schrödinger equation of motion, Dirac's *principle of superposition*, his *axiom of measurement* (now including the actual information "bits" measured), and - most importantly - Dirac's *projection postulate*, the "collapse" of ψ that so many interpretations of quantum mechanics deny.

And Einstein's quantum theory does not need the "conscious observer" of the Copenhagen Interpretation thought to be required for a projection, for the wave-function to "collapse," for one of the possibilities to become an actuality. All the collapse does require is an interaction between systems that creates irreversible and observable, but not necessarily observed, information.

Einstein's quantum theory denies that particles have no properties until measurements are made by these "conscious observers.

Among the founders of quantum mechanics, almost everyone agreed that irreversibility is a key requirement for a measurement. As Einstein appreciated, irreversibility introduces statistical mechanics and thermodynamics into a proper formulation of quantum mechanics.

Information is not a conserved quantity like energy and mass, despite the view of many mathematical physicists, who generally accept the determinist idea that information too is conserved.

The universe began in a state of equilibrium with minimal information, and information is being created every day, despite the second law of thermodynamics. Classical interactions between large macroscopic bodies do not generate new information. Newton's laws of motion are thought to be deterministic so that the information in any configuration of bodies, motions, and force is enough to know all past and future configurations (Laplace's intelligent demon). Classical mechanics does, in principle, conserve information.

In the absence of interactions, an isolated quantum system evolves according to the unitary Schrödinger equation of motion. Just like classical systems. The deterministic Schrödinger equation also conserves information.

Unlike classical systems however, when there is an interaction between quantum systems, the two systems become entangled and there may be a change of state in either or both systems. This change of state may create new information.

If that information is instantly destroyed, as in most interactions, it may never be observed macroscopically. If, on the other hand, the information is stabilized for some length of time, it may be seen by an observer and considered to be a "measurement." But it need not be seen by anyone to become new information in the universe. The universe is its own observer!

For the information (negative entropy) to be stabilized, the second law of thermodynamics requires that an amount of positive entropy greater than the negative entropy must be transferred away from the new information structure.

Exactly how the universe allows pockets of negative entropy to form as "information structures" we describe as the "cosmic creation process." This core two-step process has been going on since the origin of the universe. It continues today as we add information to the sum of human knowledge. We'll discuss it further briefly in chapter 41.

Note that despite the Heisenberg uncertainty principle, quantum mechanical measurements are not always uncertain. When a system is measured (prepared) in an eigenstate, a subsequent measurement (Pauli's measurement of the first kind) will find it in the same state with perfect certainty.

What are the normal possibilities for new quantum states? The transformation theory of Dirac and Jordan lets us represent ψ in a set of basis functions for which the combination of quantum systems (one may be a measurement apparatus) has eigenvalues (the *axiom of measurement*). We represent ψ as in a linear combination (the *principle of superposition*) of those "possible" eigenfunctions. Quantum mechanics lets us calculate the probabilities of each of those "possibilities."

Interaction with the measurement apparatus (or indeed interaction with any other system) may select out (the *projection postulate*) one of those possibilities as an actuality. But for this event to be an "observable" (a John Bell "beable"), information must be created and positive entropy must be transferred away from the new information structure, in accordance with our two-step information creation process.

All interpretations of quantum mechanics predict the same experimental results. Einstein's "objectively real" quantum theory is no exception, because the experimental data from quantum experiments is the most accurate in the history of science.

Where interpretations differ is in the picture (the *visualization*) they provide of what is "really" going on in the microscopic world - so-called "quantum reality." Schrödinger called it *Anschaulichkeit*. He and Einstein were right that we should be able to picture "quantum reality."

However, the Copenhagen Interpretation of Bohr and Heisenberg discourages all attempts to visualize the nature of the "quantum world," because they say that all our experience is derived from the "classical world" and should be described in ordinary language. This is why Bohr and Heisenberg insisted on some kind of "cut" between the quantum event and the mind of an observer.

Copenhageners were proud of their limited ability to know what is going on in "quantum reality." Bohr actually claimed...:

> There is no quantum world. There is only an abstract quantum physical description. It is wrong to think that the task of physics is to find out how nature is. Physics concerns what we can say about nature.

Einstein's "objective reality" is based on things we can visualize, without being able to measure them directly. (See our on-line animation of the two-slit experiment[3], our EPR experiment visualizations[4], and Dirac's three polarizers[5] to visualize the superposition of states and the projection or "collapse" of a wave function.)

Einstein and Schrödinger made fun of superposition, but Einstein never doubted the validity of any of Dirac's "principles" of quantum mechanics. What Einstein attacked was the nonsense of assuming that real objects could be in such a superposition, both here and there, both dead and alive. etc.

Bohr was of course right that classical physics plays an essential role. His Correspondence Principle allowed him to recover some important physical constants by assuming that the discontinuous quantum jumps for low quantum numbers (low "orbits" in his old quantum theory model) converged in the limit of large quantum numbers to the continuous radiation emission and absorption of classical electromagnetic theory.

In addition, we know that in macroscopic bodies with enormous numbers of quantum particles, quantum effects are averaged over, so that the uncertainty in position and momentum of a large body still obeys Heisenberg's indeterminacy principle, but the uncertainty is for all practical purposes unmeasurable and the body can be treated classically.

We can say that the quantum description of matter also converges to a classical description in the limit of large numbers of quantum particles. We call this "adequate" or statistical determinism. It is the apparent determinism we find behind Newton's laws of motion for macroscopic objects. The statistics of averaging over many

3   informationphilosopher.com/solutions/experiments/two-slit_experiment/
4   informationphilosopher.com/solutions/experiments/EPR/
5   www.informationphilosopher.com/solutions/experiments/dirac_3-polarizers/

Chapter40

independent quantum events then produces the "quantum to classical transition" for the same reason as the "law of large numbers" in probability theory approaches a continuous function..

Note that the macromolecules of biology are large enough to stabilize their information structures. DNA has been replicating its essential information for billions of years, resisting equilibrium despite the second law of thermodynamics. The creation of irreversible new information also marks the transition between the quantum world and the "adequately deterministic" classical world, because the information structure itself must be large enough (and stable enough) to be seen. Biological entities are macroscopic, so the quantum of action h becomes small compared to the mass m and h/m approaches zero.

Decoherence theorists say that our failure to see quantum superpositions in the macroscopic world *is* the measurement problem Einstein's "objective reality" interpretation thus explains why quantum superpositions like Schrödinger's Cat are not seen in the macroscopic world. Stable new information structures in the dying cat reduce the quantum possibilities (and their potential interference effects) to a classical actuality. Upon opening the box and finding a dead cat, an autopsy will reveal that the time of death was observed/recorded. *The cat is its own observer.*

The nadir of interpretation was probably the most famous interpretation of all, the one developed in Copenhagen, the one Niels Bohr's assistant Leon Rosenfeld said was not an interpretation at all, but simply the "standard orthodox theory" of quantum mechanics.

It was the nadir of interpretation because Copenhagen wanted to put a stop to "interpretation" in the sense of understanding or "visualizing" an underlying reality. The Copenhageners said we should not try to "visualize" what is going on behind the collection of observable experimental data. Just as Kant said we could never know anything about the "thing in itself," the *Ding-an-sich*, so the positivist philosophy of Auguste Comte, Ernst Mach, Bertrand Russell, Rudolf Carnap, as well as the British empiricist thinkers John Locke and David Hume, claim that knowledge stops at the "secondary" sense data or perceptions of phenomena, preventing access to the primary "objects."

Einstein's views on quantum mechanics have been seriously distorted (and his early work largely forgotten), perhaps because of his famous criticisms.

Though its foremost critic, Einstein frequently said that quantum mechanics was a most successful theory, the very best theory so far at explaining microscopic phenomena, but that he hoped his ideas for a continuous field theory would someday add to the discrete particle theory and its "nonlocal" phenomena. It would allow us to get a deeper understanding of underlying reality, though at the end he despaired any his continuous field theory compared to particle theories.

Many if not most of the "interpretations" of quantum mechanics deny a central element of quantum theory, one that Einstein himself established in 1916, namely the role of indeterminism, or "chance," to use its traditional name, as Einstein did in physics (in German, *Zufall*) and as William James did in philosophy in the 1880's. These interpretations all hope to restore the determinism of classical mechanics.

Many interpretations even deny the existence of particles. They admit only waves that evolve unitarily under the Schrödinger equation.. They like to regard the wave function as a real entity rather than an abstract possibilities function.

We can therefore classify various interpretations by whether they accept or deny chance, especially in the form of the so-called "collapse" of the wave function, also known as the "reduction" of the wave packet or what Paul Dirac called the "projection postulate." Most "no-collapse" theories are deterministic. "Collapses" in standard quantum mechanics are irreducibly indeterministic.

Einstein's criticisms of quantum mechanics, in the form of many attempts to visualize what is going on in "quantum reality," led him to make many mistakes, as we shall see in chapter 42

But behind almost every Einstein "mistake" was an extraordinary insight that has led to some of today's most fascinating and puzzling aspects of quantum mechanics. Einstein's "objective reality" is our best hope for resolving some of those puzzles.

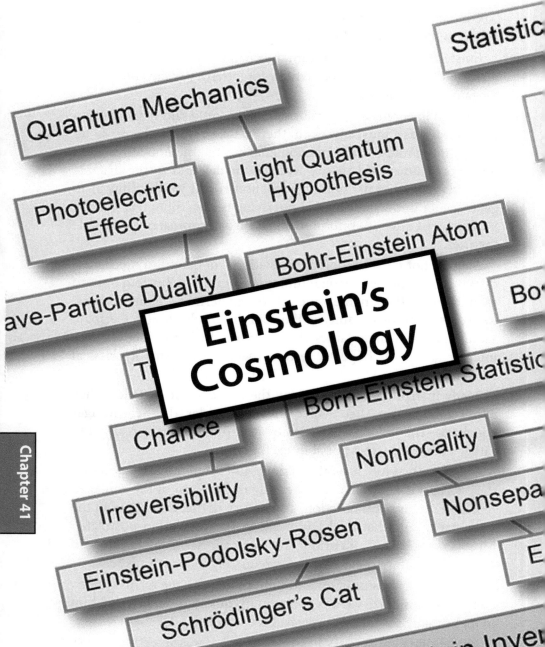

Quantum Mechanics

Statistic

Photoelectric Effect

Light Quantum Hypothesis

ave-Particle Duality

Bohr-Einstein Atom

Bo

Einstein's Cosmology

T

Chance

Born-Einstein Statistic

Nonlocality

Irreversibility

Nonsepa

Einstein-Podolsky-Rosen

E

Schrödinger's Cat

Did Albert Einstein Inver

Einstein's Cosmology

The Cosmological Constant

When ALBERT EINSTEIN was completing his work on general relativity in 1916, it was said that he asked some astronomers whether the stars were falling towards us or perhaps expanding away from us. "Oh, Dr. Einstein, it is well known that the stars are 'fixed,' in the celestial sphere." Since his new equations suggested otherwise, Einstein added a small term called the cosmological constant that would prevent expansion or contraction.

One very simple way to understand expansion in non-relativistic terms is to compare the amount of gravitating matter in the universe, whose mutual attraction would collapse the universe, to the motion energy seen in the distant galaxies.

The positive "kinetic" energy of the motion is either larger or smaller than the negative "potential" binding energy. We can distinguish three cases.

K.E. < P.E. The universe is said to be *positively* curved. The self-gravitating force will eventually slow down and stop the expansion. The universe will then collapse in a reverse of the "Big Bang" origin.

K.E. > P.E. The universe is said to be *negatively* curved. The self-gravitating force will be overcome by the motion energy. The universe will expand forever. When galaxies are infinitely apart, they will still be moving.

K.E. = P.E. The universe is flat. Average curvature is *zero*. The geometry of the universe is Euclidean. The expansion will stop, but only when the distances between remote galaxies approaches infinity after an infinite time.

By just adding a cosmological constant to achieve a result, Einstein masked the underlying physics for time.

The Flatness Problem

The universe is very likely flat because it was created flat. A flat universe starts with minimal information, which is fine since our cosmic creation process can create all the information that we have today. Leibniz' question, "Why is there something rather

Chapter 41

than nothing?" might be "the universe is made out of something (matter energy) and the opposite of that something (motion energy)."

When I was a first-year graduate student in astrophysics at Harvard University in 1958, I encountered two problems that have remained with me all these years. One was the fundamental problem of information philosophy - *"What creates the information structures in the universe?"* The other was the flat universe.

At that time, the universe was thought to be positively curved. EDWIN HUBBLE's red shifts of distant galaxies showed that they did not have enough kinetic energy to overcome the gravitational potential energy. Textbooks likened the universe to the surface of an expanding balloon decorated with galaxies moving away from one another.

That balloon popped for me when WALTER BAADE came to Harvard to describe his work at Mount Wilson. Baade took many images with long exposures of nearby galaxies and discovered there are two distinct populations of stars. And in each population there was a different kind of Cepheid variable star. The period of the Cepheid's curve of light variation indicated its absolute brightness, so they could be used as "standard candles" to find the distances to star clusters in the Milky Way.

Baade then realized that the Cepheids being used to calculate the distance to Andromeda were 1.6 magnitudes brighter than the ones used in our galaxy. Baade said Andromeda must be twice as far away as Hubble had thought.

As I listened to Baade, for me the universe went from being positively curved to negatively curved. It jumped right over the flat universe! I was struck that we seemed to be within observational error of being flat. Some day a physicist will find the reason for perfect flatness, I thought.

I used to draw a line with tick marks for powers of ten in density around the critical density ρ_c to show how close we are to flat. Given so many orders of magnitude of possible densities, it seemed improbable that we were just close by accident We could increase the density of the universe by thirty powers of ten before it would have the same density as the earth (too dense!). But on the lighter side, there are an infinite number of powers of ten. We can't exclude a universe with average density zero, which still allows us to exist, but little else in the distance.

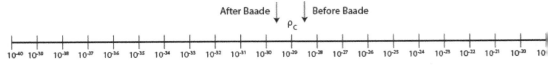

After Baade   Before Baade

ρ_C

| 10^{-40} | 10^{-39} | 10^{-38} | 10^{-37} | 10^{-36} | 10^{-35} | 10^{-34} | 10^{-33} | 10^{-32} | 10^{-31} | 10^{-30} | 10^{-29} | 10^{-28} | 10^{-27} | 10^{-26} | 10^{-25} | 10^{-24} | 10^{-23} | 10^{-22} | 10^{-21} | 10^{-20} | 10 |

In the long run we are approaching a universe with average density zero. All the non-gravitationally bound systems will slip over our light horizon as the expansion takes them higher than the velocity of light. At that time, we will be alone in the universe with the nearby, gravitationally bound members of our "local group" of galaxies, the Milky Way, Andromeda, the Large and Small Magellanic Clouds, and a few dozen dwarf galaxies.

Beyond them will be ghostly images of galaxies, quasars, supernovae, and other objects with whom communication will never be possible at the speed of light.

But note that we may always be able to see back to the cosmic microwave background, all the same contents of the universe that we see today, all extremely red-shifted to the point of no visible energy in the photons!

The Problem of Missing Mass (Dark Matter)

Given our assumption that the universe is exactly flat, the missing mass problem is that there is not enough observable material so that in Newtonian cosmology the gravitational binding energy can exactly balance the kinetic energy. The visible (luminous mass) accounts for only about 4-5 percent of the needed mass. Studying the rotation curves of galaxies and galaxy clusters reveals an invisible mass (called dark matter) contained inside the galaxies and clusters that amounts to perhaps 6 times the visible matter, which accounts for about 30 percent of the critical mass density needed to make the universe exactly flat. Current theory accounts for the balance by "dark energy," an interpretation of the cosmological constant Einstein considered adding to his equations as a pressure to keep it from collapsing (known as "vacuum energy"). But the missing mass could just be more dark matter between the galaxies and clusters. About 3 times the estimated dark matter would do.

Chapter 41

And I am delighted that observations are within a factor of three of the critical density ρ_c.

When Baade showed the universe was open in the 1950's, we needed ten times more matter for a flat universe. Now we need only three times more. More than ever, we are obviously flat!

Dark Energy (Is the Expansion Accelerating?)

Finding the missing mass can close the universe and explain its flatness. But it would not explain the apparent accelerating expansion seen in Type 1a supernovae. This might be an artifact of the assumption they are perfect "standard candles." Recent evidence suggests that distant Type 1a supernovae are in a different population than those nearby, something like Baade's two populations.

It seems a bit extravagant to assume the need for an exotic form of vacuum energy on the basis of observations that could have unknown but significant sources of error. Fortunately, the size of this problem is only another factor of between 3 and 4, well within observational error.

String theorists claim conditions at the universe origin must have been "fine tuned" to within 120 orders of magnitude to produce our current universe. This seems to be nonsense.

The Horizon Problem

The horizon problem arises from the perfect synchronization of all the parts of our visible universe, when there may never have been a time in the early universe that they were close enough together to exchange synchronization signals.

We propose a solution to the horizon problem based on Einstein's (mistaken) insight that in the wave-function collapse of entangled particles, something is "traveling" faster than the speed of light.

Einstein said that events in a spacelike separation cannot interact, That would violate his special theory of relativity. He described it as the "impossibility of simultaneity." But something can simultaneously change great disstances. That something is *information about possibilities.*

When the "universal wave function" Ψ collapsed at t = 0, parts of the universe that are outside our current light horizon may have been "informed" that it was time to start, no matter the distance.

This radical idea is consistent with RICHARD FEYNMAN's path integral (or "sum-over-histories") formulation of quantum mechanics. In calculating the probability of a quantum event, the path integral is computed over all the possible paths of virtual photons, many traveling faster than the speed of light.

The Information Paradox

Can we speculate about what Einstein might have thought about the black-hole information paradox?

Perhaps not. For Einstein, entropy is defined by Boltzmann's principle. $S = k \log W$, where W is the number of phase-space cells.

Since the size of the black hole is smaller when matter is added, we can see that STEPHEN HAWKING and Jakob Bekenstein were correct that the information content of physical objects falling into a black hole will be lost forever. More particles are now distributed in a smaller number of cells

In 1997, John Preskill made a bet with Hawking, claiming that information must be preserved, according to quantum theory.

In fact, neither quantum nor classical theory requires the conservation of information. Being simply the arrangement of material particles in phase space, information is not a conserved quantity like energy and momentum, as Einstein would have known.

The idea of conserved information comes from mathematical physicists who want a deterministic universe in which all the information existing today was present at the origin of the universe.

In 2004, Hawking published a paper showing how some information might escape from a black hole, and he conceded his loss of Preskill's bet. Hawking is right that particles emerge from pair production at the black hole horizon, but the idea that it is the same information that was destroyed when information structures fell into the black hole is simply absurd.

Hawking may have told us this when he quipped that he should have burned the baseball encyclopedia he gave to Preskill and pay off the lost bet by sending him the ashes!

Once again, it was Einstein's phenomenal imagination that first conceived of extraordinary ideas only recently confirmed, like gravitational waves, gravitational lensing, and of course black holes.

Chapter 41

Statistic

Quantum Mechanics

Light Quantum
Hypothesis

Photoelectric
Effect

Bohr-Einstein Atom

ave-Particle Duality

Bo

Einstein's Mistakes

Tra

Born-Einstein Statistic

Chance

Nonlocality

Irreversibility

Nonsepa

Einstein-Podolsky-Rosen

E

Schrödinger's Cat

Did Albert Einstein Inver

Einstein's Mistakes

We must first acknowledge that Einstein's mistakes have given us in general more important theoretical insights than those of all but a handful of great physicists's successes. Einstein's mistakes lie behind the greatest puzzles and mysteries in physics today.

While Einstein did not solve these mysteries, in most of them so far neither has any other scientist provided convincing explanations. That his phenomenal mind saw them at all is his great gift to science.

When we see his mistakes for what they are, and when we add them to his extraordinary successes, Einstein emerges as the single greatest force behind both of the leading fields of physics today, relativity and quantum mechanics.

Fields and Particles

Unified Field Theory

In terms of effort spent and results achieved, surely his unified field theory was Einstein's greatest mistake, first because it was deterministic, second because there are now so many fields.

He wrote his friend Michele Besso the year before he died,

> "I consider it quite possible that physics cannot be based on the field concept, i.e:, on continuous structures. In that case, nothing remains of my entire castle in the air, gravitation theory included, [and of] the rest of modern physics." [1]

Space and Time

Einstein is said to have combined space and time into a single four-dimensional continuum. This was first done by Hermann Minkoswki, but Einstein deserves credit for developing the four-dimensional energy-momentum tensor that describes his theory of general relativity.

---

1   Pais, 1982, p.467

Chapter42

In his later years Einstein had many doubts about the reality of space and time, wondering if they may be just convenient fictions, "free creations of the human mind," which just happen to describe accurately the "real" things, the material particles.

Quantum Physics

Ontological Chance

Without a doubt, it was Einstein's two papers in 1916 and early 1917 that established chance in the emission and absorption of his light quanta. The times and directions of light interactions with matter are completely indeterminate. Einstein gave credit to Ernest Rutherford for discovering a similar indeterminacy in radioactive decay.

Einstein said chance must be considered a "weakness in the theory."

But it was Einstein's proof that thermal equilibrium between Planck's radiation distribution and the Maxwell-Boltzmann velocities distribution of matter could not be maintained without the emission of photons going off in all directions at random.

Einstein's canonical paper on the A and B coefficients for emission and absorption is a foundational element of the statistical nature of quantum mechanics, and it predicted the stimulated emission of radiation that underlies the working of lasers.

Einstein's mistake was to not accept for many years the conclusion that natural processes involve chance. "God does not play dice."

This one "mistake" explains how the universe can create unpredictable new information structures like atoms, stars, galaxies, living things, minds, and new ideas! See chapter 43.

The Statistical Interpretation

MAX BORN's interpretation of the quantum mechanical wave function of a material particle as the probability (amplitude) of finding a material particle was a direct extension of Einstein's interpretation of light waves as giving probability of finding photons.

To be sure, Einstein's interpretation may be considered only qualitative, where Born's was quantitative. He made it the squared modulus of the probability amplitude $|\psi|^2$. The new quantum mechanics gives us exact calculations - of statistics!

As with his dislike of chance, Einstein was happy to give Born all the credit, including a Nobel Prize, for the statistical interpretation.

Nonlocality

When Einstein first thought about a light wave spreading out in space, only to collapse to a point when all the light was collected into a single atom in metal to eject a single electron, he briefly thought distributed energy must have moved faster than light to collect itself together.

To be sure, Einstein hypothesized that perhaps light is not continuously distributed over an increasing space but consists of a finite number of energy quanta which are localized at points in space. But this did not stop him from worrying about nonlocality.

Einstein saw spacelike separated events occurring *simultaneously*, an apparent violation of his special theory of relativity, which claims that simultaneity is impossible in an absolute sense

Symmetry and Conservation

EPR and Entanglement

As we mentioned in the EPR chapter 26, Einstein's greatest scientific biographer, Abraham Pais, concluded in 1982 that the EPR paradox "had not affected subsequent developments in physics, and it is doubtful that it ever will." [2] Einstein had drawn attention for decades to the appearance of nonlocality and in the 1935 EPR paper added his *separation principle*, but his orthodox physicist colleagues could make no sense of his paper.

Einstein's mistake was to say we should absolutely agree that the real factual situation of one system is independent of what is done with another which is spatially separated. [3] ERWIN SCHRÖDINGER immediately pointed out that the two-particle wave function would not separate without an interaction or measurement.t

2   Pais, 1982, p.456
3   Einstein, 1949a, p.85

But it was Einstein himself who first imagined two events in a spacelike separation occurring simultaneously, an impossibility according to his own special theory of relativity. Without this mistake of Einstein, we might never have discovered entanglement!

Spooky Action-at-a-Distance

Einstein described spooky action as one particle acting "telepathically" on another particle spatially separated.[4] It may be no exaggeration to say that spooky action is one of Einstein's greatest original ideas.

Adding "spooky" in 1949 to his decades of complaints about non-locality and nonseparability did catch the world's attention.

But Einstein should have seen that all these cases were not "actions" by one particle on a distant particle. Einstein added a *false asymmetry* into a symmetric situation.

Schrödinger's Cat

This famous cat began with Einstein criticizing the implication of Schrödinger's wave equation. He told Schrödinger to imagine a charge of gunpowder that can spontaneously combust, on average once a year. Then "your ψ-function describes a sort of blend of not-yet and already exploded systems." Schrödinger famously adapted Einstein's idea to his cat in a "superposition" of dead and alive.

Both Einstein and Schrödinger were making fun of superposition, but Einstein should have known it was just a mathematical tool to calculate statistical probabilities.

Schrödinger switched from joking about superposition to claiming that entanglement is the "characteristic trait" of quantum mechanics. He and Einstein parted ways.

Cosmology

The Cosmological Constant

Einstein himself described the addition of a constant to his equations of general relativity, in order to produce a static universe, his "biggest blunder," in conversation with George Gamow.[5]

---

4   Schilpp, 1949, p. 85
5   Gamow, 1970, p.44

The Expansion of the Universe

Had Einstein not forced his theory to match the poor observational data of his time, he might have speculated that the universe was adding space by expanding or contracting, over a decade before Edwin Hubble found the expansion of external galaxies in 1927.

The Flat Universe

As Einstein's field equations for general relativity improved in the early years, he might have noted that when the expansion rate - the motion energy gets near the gravitational binding energy, the overall curvature approaches zero and the "radius" of the observable universe approaches infinity.

As observations have improved, the universe now appears within a factor of three of having enough matter to make the universe "flat" and its geometry Euclidean.

Einstein might have appreciated this symmetry between energy and matter,

Thermodynamics and Statistical Mechanics

Gibbs-Liouville

The conservation of any particular volume of phase space (the Liouville theorem) led J. WILLARD GIBBS to claim that information is also conserved. Einstein claimed that he did little or nothing more than Gibbs. But this was a mistake. Gibbs' statistical mechanics is a formal theory that does not even mention material particles. Einstein's work led to the proof of the existence of atoms!

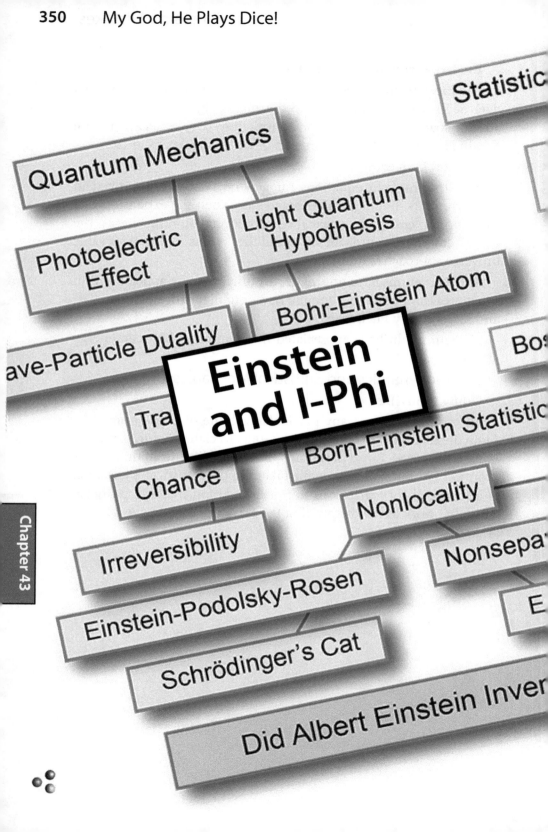

Statistic

Quantum Mechanics

Light Quantum Hypothesis

Photoelectric Effect

Bohr-Einstein Atom

ave-Particle Duality

Bos

Einstein and I-Phi

Tra

Born-Einstein Statistic

Chance

Nonlocality

Irreversibility

Nonsepar

Einstein-Podolsky-Rosen

E

Schrödinger's Cat

Did Albert Einstein Inver

Albert Einstein and Information Philosophy

On Information Philosophy

Information is neither matter nor energy, although it needs matter to be embodied and energy to be communicated. Why should information become the new basis for understanding and solving so many problems in philosophy and science?

It is because everything in the universe that is distinguishable from chaos and disorder is an *information structure* that was created since the structureless, pure energy origin of the universe.

As most all of us know, matter and energy are *conserved*. This means that there is just the same total amount of matter and energy today as there was at the universe origin. Einstein showed us that matter can be converted into energy with his equation $E=mc^2$, so there is just one unchanging total of "stuff" in the universe.

But then what accounts for all the *change* that we see, the new things under the sun? It is information, which is not conserved and has been increasing since the beginning of time, despite the second law of thermodynamics, with its increasing entropy, which destroys order.

What is changing is the *arrangement* of the existing matter in what we call information structures. What is emerging is new information. What idealists and holists see is that emergence of immaterial information embodied in material structures.

Living things, you and I, are dynamic growing information structures, forms through which matter and energy continuously flow. And it is information processing that controls those flows!

At the lowest levels, living information structures blindly replicate their information. At higher levels, natural selection adapts them to their environments. At the highest levels, they develop behaviors, intentions, goals, and agency, introducing purpose into the universe.

Chapter 43

Information is the modern spirit, the ghost in the machine, the mind in the body. It is the soul, and when we die, it is our information that perishes, unless the future preserves it. The matter remains.

Information can explain the fundamental metaphysical connection between materialism and idealism. Information philosophy replaces the determinism and metaphysical necessity of eliminative materialism and reductionist naturalism with metaphysical possibilities. Alternative possibilities can not exist without ontological chance. Determinism says there is but one possible future.

Many mathematical physicists like the idea of a completely deterministic universe. The Bohmians, Everett's many worlders, John Bell, and the Decoherence theorists are all determinists. They believe that the "wave function of the universe" evolves deterministically, and it does. But they deny the many "collapses of the wave function" which are indeterministic and are the creative source of all new information.

Einstein saw chance as a "weakness in the theory." But the important thing is that he was the first person to see ontological "objectively real" chance in physics. Chance in classical physics had been regarded as epistemological, merely human ignorance.

Perhaps the most amazing thing about information philosophy is its discovery that abstract and immaterial information (the quantum wave field) can exert an influence over concrete matter, perhaps explaining how mind can move body, how our thoughts can control our actions, deeply related to the way the quantum wave function controls the probabilities of locating quantum particles, as first seen, but never understood, by Einstein.

Einstein did not like probabilities but clearly saw that quantum physics is a *statistical* theory.

How abstract *probability* amplitudes Ψ control the *statistics* of experiments remains the *one deep mystery of quantum mechanics*.

Knowledge is information in minds that is a partial isomorphism (mapping) of the information structures in the external world. Information philosophy is a correspondence theory.

Sadly, there is no isomorphism, no information in common, between words and objects. This accounts for much of the failing of analytic language philosophy in the past century. The arbitrary and conventional connections between words and objects is the source of confusion in Niels Bohr's Copenhagen Interpretation of quantum mechanics.

Although language is a fine tool for human communication, it is arbitrary, ambiguous, and ill-suited to represent the world directly. Human languages do not picture reality. Information is the true *lingua franca* of the universe.

The extraordinarily sophisticated connections between words and objects are "free creations of human minds," mediated by the brain's experience recorder and reproducer (ERR). Words stimulate wired neurons to start firing and to play back those experiences that include related objects.

Neurons that were wired together in our earliest experiences fire together at later times, contextualizing our new experiences, giving them meaning. And by replaying emotional reactions to similar earlier experiences, it makes then "subjective experiences," giving us the feeling of "what it's like to be me" and solving the "hard problem" of consciousness.

Without words and related experiences previously recorded in our mental experience recorders, we could not comprehend words. They would be mere noise, with no meaning.

Far beyond words, a dynamic information model of an information structure in the world is presented immediately to the mind as a simulation of reality experienced for itself.

This is why we are creating *animations* of mysterious quantum phenomena to show you the two-slit experiment, entanglement, and the interaction of radiation with microscopic matter that leads to the macroscopic irreversibility underlying the second law of thermodynamics.

We will analyze all the quantum "mysteries" we hope to solve in terms of information structures and the communication of information between information structures. We will look to find the information in each of the quantum mysteries

Chapter 43

Where's the Information in Entangled Particles?

The central mystery in entanglement for eighty years has been how Alice's measurement of a property can be "transmitted," presumably faster than the speed of light, to Bob at a remote space-like separation, so that Bob's measurement of a related property can be perfectly correlated with Alice's measurement.

The information needed is the electron spin or photon polarization direction (up or down) for each particle. The Copenhagen Interpretation says we cannot know those spin values, that they do not even exist until the measurements are made.

Einstein's "objective reality" says that they do have values, independent of our measurements. When we prepare the experiment, we know that one particle is up and the other down, but we don't know which is which.

Because we lack that knowledge, quantum mechanics assumes they are best described by a linear superposition of up-down and down-up. Objective reality, however, says they always will be found in one of those states, *either* up-down or down-up.

Now Einstein's principles of conservation say that the initial properties are conserved as long as there is no external interaction with the two particles. The information is therefore carried along in each particle. Whichever particle starts out with spin up will be measured with spin up at any later time, the other will be found spin down.

We have shown that the opposite spins can be regarded as "hidden constants" of the motion traveling locally from their creation, consistent with Einstein's picture of an "objective reality." When Alice exercises her "free choice" of a spin direction in which to measure, she adds new information to the universe, she "creates" properties that could not have been know at the start of the experiment.

To a quantum physicist of the Copenhagen school, who thinks the particles lack properties simply because we don't know them, it will *appear* as if the particles are communicating the needed correlation information instantly over large distances. See chapter 29.

But the information moves *locally*, only as fast as the particles.

Where's the Information in the Two-Slit Experiment?

Is it in the particles themselves as we found for entanglement? No. Here the Copenhagen physicist is closer to the truth. We know nothing about the current path. We only know particles were fired from a distance away from the two slits.

Once a particle hits the screen, we know the beginning and ending of the path, as we do for entanglement, but we do not know which slit the particle went through if both slits are open.

So where is the information that produces one interference pattern when both slits are open, and two distinctly different patterns when either slit 1 or slit 2 is open?

In this case the information is in the wave function, and as Einstein first knew, that information is only statistical information. It gives us only probabilities of finding particles, which we will confirm for very large numbers of particles. We know nothing about an individual path.

Nevertheless, Einstein's "objective reality" says the particle has a path. And his principles of conservation tell us that the particle never splits in two, so it must travel through just one of the slits.

We saw in chapter 33 that the wave patterns are different when one slit is open or both slits are open.

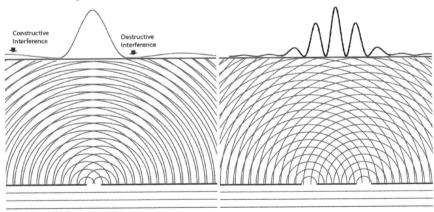

So the ulimate source of the information in the wave field is in the boundary conditions, the distribution of local material, just the way the gravitational field is determined by material nearby.

How abstract *probability* amplitude wave function can influence the motions of the particles so that

they produce the *statistics* of many experiments remains the *one mystery of quantum mechanics.*

The mystery is not, as RICHARD FEYNMAN thought, how the particle can go through both slits. It is somewhat deeper. How the wave function can influence particle motions. The information needed to generate interference patterns is in the wave function.

Where's the Information in Microscopic Irreversibility?

In 1874, JOSEF LOS-CHMIDT criticized his younger colleague LUDWIG BOLTZMANN's attempt to derive from basic classical dynamics the increasing entropy required by the second law of thermodynamics. Loschmidt said that the laws of classical dynam-ics are time reversible.

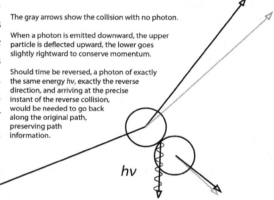

The gray arrows show the collision with no photon.

When a photon is emitted downward, the upper particle is deflected upward, the lower goes slightly rightward to conserve momentum.

Should time be reversed, a photon of exactly the same energy *hv*, exactly the reverse direction, and arriving at the precise instant of the reverse collision, would be needed to go back along the original path, preserving path information.

hv

Consequently, if we just turn the time around, the time evolution of the system should lead to decreasing entropy.

Boltzmann investigated the classical paths of particles in collision to develop his "transport equation." He wondered if after a collision a particle might lose some of the information from a particular collision after colliding with a few more particles. He called this "molecular disorder."

Now Einstein has shown us how information about a path before a collision will be lost during the collision if the collision emits or absorbs a photon. The interaction of radiation with the particles is *irreversible*. Einstein says radiation interactions are not "invertible."

In this case we cannot know the information, but we can say that information needed to reverse collisions has been lost.

Where's the Information in the Measurement Problem?

Some define the problem of measurement simply as the logical contradiction between two laws describing the motion of quantum systems; the unitary, continuous, and deterministic time evolution of the Schrödinger equation versus the non-unitary, discontinuous, and indeterministic collapse of the wave function. JOHN VON NEUMANN saw a problem with two distinct (indeed, opposing) processes. See chapter 25.

The mathematical formalism of quantum mechanics provides no way to predict exactly when the wave function stops evolving in a unitary fashion and collapses. If it could predict this perfectly, it would no longer be quantum mechanics. Experimentally and practically, however, we can say that this occurs when the microscopic system interacts with a macroscopic measuring apparatus.

It takes energy to record the information about the measurement in the material of the apparatus. for example by moving a pointer, marking a chart recorder, or storing data in computer memory.

New information creation requires a local reduction in the entropy. And in order for that new information to remain stable for a observer to read it, the overall global entropy must increase by a larger amount to satisfy the second law. Waste energy is carried away from the measurement apparatus.

Where's the Information in a Deterministic World?

PIERRE-SIMON LAPLACE imagined a super-intelligence that could know the positions, velocities, and forces on all the particles in the universe at one time, together with the deterministic laws of motion, and thus know the universe for all times, past and future. The concept has been criticized for the vast amount of information that would be required, impractical if not impossible to collect instantaneously. And where would the information be kept? If in some part of the universe, there would be an infinite regress of information storage.

Determinists, especially mathematical physicists and compatibilist philosophers, are comfortable with this idea.

A moment's thought tells us that information is being created in the universe at every moment. Which leads us to the question...

How Did All the Information in the Universe Get Created?

Information philosophy has solved this great problem, perhaps the greatest of all problems in physics and philosophy.

And our solution depends on Einstein's expansion of the universe. If the universe were static, it would have come to thermal equilibrium, the "heat death", ages ago.

Many scientists think the universe must have started in a state of very high information. Since information is destroyed by the entropy increase of the second law, they argue there must have been even more information at the beginning than we see today.

But the reverse is true. The early universe was far denser than today. Particles were jammed together at an extraordinarily high temperature which prevented even elementary particles like protons and neutrons from forming, let alone atoms (which did not become stable for the first 380,00 years) or the galaxies, stars, and planets (which had to wait over 400 million years for the gas to cool down enough for gravity to overcome the high pressure and temperature, and the radiation to cool to a black sky everywhere).

The expansion opened up space between the gas particles. As Boltzmann's and Einstein's statistical mechanics would have described it, there appeared many more phase-space cells for the fixed number of particles to arrange themselves in.

And the *arrangement* of particles is their *information structure.*

The early universe was at nearly maximum entropy and minimal information. The expansion increased the maximum possible entropy, and it did it faster than the gas and radiation could approach a new equilibrium with that new maximum entropy.[1] The difference between the maximum and the actual entropy we call negative entropy, or potential information.

Now each new bit of information created has to go through the same two steps we have identified as necessary to create any information structure, from a quantum measurement to a nucleotide position in a strand of DNA.

Similar steps are the basis of our two-stage model of free will. First quantum chance allows *alternative possibilities* to exist Then a "free choice" , *adequately determined* to make us responsible for our actions, creates the new information in our decision.

1   See Layzer, 1991

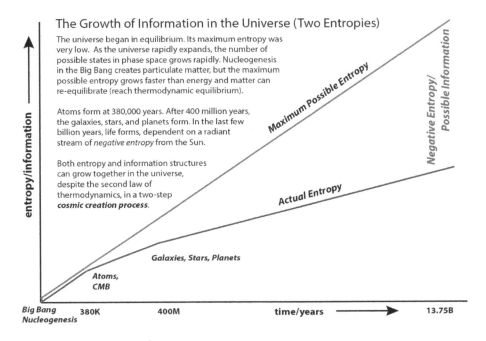

The Growth of Information in the Universe (Two Entropies)

The universe began in equilibrium. Its maximum entropy was very low. As the universe rapidly expands, the number of possible states in phase space grows rapidly. Nucleogenesis in the Big Bang creates particulate matter, but the maximum possible entropy grows faster than energy and matter can re-equilibrate (reach thermodynamic equilibrium).

Atoms form at 380,000 years. After 400 million years, the galaxies, stars, and planets form. In the last few billion years, life forms, dependent on a radiant stream of *negative entropy* from the Sun.

Both entropy and information structures can grow together in the universe, despite the second law of thermodynamics, in a two-step *cosmic creation process*.

entropy/information

Maximum Possible Entropy

Negative Entropy/ Possible Information

Actual Entropy

Galaxies, Stars, Planets

Atoms, CMB

Big Bang Nucleogenesis 380K 400M **time/years** 13.75B

1. The Quantum Step. Whenever matter is rearranged to create a new information structure, the quantum processes involve a collapse of the wave function that introduces an element of chance. Without chance and alternative possibilities, no new information is possible. With those possibilities, things could have been otherwise.

2) The Thermodynamic Step. A new information structure reduces the local entropy. It cannot be stable unless it transfers away enough positive entropy to satisfy the second law of thermodynamics, which says that the total entropy (disorder) must always increase.

Information philosophy tells a story of cosmic and biological evolution that is but one creation process all the way from the original cosmic material to life on earth to the immaterial minds that have now discovered the cosmic creation process itself!

These same two steps are involved in our minds whenever we freely create a new idea! Most of our ideas are simply inherited as the traditional knowledge of our culture. This book emphasizes how many of our ideas about quantum physics we owe to Albert Einstein. But many new thoughts are the work of our creative imaginations. And in that sense, we are all co-creators of our universe.

Chapter 43

Statistic

Quantum Mechanics

Light Quantum Hypothesis

Photoelectric Effect

Bohr-Einstein Atom

ave-Particle Duality

Quantum Information

Bo

T

Born-Einstein Statistic

Chance

Nonlocality

Irreversibility

Nonsepa

Einstein-Podolsky-Rosen

E

Schrödinger's Cat

Did Albert Einstein Inve

Quantum Information

Quantum information, quantum computing, quantum encryption with key distribution, and quantum teleportation, are all described as using *entanglement as a resource*.

So the key question for Einstein's "objective reality" view is whether its "objective" form of entanglement is identical to the concept of quantum entanglement, so as to be useful.

In Einstein's first description of a two-particle system that might be *nonseparable* (he of course mistakenly hoped they could be separable), it was the linear momentum that exhibited "action-at-a-distance." We now understand linear momentum as a "hidden constant" of the motion, giving us "knowledge-at-a-distance."

In our extension of Einstein's "objective reality," all other properties of the two-particle entangled system (angular momentum, spin, polarization) travel along with the particles, conserved as "hidden constants," from their initial entanglement in the center of their "special frame."

The angular momentum, spin, and polarization vectors have not been "measured" at their entanglement. Entanglement is not a "state preparation." Angular spin components are undefined.

It is thus the projections of some properties by "Alice" in specific directions that are instantly correlated with Bob's particle at all spacelike separations.

We start with the two-particle quantum wave function, which in standard quantum mechanics is described as a *superposition* of two-particle states,

$$\psi = |+->-|-+>.$$

PAUL DIRAC tells us that superposition is just a "manner of speaking" and that an individual system is in just one of the superposed states, although there is no way to know which, so say it is

$$\psi = |+->.$$

Upon disentanglement by any external interaction, say by a measurement/collapse of the two-particle wave function, this becomes the product of two single-particle wave functions,

$$\psi = |+>|->.$$

We can visualize the | + > state as keeping the + spin or polarization of the *directionless* spin, but still without that state having a specific spatial component, e.g., z+. It is when a measurement is made that two things happen. 1) the wave function is factorized. 2) The single-particle wave functions both acquire a spatial component direction. One will be a projection of | + >, the other of | - >. These two must be in opposite spatial directions in order to maintain the conservation of total spin zero!

These will be acquired *simultaneously*, in apparent violation of special relativity. But nothing is traveling between them. Whoever measures first, Alice or Bob, breaks the symmetry of the directionless spins in the two-particle wave function and forces the two spins into opposite spatial directions, say z+ and z-.

Subsequent examination of the pairs of measurements by Alice and Bob in the same direction will reveal their perfect correlations. There is no way this can be used for faster-than-light communications.

Notice that if Bob makes a measurement after Alice, it has no effect on Alice's particle. They have been decohered, disentangled, and finally separated. For example, if Bob measures at a different angle α, he will get weaker correlations proportional to $(\cos \alpha)^2$, as predicted by quantum mechanics. [1]

John Bell's claim that "hidden variables" would produce straight-line correlations has no physical foundation whatever. When Bell says that "the Einstein program fails," it is Bell's physically absurd straight line correlation, with "kinks," that fails. See chapter 32.

Objectively real "hidden constants" are not mysteriously transmitted instantaneously, which is impossible. They are carried along at the particles' speed as "constants of the motion." The spatial components in a particular direction are not carried along, they are *created* by the measurement, with the direction a "free choice" of the experimenter.

The most obvious "hidden constant" is the particle momentum, whose conservation was used in the 1935 EPR paper.

---

[1]   See Dirac's discussion of polarizers in chapter 19.

Entangled Qubits

In order to decide if this entanglement is good enough for quantum computing, we need to know how the qubits in a particular quantum computer get entangled. And then we need to understand the type of directional *measurement* that *creates* the perfectly correlated (or anti-correlated) states at any distance.

There are at least a dozen physical realizations of a quantum computer. They all involve a number of entangled qubits, arranged in a sequence. They are typically very close together, for example arranged in a vertical (z) column in an ion trap that constrains their x and y positions. An array of ion traps can be arranged in a quantum charge-coupled device (a QCCD chip). A large array has areas for memory storage and interaction areas for implementing algorithmic computations.

Qubits are initialized, stored as computer memory, then manipulated to communicate (teleport) data from qubit to qubit.

The qubits are initialized by a laser that optically "pumps" the ion from its ground state, either into a hyperfine state (the electron spin flips to be parallel with the nuclear spin), or the electron is pumped up into an "excited" but "metastable" state (one of the atom's optical energy levels that cannot drop back to the ground state with a single-photon quantum jump).

Pairs of qubits can now be entangled by the application of quantum logic gates like the "controlled not" (C-NOT). Qubits can then be teleported between different ion traps in the array. They can also be converted to light and sent through photonic channels, locally or out over fiber optic cables or free space transmission to satellites and beyond.

"Objectively real" qubits in the form of "hidden constants" have values that were determined at the time of entanglement. But they are fully correlated and perfectly random bit sequences.

The fully correlated "Bell states" or "EPR pairs" that appear at an arbitrary angle decided by Alice's "free choice" may also have been hidden in directionless spin states. Whether they are adequate for quantum information systems remains to be decided.

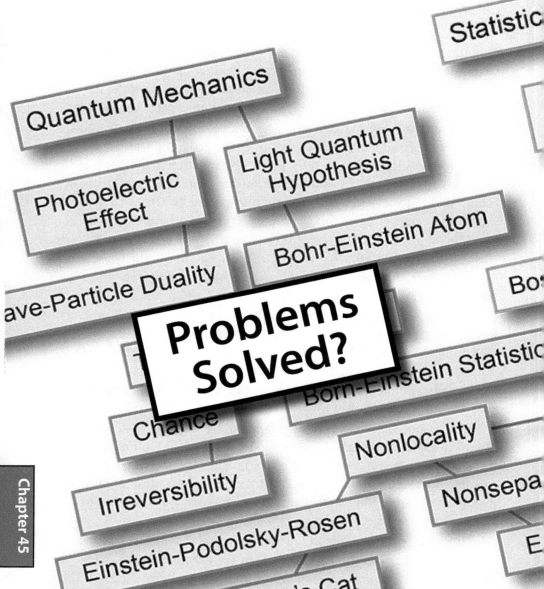

Problems Solved?

In the preface we posed thirteen problems for which a deep analysis of Einstein's thinking, especially his idea of an "objective reality," might lead to plausible solutions.

1. The 19th-century problem of microscopic irreversibility
2. Nonlocality, first seen by Einstein in 1905
3. Wave and particle "duality" (1909)
4. The metaphysical question of ontological chance (1916)
5. Nonlocality and "action-at-a-distance" (1927)
6. The "one mystery" of the two-slit experiment
7. The measurement problem (1930)
8. The role of a "conscious observer" (1930)
9. Entanglement and "spooky" action-at-a-distance (1935)
10. Schrödinger's Cat - dead *and* alive? (1936)
11. No "hidden variables," but hidden constants (1952)
12. Conflict between relativity and quantum mechanics?
13. Is the universe deterministic or indeterministic?

Our proposed solutions are *radical*, if only compared to decades of confusion and mystery surrounding quantum mechanics, but we hope that you find most of them visualizable and intuitive, not characteristics normally associated with the quantum.

Microscopic Irreversibility

Problem: In classical mechanics, microscopic particle collisions are time reversible, conserving entropy and information. Neither entropy, nor more importantly information, can increase in a deterministic, classical world. Ludwig Boltzmann showed that random collisions could increase the macroscopic entropy, but reversing the time would decrease it again.. Thus the puzzle, how to reconcile macroscopic entropy with microscopic reversibility.

Solution: Reversibility fails when any matter interacts with radiation, e.g., emission of a photon during the collision, or changes (quantum jumps) between internal energy levels, are

Chapter 45

taken into account. Any quantum process with such transitions involves ontological chance as discovered by Einstein in 1916. Interaction with light introduces random changes in the energy and momentum of either or both particles. If all particle motions could be reversed, the absorption of a photon with the same energy in the opposite direction at the correct moment is not impossible, but statistically very unlikely to occur.

Comment: As Einstein noted in 1909, emission processes are not "invertible." There are outgoing spherical waves, but incoming spherical waves are never seen. Josef Loschmidt's reversibility paradox is removed. Ernst Zermelo's recurrence objection is also eliminated because the recurrence of original, low entropy states is prevented by the expansion of the universe. The environment is always different. See chapters 11 and 12.

Nonlocality

Problem: When a light wave, possibly carrying energy, spreads out in all directions, how can that energy be suddenly collected together at one point to eject an electron in the photoelectric effect? In 1909 Einstein feared this instantaneous "collapse" of the light wave was a violation of his special theory of relativity?

Solution: It took Einstein some years to see that the light wave is really just the abstract probability of finding his light quanta or material particles. One can think of the probability of finding a particle somewhere other than where it is actually found as suddenly going to zero, which gives the appearance of a "collapse." In any case no matter, energy, or even abstract information is moving when a particle is found somewhere. Nonlocality is only the *appearance* of change in spatially separated places. Nothing objectively real is moving.

Comment: Probabilities are solutions to the Schrödinger equation, determined by the boundary conditions of the experiment and the wavelength of incoming particles. Probabilities for other particles in the space do not change when one particle is detected. See chapters 6, 9, and 23.

Wave-Particle Duality

Problem: Popular interpretations of quantum mechanics describe quantum objects as sometimes waves and sometimes particles, or perhaps both at the same time?

Solution: Particles are real objects. Einstein was first to see waves as imaginary, mathematical fictions, "ghostly" and "guiding" fields, that allow us to calculate probabilities for finding particles. These waves have a statistical power over the location of particles that is the one deep mystery of quantum mechanics.

Particles are *discrete* discontinuous localized quanta of matter or energy. It was Einstein in 1905 who proved the existence of matter particles and hypothesized light particles, the prototypes of the two families of elementary particles in the "standard model" - fermions and bosons. Twenty years later, he discovered their different quantum statistics!

Waves, or wave functions, are mathematical solutions to the Schrödinger equation, with *continuous* values in all space, which provide probabilities for finding particles in a given place and in a specific quantum state.

Comment: The time evolution of the wave function is not the motion of the particle. It is only the best estimate of where the particle might be found. *Continuous* wave functions evolve deterministically. Particles are *discrete* and change their quantum states indeterministically.

As MAX BORN described it "The motion of the particle follows the laws of probability, but the probability itself propagates in accord with causal laws."

Particles are physics. Waves, and fields, are metaphysics.

See chapter 9.

Ontological Chance

Problem: If every collision between material particles is controlled completely by the distribution and motions of all other particles together with the natural force laws of classical physics, then there is only one possible future.

Solution: In modern physics, all interactions between material particles are mediated by the exchange of energy particles. Einstein's light quanta (photons) are the mediating particles for electromagnetic radiation. In 1916, Einstein showed that these energy particle exchanges always involve chance. Quantum mechanics is statistical, opening the possibilities needed for free will, the "free choice" of the experimenter, and "free creations of the human mind."

Comment: The emergence of classical laws and apparent deterministic causality occurs whenever the number of particles grows large so quantum randomness can be averaged over. Bohr's "correspondence principle" claims classicality also occurs when quantum numbers are large.

The "quantum-to-classical transition" occurs when the mass of an object m is very large compared to Planck's constant h, so the uncertainty $\Delta v \, \Delta x \geq h / m$ is very small. See chapters 1 and 11.

Nonlocality and Action-at-a-Distance

Problem: Einstein's 1927 presentation at the fifth Solvay conference was his first public description of an issue that had bothered him since 1905. He thought he saw events at two places in a spacelike separation happening simultaneously. His special theory of relativity claims to show the *impossibility of simultaneity*.

Solution: Einstein's blackboard drawing shows us that the electron's wave function propagates in all directions, but when the particle appears, all of it is found at a single point.

Using Einstein's idea of "objective reality," without any interactions that could change the momentum, the particle must have traveled in a straight line from the origin to the point where it is found. The properties of the particle considered by Einstein in 1927 could have evolved *locally* from the start of the experiment as what we called "hidden constants" of the motion.

Comment: There was no "action" by either particle on the other in this case, so we call it "knowledge-at-a-distance." See chapters 9, 17, 18, and 23.

Chapter 45

Two-Slit Experiment

Problem: In experiments where a single particle travels to the screen at a time, large numbers of experiments show interference patterns when both slits are open, suggesting that a particle must move through both slits in order to "interfere with itself."

Solution: Solutions to the time-independent Schrödinger equation for the given boundary conditions - two open slits, screen, particle wavelength - are different for the case of one slit open. In Einstein's "objective reality," the particle conserves all its properties and goes through only one slit. Probability amplitudes of the wave function are different when two slits are open, explaining interference.

Comment: Feynman's path integral formulation of quantum mechanics suggests the same solution. His "virtual particles" explore all space (the "sum over paths") as they determine the variational minimum for least action, thus the resulting probability amplitude wave function can be said to "know" which holes are open. How abstract probabilities influence the particles' motions is the one remaining mystery in quantum mechanics.

Bohmian mechanics also defends a particle that goes through one slit reacting to probabilities that are based on two slits being open. See chapter 33.

Measurement Problem

Problem: JOHN VON NEUMANN saw a *logical* problem with two distinct (indeed, opposing) processes, the unitary, *continuous*, and deterministic time evolution of the Schrödinger equation versus the non-unitary, *discontinuous*, and indeterministic "collapse of the wave function." Decoherence theorists and many-worlders are convinced that quantum mechanics should be based on the wave function alone. There are no particles, they say. Schrödinger agreed.

Solution: We can think of the time evolution of a system as involving these two processes, but one after the other. First, the system evolves as a probability amplitude wave function according to the time-dependent Schrödinger equation. Then, at an unknown time (which bothers the critics), the particle appears somewhere.

The time of collapse may simply be the moment an experimenter makes a measurement. Measurement requires the recording of *irreversible* information about the location of the particle, as von Neumann knew. It does not have to be in the mind of a conscious observer.

Comment: This problem shows why we need to get "beyond logic" in the philosophy of science.

Conscious Observer

Problem: The Copenhagen Interpretation and many of its supporters, e.g., Werner Heisenberg, John von Neumann, Eugene Wigner, considered a measurement not complete until it reaches the mind of the observer. They asked where is the "cut" (*Schnitt*) between the experiment and the mind?

Solution: Information must be recorded *irreversibly* before any observer can know the results of a measurement. Data recorded (ontologically) by a measuring instrument creates new information in the universe. But so does any newly created information structure in nature without an observer. Einstein wanted objective reality to be independent of observers, but there are measurements that are a "free choice" of the experimenter, creating a new part of reality.

Comment: We might say that information becomes known (epistemological) when it is recorded in the world and then seen by a human observer. But most new information created is ontological, the universe is *observing itself*. See chapter 25.

Entanglement and "Spooky" Action-at-a-Distance

Problem: In his 1935 EPR paper, Einstein discussed two particles traveling away from the center. He used conservation principles to show that measuring one particle gives information about the other without measuring it directly. We have shown the two particles' properties could have evolved *locally* from their original values at the center no matter how far the particles are apart, as long as no interaction with the environment has altered their values and destroyed their "coherence." But a true *nonlocality* appears in David Bohm's 1952 version the EPR experiment, in which electron spin components are measured instead of linear momenta.

Solution: As the electrons travel apart, each one stays in its state by conservation laws. Their spins and linear momenta are conserved.

The left-moving particle electron is say -p. The other is p. The total linear momentum is zero. Similarly their total spin is zero. If one electron is spin $\hbar/2$, the other is exactly opposite. But the original process of entanglement has not left the electron spins with a definite spatial direction.

When Alice uses her "free choice" of which angle to measure the spin (or polarization) component, she adds new information which was not present at the original entanglement. Alice's measurement decoheres and disentangles the two-particle wave function. The particles now appear in a spacelike separation equidistant from the origin. The *directionless* and opposite spins are projected by her measurement into spin components, say z+ and z-. If Bob then measures at the same angle, he gets the perfectly correlated opposite value.

Comments: It is part of the deep mystery of quantum mechanics how the spatial directions of the two spins, created by a measurement of the two-particle wave function anywhere, come out in perfectly correlated directions. But had they not, something even worse would have happened. Symmetry and conservation laws would have been violated.

Schrödinger's Cat

Problem: Erwin Schrödinger imagined that the time evolution of his equation could start with a microscopic radioactive nucleus in a superposition of decayed and undecayed state, leading to a macroscopic cat in a similar superposition. When he suggested it, he was criticizing, really ridiculing, what he thought was an absurd consequence of Paul Dirac's *principle of superposition,* with its probabilities for a system to be in different states

Solution: Schrödinger was just criticizing superposition and its probabilities. There is never an *individual* cat simultaneously dead and alive. What the superposition of possible states in quantum mechanics gives us are only *probabilities* for the cat being dead *or* alive. The predicted *probabilities* are empirically confirmed by the *statistics* in large numbers of identical experiments, each one of which ends up with either a live or dead cat.

Comment: The individual radioactive nucleus is never in a superposition of decayed and not decayed. Quantum mechanics gives us the *probabilities* of a decay or remaining undecayed. Once there is a decay, the evolution results in a dead cat. If no decay, then a live cat. Indeed, not only do macroscopic superpositions of cats not exist, the radioactive nucleus is not in a superposition. There are no macroscopic superpositions because there are no microscopic superpositions either.

No "Hidden Variables," but Hidden Constants

Problem: DAVID BOHM suggested that "hidden variables" could instantaneously communicate information between entangled particles to perfectly correlate their properties at great distances, specifically the opposite $+1/2$ and $-1/2$ electron spins of a two-electron system with total spin zero.

Solution: In our adaptation of Einstein's "objective reality," the particles are generated with individual properties, momenta, angular momenta, spins, and they conserve these properties until they are measured. These properties are carried along "locally" with the particles, so do not violate special relativity as Einstein feared.

While there might not be Bohmian "hidden variables," we can call these conserved quantities "hidden constants" ("constants of the motion," hidden in plain sight). They explain the *appearance* of Einstein's "spooky" action-at-a-distance. Our hidden constants can explain the original EPR results, but they cannot explain the measurements of electron spin components, which are *created* by Alice's measurement.

Comment: The two spin components, say z+ and z-, are Alice's *nonlocal* projections of the opposing spins that traveled *locally* from the origin. The *nonlocal* aspect is that these spin components have perfectly opposing directions even though they are about to be greatly separated, once the two-particle wave function has collapsed into the product of two single-particle wave functions.

Of course if the opposing spins of the electrons that travel *locally* from the origin did not remain perfectly anti-correlated when measured and projected into a specific direction, that would be a violation of the conservation laws.

Is the Universe Deterministic or Indeterministic?

Problem: Einstein was well known, especially in his younger years, for hoping quantum physics could be found to be a deterministic theory. When in 1916 he discovered the randomness in quantum physics, he called chance a "weakness in the theory." And many times he insisted that "God does not play dice." Many of the alternative "interpretations" of quantum mechanics are deterministic. See chapters 30, 31, 32, and 34.

Solution: Einstein had fully accepted the indeterministic nature of quantum mechanics by some time around 1930. But his colleagues paid little attention to his concerns, which had turned entirely to the *nonlocal* aspects of quantum mechanics.

Comment: Without indeterminism, we could not have a creative universe and Einstein's "free creations of the human mind."

What Is Quantized?

The "quantum condition" describes the underlying deep reason for the existence of discrete objects.

For Bohr in 1913, it was the angular momentum of electrons in their orbits, as suggested by J.W.Nicholson. For Louis de Broglie in 1924 it was that the linear momentum $p = h/\lambda$ and that an integer number of wavelengths fits around an electron orbit. For Heisenberg in 1925, it was the non-commutation of momentum and position operator matrices, and in 1927 his resulting uncertainty principle $\Delta p \Delta x = h$. In Bohr's otherwise obscure Como lecture of 1927, he showed that $\Delta v \Delta t = 1$, thus deriving the uncertainty principle with no reference to measurements as "disturbances," and embarrassing Heisenberg.

Multiplying $\Delta v \Delta t = 1$ by Max Planck's constant h, and noting $E = hv$, we have $\Delta E \Delta t = \Delta p \Delta x = \Delta J \Delta \varphi = h$. All of these expressions have the same physical dimensions as angular momentum J.

As Erwin Schrödinger explained, it is always *action, or angular momentum, that is being quantized.* Momentum *p*, position *x*, energy *E*, and time *t*, all take on continuous values. It is the angular momentum or spin *J* that comes in integer multiples of *h*.

Any interaction of radiation and matter involves at least one unit of Planck's quantum of action *h*, which first appeared in 1900, though only as a heuristic mathematical device, not the radical core idea of a new physics. That was seen first by Einstein, like so many of the quantum mechanical concepts he saw long before the "founders" developed their powerful quantum calculation methods.

The Bottom Line

There is no microscopic reversibility.

There is no nonlocality in the form of one event *acting* on another in a spacelike separation. There are simultaneous synchronized events in a spacelike separation, which Einstein feared violated his special theory of relativity. They do not.

Particles are real physics. Waves are imaginary. Fields are metaphysics.

Ontological chance exists. Without it, nothing ever happens.

Nothing physically "collapses" when a possibility is actualized.

The "one mystery" of quantum mechanics is how probability waves control the statistical motions of particles to produce interference effects.

The measurement problem is explained as when new information is irreducibly recorded in the measurement apparatus. Local entropy is reduced. Global entropy increases.

There is no nonseparability. Particles separate as soon one leaves the other's light cone. But two entangled particles retain their perfect correlation of properties as required by the conservation laws, until one interacts with something in the environment and decoheres. A measurement begins with the properties of the particles still correlated. It ends with decorrelation and disentanglement. The mysterious power of the two-particle wave function separates into single-particle functions with their new spatial spin direction also perfectly correlated. But the particular spin component direction chosen by Alice was not known at the origin. It can be viewed as new information appearing nonlocally, i.e. simultaneously in a spacelike separation.

"Spooky action-at-a-distance" is just the *appearance* of communication or interaction when entangled particles are measured at separation and found to remain perfectly correlated. There is no "action" by one particle on the other. It is simply "knowledge-at-a-distance."

There is no conflict between special relativity and quantum mechanics, though there would have been if the probability waves had been carrying energy or matter.

Schrödinger's cat will always be found as alive, dead, or dying if the nuclear decay has occurred. This is just as individual objects are never in a superposition, never in two places at the same time.

There is one world. It is a quantum world. The world *appears* classical for objects with large mass. And it is indeterministic, which opens *alternative possibilities* for an open, free, and creative future, for Einstein's "free creations of the human mind."

Einstein's "objective reality" can explain the world with standard quantum mechanics, so much of which he discovered or created.

His many criticisms and objections did not prevent him from seeing the truly mysterious aspects of quantum physics well before his colleagues, who often get the credit that belongs to him.

How to Restore Credit to Einstein

To correct this problem, historians of physics and especially teachers of quantum mechanics must change the way they discuss and especially to *teach* Einstein's contributions to physics.

His paper explaining Brownian motion should be taught as the first proof that matter is not continuous, but discrete. It consists of quanta. He thought he had proved Boltzmann's controversial hypothesis of atoms.

His paper explaining the "photoelectric effect," for which he was awarded the Nobel Prize, should be taught as the revolutionary hypothesis that light energy also comes in discrete quanta hv.

In these two 1905 papers, Einstein was the first to see the elements in today's "standard model" of particle physics - the fermions (matter) and the bosons (energy). For this work alone, Einstein should be seen as the true founder of quantum mechanics

His third paper in 1905, explaining relativity, should not overshadow his quantization of matter and energy and his fourth paper that year, showing their interchangeability - $E = mc^2$.

His 1907 paper explaining the anomalous specific heat of certain atoms should be taught as the discovery of energy levels in atoms and the "jumps" between them, six years before Niels Bohr's quantum jumps between his postulated energy levels in the atom.

Einstein's 1909 paper explaining wave-particle duality should be taught as the continuous wave (and later the wave function ψ) giving us the *probability* of finding a discrete particle. Quantum mechanics is *statistical*!

His 1916 paper on transition probabilities between energy levels, which discovered the stimulated emission of radiation behind today's lasers, should be taught as the discovery of ontological *chance* in nature whenever matter and radiation interact. The interactions always involve at least one quantum of action h. They introduce statistics and indeterminacy a decade before Werner Heisenberg's uncertainty principle.

Arthur Holly Compton's 1923 explanation of the "Compton effect," which confirmed Einstein's 1916 prediction that particles of light have momentum as well as energy, should be taught as Einstein's deep confidence in conservation principles, so that the motions and paths of quantum particles *objectively* exist and at all times are obeying those conservation laws for momentum and energy. Einstein had used these fundamental principles to invalidate Niels Bohr's final attempt to deny Einstein's light quantum hypothesis in 1924, in the Bohr-Kramers-Slater paper. This work should be taught as the basis for Einstein's belief in an "objective reality."

Particles don't cease to exist, or appear simultaneously at multiple places, as claimed by the Copenhagen Interpretation of quantum mechanics. Just because we can't continuously measure paths does not mean that particles do not exist until we observe them.

Einstein's 1925 papers based on Satyendra Nath Bose's very simple quantum derivation of the Planck law in 1924, should be taught as Einstein's discovery of the *indistinguishability* of elementary particles and their consequent strange and different statistics for half-spin "fermions"and unit-spin "bosons."

Einstein's misunderstood and ignored presentation at the Solvay conference of 1927 showing the *nonlocal* behavior in a single particle passing through a slit should be taught as the beginning of his 1935 EPR paper, when he showed that two particles a great distance apart can acquire perfectly correlated properties instantaneously, his discovery of nonseparability and entanglement.

Poincaré and Einstein

Some historians of science have pointed out how much Einstein was inspired by Henri Poincaré's great book *Science and Hypothesis*.

Many of Einstein's biographers have described the young Einstein's colleagues who met frequently to discuss new ideas in philosophy and physics. They called themselves the Olympia Academy. After a frugal evening meal of sausage, cheese, fruits, honey, and tea, they read and discussed the great works of David Hume, John Stuart Mill, Ernst Mach, and Karl Pearson. Several weeks were spent on Henri Poincare's *La Science et l'Hypothèse*.

Recently a few scholars have shown that in his "miracle year" of 1905 Einstein solved three great problems described by Poincaré, just one year after his book had been translated into German. Arthur I. Miller cited three problems he thought Poincaré felt were "pressing;" the failed attempts to detect the motion of Earth through the "ether," the photoelectric effect, and Brownian motion.[1] A close reading of Poincaré's book shows that great thinker suggested several more problems to Einstein, most importantly the principle of relativity, but also the one-way increase of entropy with its problem of irreversibility, Maxwell's demon, the question of determinism or indeterminism, and amazingly "action-at-a-distance." We now realize that in quantum mechanics what Einstein discovered is only "knowledge-at-a-distance."

We hope to have shown that the far-seeing Einstein grappled with all these problems, a few unsuccessfully but always creatively, between reading Poincaré in 1904 and his death five decades later.

Chapter 45

---

1   Miller, 2002, p.185. Rigden 2005, p.8, Holt, 2018, p.5

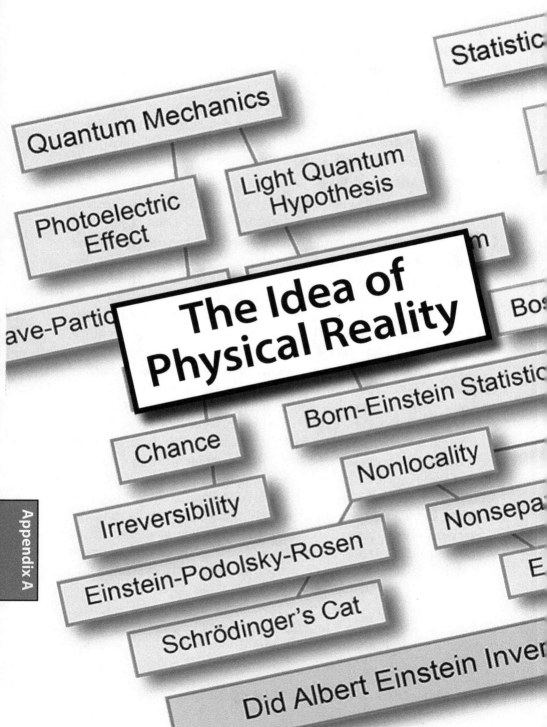

Quantum Mechanics

Statistic

Photoelectric
Effect

Light Quantum
Hypothesis

ave-Partic

m

The Idea of
Physical Reality

Bos

Born-Einstein Statistic

Chance

Nonlocality

Irreversibility

Nonsepa

Einstein-Podolsky-Rosen

E

Schrödinger's Cat

Did Albert Einstein Inver

The Idea of Physical Reality

"Maxwell's Influence on the Evolution of the Idea of Physical Reality"

On the one hundredth anniversary of Maxwell's birth.

Published, 1931, in *James Clerk Maxwell: A Commemoration Volume*, Cambridge University Press,

The belief in an external world independent of the perceiving subject is the basis of all natural science. Since, however, sense perception only gives information of this external world or of "physical reality" indirectly, we can only grasp the latter by speculative means. It follows from this that our notions of physical reality can never be final. We must always be ready to change these notions—that is to say, the axiomatic basis of physics—in order to do justice to perceived facts in the most perfect way logically. Actually a glance at the development of physics shows that it has undergone far-reaching changes in the course of time.

The greatest change in the axiomatic basis of physics—in other words, of our conception of the structure of reality—since Newton laid the foundation of theoretical physics was brought about by Faraday's and Maxwell's work on electromagnetic phenomena. We will try in what follows to make this clearer, keeping both earlier and later developments in sight.

According to Newton's system, physical reality is characterized by the concepts of space, time, material point, and force (reciprocal action of material points). Physical events, in Newton's view, are to be regarded as the motions, governed by fixed laws, of material points in space. The material point is our only mode of representing reality when dealing with changes taking place in it, the solitary representative of the real, in so far as the real is capable of change. Perceptible bodies are obviously responsible for the concept of the material point; people conceived it as an analogue of mobile bodies, stripping these of the characteristics of extension, form, orientation in space, and all "inward" qualities, leaving only inertia and translation and adding the concept of force. The material bodies, which had led psychologically to our formation of the concept of the "material point," had now themselves to be regarded as systems of material points. It should be noted that this theoretical scheme is in essence an atomistic and mechanistic one. All happenings were to be interpreted purely mechanically— that is to say, simply as motions of material points according to Newton's law of motion.

The most unsatisfactory side of this system (apart from the difficulties involved in the concept of "absolute space" which have been raised once more quite recently) lay in its description of light, which Newton also conceived, in accordance with his system, as composed of material points. Even at that time the question, What in that case becomes of the material points of which light is composed, when the light is absorbed?, was already a burning one. Moreover, it is unsatisfactory in any case to introduce into the discussion material points of quite a different sort, which had to be postulated for the purpose of representing ponderable matter and light respectively. Later on, electrical corpuscles were added to these, making a third kind, again with completely different characteristics. It was, further, a fundamental weakness that the forces of reciprocal action, by which events are determined, had to be assumed hypothetically in a perfectly arbitrary way. Yet this conception of the real accomplished much: how came it that people felt themselves impelled to forsake it?

In order to put his system into mathematical form at all, Newton had to devise the concept of differential quotients and propound the laws of motion in the form of total differential equations— perhaps the greatest advance in thought that a single individual was ever privileged to make. Partial differential equations were not necessary for this purpose, nor did Newton make any systematic use of them; but they were necessary for the formulation of the mechanics of deformable bodies; this is connected with the fact that in these problems the question of how bodies are supposed to be constructed out of material points was of no importance to begin with.

Thus the partial differential equation entered theoretical physics as a handmaid, but has gradually become mistress. This began in the nineteenth century when the wave-theory of light established itself under the pressure of observed fact. Light in empty space was explained as a matter of vibrations of the ether, and it seemed idle at that stage, of course, to look upon the latter as a conglomeration of material points. Here for the first time the partial differential equation appeared as the natural expression of the primary realities of physics. In a particular department of theoretical physics the continuous field thus appeared side by side with the material point as the representative of physical reality. This dualism remains even today, disturbing as it must be to every orderly mind.

If the idea of physical reality had ceased to be purely atomic, it still remained for the time being purely mechanistic; people still tried to explain all events as the motion of inert masses; indeed no other way of looking at things seemed conceivable. Then came the great change, which will be associated for all time with the names of Faraday, Maxwell, and Hertz. The lion's share in this revolution fell to Maxwell. He showed that the whole of what was then known about light and electromagnetic phenomena was expressed in his well-known double system of differential equations, in which the electric and the magnetic fields appear as the dependent variables. Maxwell did indeed, try to explain, or justify, these equations by the intellectual construction of a mechanical model.

But he made use of several such constructions at the same time and took none of them really seriously, so that the equations alone appeared as the essential thing and the field strengths as the ultimate entities, not to be reduced to anything else. By the turn of the century the conception of the electromagnetic field as an ultimate entity had been generally accepted and serious thinkers had abandoned the belief in the justification, or the possibility, of a mechanical explanation of Maxwell's equations.

Before long they were, on the contrary, actually trying to explain material points and their inertia on field theory lines with the help of Maxwell's theory, an attempt which did not, however, meet with complete success.

Appendix A

Neglecting the important individual results which Maxwell's life-work produced in important departments of physics, and concentrating on the changes wrought by him in our conception of the nature of physical reality, we may say this: before Maxwell people conceived of physical reality—in so far as it is supposed to represent events in nature—as material points, whose changes consist exclusively of motions, which are subject to total differential equations. After Maxwell they conceived physical reality as represented by continuous fields, not mechanically explicable, which are subject to partial differential equations. This change in the conception of reality is the most profound and fruitful one that has come to physics since Newton; but it has at the same time to be admitted that the program has by no means been completely carried out yet. The successful systems of physics which have been evolved since rather represent compromises between these two schemes, which for that very reason bear a provisional, logically incomplete character, although they may have achieved great advances in certain particulars.

The first of these that calls for mention is Lorentz's theory of electrons, in which the field and the electrical corpuscles appear side by side as elements of equal value for the comprehension of reality. Next come the special and general theories of relativity, which, though based entirely on ideas connected with the field-theory, have so far been unable to avoid the independent introduction of material points and total differential equations. The last and most successful creation of theoretical physics, namely quantum-mechanics, differs fundamentally from both the schemes which we will for the sake of brevity call the Newtonian and the Maxwellian. For the quantities which figure in its laws make no claim to describe physical reality itself, but only the *probabilities* of the occurrence of a physical reality that we have in view. Dirac, to whom, in my opinion, we owe the most perfect exposition, logically, of this theory, rightly points out that it would probably be difficult, for example, to give a theoretical description of a photon such as would give enough information to enable one to decide whether it will pass a polarizer placed (obliquely) in its way or not.

I am still inclined to the view that physicists will not in the long run content themselves with that sort of indirect description of the real, even if the theory can eventually be adapted to the postulate of general relativity in a satisfactory manner. We shall then, I feel sure, have to return to the attempt to carry out the program which may be described properly as the Maxwellian—namely, the description of physical reality in terms of fields which satisfy partial differential equations without singularities.

Analysis

Here Einstein explains how physical reality came to be conceived as *continuous* fields not mechanically explainable in terms of material objects.

He describes Paul Dirac's formulation of quantum mechanics as "the most perfect exposition," in which there is not enough *information* to know in which of two states a particle will be found.

Einstein's "objective reality" is simply "an external world independent of the perceiving subject."

Quantum mechanics, he says, "make no claim to describe physical reality itself, but only the *probabilities* of the occurrence of a physical reality that we have in view."

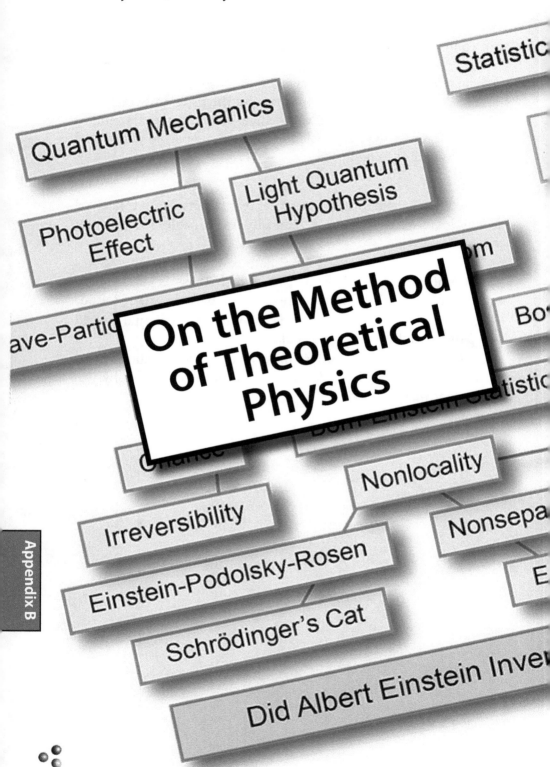

**On the Method
of Theoretical
Physics**

On the Method of Theoretical Physics

The Herbert Spencer Lecture, delivered at Oxford, June 10, 1933.

IF YOU wish to learn from the theoretical physicist anything about the methods which he uses, I would give you the following piece of advice: Don't listen to his words, examine his achievements. For to the discoverer in that field, the constructions of his imagination appear so necessary and so natural that he is apt to treat them not as the creations of his thoughts but as given realities.

This statement may seem to be designed to drive my audience away without more ado. For you will say to yourselves, 'The lecturer is himself a constructive physicist; on his own showing therefore he should leave the consideration of the structure of theoretical science to the epistemologist'.

So far as I personally am concerned, I can defend myself against an objection of this sort by assuring you that it was no suggestion of mine but the generous invitation of others which has placed me on this dais, which commemorates a man who spent his life in striving for the unification of knowledge.

But even apart from that, I have this justification for my pains, that it may possibly interest you to know how a man thinks about his science after having devoted so much time and energy to the clarification and reform of its principles.

Of course his view of the past and present history of his subject is likely to be unduly influenced by what he expects from the future and what he is trying to realize to-day. But this is the common fate of all who have adopted a world of ideas as their dwelling-place.

He is in just the same plight as the historian, who also, even though unconsciously, disposes events of the past around ideals that he has formed about human society.

I want now to glance for a moment at the development of the theoretical method, and while doing so especially to observe the relation of pure theory to the totality of the data of experience.

Appendix B

Here is the eternal antithesis of the two inseparable constituents of human knowledge, Experience and Reason, within the sphere of physics. We honour ancient Greece as the cradle of western science. She for the first time created the intellectual miracle of a logical system, the assertions of which followed one from another with such rigor that not one of the demonstrated propositions admitted of the slightest doubt-Euclid's geometry. This marvellous accomplishment of reason gave to the human spirit the confidence it needed for its future achievements. The man who was not enthralled in youth by this work was not born to be a scientific theorist. But yet the time was not ripe for a science that could comprehend reality, was not ripe until a second elementary truth had been realized, which only became the common property of philosophers after Kepler and Galileo. Pure logical thinking can give us no knowledge whatsoever of the world of experience; all knowledge about reality begins with experience and terminates in it.

Conclusions obtained by purely rational processes are, so far as Reality is concerned, entirely empty. It was because he recognized this, and especially because he impressed it upon the scientific world that Galileo became the father of modern physics and in fact of the whole of modern natural science.

But if experience is the beginning and end of all our knowledge about reality, what role is there left for reason in science? A complete system of theoretical physics consists of concepts and basic laws to interrelate those concepts and of consequences to be derived by logical deduction. It is these consequences to which our particular experiences are to correspond, and it is the logical derivation of them which in a purely theoretical work occupies by far the greater part of the book. This is really exactly analogous to Euclidean geometry, except that in the latter the basic laws are called 'axioms'; and, further, that in this field there is no question of the consequences having to correspond with any experiences. But if we conceive Euclidean geometry as the science of the possibilities of the relative placing of actual rigid bodies and accordingly interpret it as a physical science, and do not abstract from its original empirical content, the logical parallelism of geometry and theoretical physics is complete.

We have now assigned to reason and experience their place within the system of theoretical physics. Reason gives the structure to the system; the data of experience and their mutual relations are to correspond exactly to consequences in the theory. On the possibility alone of such a correspondencer ests the value and the justification of the whole system, and especially of its fundamental concepts and basic laws. But for this, these latter would simply be free inventions of the human mind which admit of no a priori justification either through the nature of the human mind or in any other way at all.

The basic concepts and laws which are not logically further reducible constitute the indispensable and not rationallyd educible part of the theory. It can scarcely be denied that the supreme goal of all theory is to make the irreducible basic elements as simple and as few as possible without having to surrender the adequater epresentation of a single datum of experience.

The conception here outlined of the purely fictitious character of the basic principles of theory was in the eighteenth and nineteenth centuries still far from being the prevailing one. But it continues to gain more and more ground because of the everwidening logical gap between the basic concepts and laws on the one side and the consequences to be correlated with our experience on the other-a gap which widens progressively with the developing unification of the logical structure, that is with the reduction in the number of the logically independent conceptual elements required for the basis of the whole system.

Newton, the first creator of a comprehensive and workable system of theoretical physics, still believed that the basic concept and laws of his system could be derived from experience; his phrase 'hypotheses non fingo' can only be interpreted in this sense. In fact at that time it seemed that there was no problematica element in the concepts, Space and Time. The concepts of mass, acceleration, and force and the laws connecting them, appeared to be directly borrowed from experience. But if this basis is assumed, the expression for the force of gravity seems to be derivable from experience; and the same derivability was to be anticipated for the other forces.

One can see from the way he formulated his views that Newton felt by no means comfortable about the concept of absolute space, which embodied that of absolute rest; for he was alive to the fact that nothing in experience seemed to correspond to this latter concept. He also felt uneasy about the introduction of action at a distance. But the enormous practical success of his theory may well have prevented him and the physicists of the eighteenth and nineteenth centuries from recognizing the fictitious character of the principles of his system.

On the contrary the scientists of those times were for the most part convinced that the basic concepts and laws of physics were not in a logical sense free inventions of the human mind, but rather that they were derivable by abstraction, i.e. by a logical process, from experiments. It was the general Theory of Relativity which showed in a convincing manner the incorrectness of this view. For this theory revealed that it was possible for us, using basic principles very far removed from those of Newton, to do justice to the entire range of the data of experience in a manner even more complete and satisfactory than was possible with Newton's principles. But quite apart from the question of comparative merits, the fictitious character of the principles is made quite obvious by the fact that it is possible to exhibit two essentially different bases, each of which in its consequences leads to a large measure of agreement with experience. This indicates that any attempt logically to derive the basic concepts and laws of mechanics from the ultimate data of experience is doomed to failure.

If then it is the case that the axiomatic basis of theoretical physics cannot be an inference from experience, but must be free invention, have we any right to hope that we shall find the correct way? Still more-does this correct approach exist at all, save in our imagination? Have we any right to hope that experience will guide us aright, when there are theories (like classical mechanics) which agree with experience to a very great extent, even without comprehending the subject in its depths? To this I answer with complete assurance, that in my opinion there is the correct path and, moreover, that it is in our power to find it. Our experience up to date justifies us in feeling sure that in Nature is actualized the ideal of mathematical simplicity. It is my conviction that pure mathematical construction enables us to discover the concepts and the laws connecting them which give us the key to the understanding of the phenomena of Nature. Experience can of course guide us in our choice of serviceable mathematical concepts; it cannot possibly be the source from which they are derived; experience of course remains the sole criterion of the serviceability of a mathematical construction for physics, but the truly creative principle resides in mathematics. In a certain sense, therefore, I hold it to be true that pure thought is competent to comprehend the real, as the ancients dreamed.

To justify this confidence of mine, I must necessarily avail myself of mathematical concepts. The physical world is represented as a four-dimensional continuum. If in this I adopt a Riemannian metric, and look for the simplest laws which such a metric can satisfy, I arrive at the relativistic gravitation-theory of empty space. If I adopt in this space a vector-field, or in other words, the antisymmetrical tensor-field derived from it, and if I look for the simplest laws which such a field can satisfy, I arrive at the Maxwell equations for free space.

Having reached this point we have still to seek a theory for those parts of space in which the electrical density does not vanish. De Broglie surmised the existence of a wave-field, which could be used to explain certain quantum properties of matter. Dirac found in the 'spinor-field' quantities of a new kind, whose simplest equations make it possible to deduce a great many of the properties of the electron, including its quantum properties. I and my colleague discovered that these 'spinors' constitute a special case of a field of a new sort which is mathematically connected with the metrical continuum of four dimensions, and it seems that they are naturally fitted to describe important properties of the electrical elementary particles.

It is essential for our point of view that we can arrive at these constructions and the laws relating them one with another by adhering to the principle of searching for the mathematically simplest concepts and their connections. In the paucity of the mathematically existent simple field-types and of the relations between them, lies the justification for the theorist's hope that he may comprehend reality in its depths.

The most difficult point for such a field-theory at present is how to include the atomic structure of matter and energy. For the theory in its basic principles is not an atomic one in so far as it operates exclusively with continuous functions of space, in contrast to classical mechanics whose most important feature, the material point, squares with the atomistic structure of matter.

The modern quantum theory, as associated with the names of de Broglie, Schr6dinger, and Dirac, which of course operates with continuous functions, has overcome this difficulty by means of a daring interpretation, first given in a clear form by Max Born:-the space functions which appear in the equations make no claim to be a mathematical model of atomic objects. These functions are only supposed to determine in a mathematical way the probabilities of encountering those objects in a particular place or in a particular state of motion, if we make a measurement. This conception is logically unexceptionable, and has led to important successes. But unfortunately it forces us to employ a continuum of which the number of dimensions is not that of previous physics, namely 4, but which has dimensions increasing without limit as the number of the particles constituting the system under examination increases. I cannot help confessing that I myself accord to this interpretation no more than a transitory significance. I still believe in the possibility of giving a model of reality, a theory, that is to say, which shall represent events themselves and not merely the probability of their occurrence. On the other hand, it seems to me certain that we have to give up the notion of an absolute localization of the particles in a theoretical model. This seems to me to be the correct theoretical interpretation of Heisenberg's indeterminacy relation. And yet a theory may perfectly well exist, which is in a genuine sense an atomistic one (and not merely on the basis of a particular interpretation), in which there is no localizing of the particles in a mathematical model. For example, in order to include the atomistic character of electricity, the field equations only need to involve that a three-dimensional volume of space on whose boundary the electrical density vanishes everywhere contains a total electrical charge of an integral amount. Thus in a continuum theory, the atomistic character could be satisfactorily expressed by integral propositions without localizing the particles which constitute the atomistic system.

Only if this sort of representation of the atomistic structure be obtained could I regard the quantum problem within the framework of a continuum theory as solved.

Analysis

Many theoreticians come to believe that their theories exist in the world, that they discovered them, whereas they were actually invented - "free creations of the human mind."

Pure thought is competent to comprehend the real.

Statistic

Quantum Mechanics

Light Quantum
Hypothesis

Photoelectric
Effect

Atom

ave-Particle

Bos

Physics and
Reality

Tra

Born-Einstein Statistic

Chance

Nonlocality

Irreversibility

Nonsepa

Einstein-Podolsky-Rosen

E

Schrödinger's Cat

Did Albert Einstein Inver

T

Physics and Reality

From The Journal of the Franklin Institute, Vol. 221, No. 3. March, 1936. Reprinted in Ideas and Opinions, p.290

It has often been said, and certainly not without justification, that the man of science is a poor philosopher. Why, then, should it not be the right thing for the physicist to let the philosopher do the philosophizing? Such might indeed be the right thing at a time when the physicist believes he has at his disposal a rigid system of fundamental concepts and fundamental laws which are so well established that waves of doubt cannot reach them; but, it cannot be right at a time when the very foundations of physics itself have become problematic as they are now. At a time like the present, when experience forces us to seek a newer and more solid foundation, the physicist cannot simply surrender to the philosopher the critical contemplation of the theoretical foundations; for, he himself knows best, and feels more surely where the shoe pinches. In looking for a new foundation, he must try to make clear in his own mind just how far the concepts which he uses are justified, and are necessities.

The whole of science is nothing more than a refinement of everyday thinking. It is for this reason that the critical thinking of the physicist cannot possibly be restricted to the examination of the concepts of his own specific field. He cannot proceed without considering critically a much more difficult problem, the problem of analyzing the nature of everyday thinking. Our psychological experience contains, in colorful succession, sense experiences, memory pictures of them, images, and feelings. In contrast to psychology, physics treats directly only of sense experiences and of the "understanding" of their connection. But even the concept of the "real external world" of everyday thinking rests exclusively on sense impressions.

Now we must first remark that the differentiation between sense impressions and images is not possible; or, at least it is not possible with absolute certainty. With the discussion of this problem, which affects also the notion of reality, we will not concern ourselves but we shall take the existence of sense experiences as given, that is to say, as psychic experiences of a special kind. I believe that the first step in the setting of a "real external world" is the formation of the concept of bodily objects and of bodily objects of various kinds. Out of the multitude of our sense experiences we take, mentally and arbitrarily, certain repeatedly occurring complexes of sense impressions (partly in conjunction with sense impressions which are interpreted as signs for sense experiences of others), and we correlate to them a concept—the concept of the bodily object. Considered logically this concept is not identical with the total-ity of sense impressions referred to; but it is a free creation of the human (or animal) mind. On the other hand, this concept owes its meaning and its justification exclusively to the totality of the sense impressions which we associate with it.

The second step is to be found in the fact that, in our thinking (which determines our expectation), we attribute to this con-cept of the bodily object a significance, which is to a high degree independent of the sense impressions which originally give rise to it. This is what we mean when we attribute to the bodily object "a real existence." The justification of such a setting rests exclusively on the fact that, by means of such concepts and mental relations between them, we are able to orient ourselves in the labyrinth of sense impressions. These notions and relations, although free mental creations, appear to us as stronger and more unalterable than the individual sense experience itself, the character of which as anything other than the result of an illusion or hallucination is never completely guaranteed. On the other hand, these concepts and relations, and indeed the postulation of real objects and, generally speaking, of the existence of "the real world," have jus-tification only in so far as they are connected with sense impres-sions between which they form a mental connection.

The very fact that the totality of our sense experiences is such that by means of thinking (operations with concepts, and the creation and use of definite functional relations between them, and the coordination of sense experiences to these concepts) it can be put in order, this fact is one which leaves us in awe, but which we shall never understand. One may say "the eternal mystery of the world is its comprehensibility." It is one of the great realizations of Immanuel Kant that the postulation of a real external world would be senseless without this comprehensibility.

In speaking here of "comprehensibility," the expression is used in its most modest sense. It implies: the production of some sort of order among sense impressions, this order being produced by the creation of general concepts, relations between these concepts, and by definite relations of some kind between the concepts and sense experience. It is in this sense that the world of our sense experiences is comprehensible. The fact that it is comprehensible is a miracle.

In my opinion, nothing can be said a priori concerning the manner in which the concepts are to be formed and connected, and how we are to coordinate them to sense experiences. In guiding us in the creation of such an order of sense experiences, success alone is the determining factor. All that is necessary is to fix a set of rules, since without such rules the acquisition of knowledge in the desired sense would be impossible. One may compare these rules with the rules of a game in which, while the rules themselves are arbitrary, it is their rigidity alone which makes the game possible. However, the fixation will never be final. It will have validity only for a special field of application (i.e., there are no final categories in the sense of Kant).

The connection of the elementary concepts of everyday thinking with complexes of sense experiences can only be comprehended intuitively and it is unadaptable to scientifically logical fixation. The totality of these connections—none of which is expressible in conceptual terms—is the only thing which differentiates the great building which is science from a logical but empty scheme of concepts. By means of these connections, the purely conceptual propositions of science become general statements about complexes of sense experiences.

We shall call "primary concepts" such concepts as are directly and intuitively connected with typical complexes of sense experiences. All other notions are—from the physical point of view—possessed of meaning only in so far as they are connected, by propositions, with the primary notions. These propositions are partially definitions of the concepts (and of the statements derived logically from them) and partially propositions not derivable from the definitions, which express at least indirect relations between the "primary concepts," and in this way between sense experiences.

Propositions of the latter kind are "statements about reality" or laws of nature, i.e., propositions which have to show their validity when applied to sense experiences covered by primary concepts. The question as to which of the propositions shall be considered as definitions and which as natural laws will depend largely upon the chosen representation. It really becomes absolutely necessary to make this differentiation only when one examines the degree to which the whole system of concepts considered is not empty from the physical point of view...

STRATIFICATION OF THE SCIENTIFIC SYSTEM

The aim of science is, on the one hand, a comprehension, as complete as possible, of the connection between the sense experiences in their totality, and, on the other hand, the accomplishment of this aim by the use of a minimum of primary concepts and relations. (Seeking, as far as possible, logical unity in the world picture, i.e., paucity in logical elements.)

Science uses the totality of the primary concepts, i.e., concepts directly connected with sense experiences, and propositions connecting them...

An adherent to the theory of abstraction or induction might call our layers "degrees of abstraction"; but I do not consider it justifiable to veil the logical independence of the concept from the sense experiences. The relation is not analogous to that of soup to beef but rather of check number to overcoat.

The layers are furthermore not clearly separated. It is not even absolutely clear which concepts belong to the primary layer. As a matter of fact, we are dealing with freely formed concepts, which, with a certainty sufficient for practical use, are intuitively connected with complexes of sense experiences in such a manner that, in any given case of experience, there is no uncertainty as to the validity of an assertion. The essential thing is the aim to represent the multitude of concepts and propositions, close to experience, as propositions, logically deduced from a basis, as narrow as possible, of fundamental concepts and fundamental relations which themselves can be chosen freely (axioms). The liberty of choice, however, is of a special kind; it is not in any way similar to the liberty of a writer of fiction. Rather, it is similar to that of a man engaged in solving a well-designed word puzzle. He may, it is true, propose any word as the solution; but, there is only *one* word which really solves the puzzle in all its parts. It is a matter of faith that nature —as she is perceptible to our five senses—takes the character of such a well-formulated puzzle. The successes reaped up to now by science do, it is true, give a certain encouragement for this faith...

First we try to get clearly in our minds how far the system of classical mechanics has shown itself adequate to serve as a basis for the whole of physics. Since we are dealing here only with the foundations of physics and with its development, we need not concern ourselves with the purely formal progresses of mechanics (equations of Lagrange, canonical equations, etc.). One remark, however, appears indispensable. The notion "material point" is fundamental for mechanics. If now we seek to develop the mechanics of a bodily object which itself can not be treated as a material point— and strictly speaking every object "perceptible to our senses" is of this category—then the question arises: How shall we imagine the object to be built up out of material points, and what forces must we assume as acting between them? The formulation of this question is indispensable, if mechanics is to pretend to describe the object *completely*.

It is in line with the natural tendency of mechanics to assume these material points, and the laws of forces acting between them, as invariable, since temporal changes would lie outside of the scope of mechanical explanation. From this we can see that classical mechanics must lead us to an atomistic construction of matter. We now realize, with special clarity, how much in error are those theorists who believe that theory comes inductively from experience. Even the great Newton could not free himself from this error ("*Hypotheses non fingo*"*)...

In my view, the greatest achievement of Newton's mechanics lies in the fact that its consistent application has led beyond this phenomenological point of view, particularly in the field of heat phenomena. This occurred in the kinetic theory of gases and in statistical mechanics in general. The former connected the equation of state of the ideal gases, viscosity, diffusion, and heat conductivity of gases and radiometric phenomena of gases, and gave the logical connection of phenomena which, from the point of view of direct experience, had nothing whatever to do with one another. The latter gave a mechanical interpretation of the thermodynamic ideas and laws and led to the discovery of the limit of applicability of the notions and laws of the classical theory of heat. This kinetic theory, which by far surpassed phenomenological physics as regards the logical unity of its foundations, produced, moreover, definite values for the true magnitudes of atoms and molecules which resulted from several independent methods and were thus placed beyond the realm of reasonable doubt. These decisive progresses were paid for by the coordination of atomistic entities to the material points, the constructively speculative character of these entities being obvious. Nobody could hope ever to "perceive directly" an atom. Laws concerning variables connected more directly with experimental facts (for example: temperature, pressure, speed) were deduced from the fundamental ideas by means of complicated calculations. In this manner physics (at least part of it), originally more phenomenologically constructed, was reduced, by being founded upon Newton's mechanics for atoms and molecules, to a basis further removed from direct experiment, but more uniform in character..

THE FIELD CONCEPT

[T]he electric field theory of Faraday and Maxwell represents probably the most profound transformation of the foundations of physics since Newton's time. Again, it has been a step in the direction of constructive speculation which has increased the distance between the foundation of the theory and sense experiences. The existence of the field manifests itself, indeed, only when electrically charged bodies are introduced into it. The differential equations of Maxwell connect the spatial and temporal differential coefficients of the electric and magnetic fields. The electric masses are nothing

Appendix C

more than places of non-vanishing divergence of the electric field. Light waves appear as undulatory electromagnetic field processes in space...

Everywhere (including the interior of ponderable bodies) the seat of the field is the empty space. The participation of matter in electromagnetic phenomena has its origin only in the fact that the elementary particles of matter carry unalterable electric charges, and, on this account, are subject on the one hand to the actions of ponderomotive forces and on the other hand possess the property of generating a field. The elementary particles obey Newton's law of motion for material points.

This is the basis on which H. A. Lorentz obtained his synthesis of Newton's mechanics and Maxwell's field theory. The weakness of this theory lies in the fact that it tried to determine the phenomena by a combination of partial differential equations (Maxwell's field equations for empty space) and total differential equations (equations of motion of points), which procedure was obviously unnatural. The inadequacy of this point of view manifested itself in the necessity of assuming finite dimensions for the particles in order to prevent the electromagnetic field existing at their surfaces from becoming infinitely large. The theory failed, moreover, to give any explanation concerning the tremendous forces which hold the electric charges on the individual particles. H. A. Lorentz accepted these weaknesses of his theory, which were well known to him, in order to explain the phenomena correctly at least in general outline.

Furthermore, there was one consideration which pointed beyond the frame of Lorentz's theory. In the environment of an electrically charged body there is a magnetic field which furnishes an (apparent) contribution to its inertia. Should it not be possible to explain the total inertia of the particles electromagnetically? It is clear that this problem could be worked out satisfactorily only if the particles could be interpreted as regular solutions of the electromagnetic partial differential equations. The Maxwell equations in their original form do not, however, allow such a description of particles, because their corresponding solutions contain a singularity. Theoretical physicists have tried for a long time, therefore, to reach the goal by a modification of Maxwell's equations. These attempts have, however, not been crowned with success. Thus it happened that the goal of erecting a pure electromagnetic field theory of matter remained unattained for the time being, although in principle no objection could be raised against the possibility of reaching such a goal. The lack of any systematic method leading to a solution discouraged further attempts in this direction. What appears certain to me, however, is that, in the foundations of any consistent field theory, the particle concept must not appear in addition to the field concept. The whole theory must be based solely on partial differential equations and their singularity-free solutions.

THE THEORY OF RELATIVITY

There is no inductive method which could lead to the fundamental concepts of physics. Failure to understand this fact constituted the basic philosophical error of so many investigators of the nineteenth century. It was probably the reason why the molecular theory and Maxwell's theory were able to establish themselves only at a relatively late date. Logical thinking is necessarily deductive; it is based upon hypothetical concepts and axioms. How can we expect to choose the latter so that we might hope for a confirmation of the consequences derived from them? ...

Probably never before has a theory been evolved which has given a key to the interpretation and calculation of such a heterogeneous group of phenomena of experience as has quantum theory. In spite of this, however, I believe that the theory is apt to beguile us into error in our search for a uniform basis for physics, because, in my belief, it is an *incomplete* reprensentation of real things, although it is the only one which can be built out of the fundamental concepts of force and material points (quantum corrections to classical mechanics). The incompleteness of the representation leads necessarily to the statistical nature (incompleteness) of the laws. I will now give my reasons for this opinion.

I ask first: How far does the Ψ function describe a real state of a mechanical system? Let us assume the Ψ_r to be the periodic solutions (put in the order of increasing energy values) of the Schrödinger equation. I shall leave open, for the time being, the question as to how far the individual Ψ_r are complete descriptions of physical states. A system is first in the state Ψ_1 of lowest energy E_1. Then during a finite time a small disturbing force acts upon the system. At a later instant one obtains then from the Schrödinger equation a Ψ function of the form

$$\Psi = \Sigma c_r \Psi_r$$

where the c_r are (complex) constants. If the c_r are "normalized," then $|c_1|$ is nearly equal to 1, $|c_2|$ etc. is small compared with 1. One may now ask: Does Ψ describe a real state of the system? If the answer is yes, then we can hardly do otherwise than ascribe to this state a definite energy E, and, in particular, an energy which exceeds E_1 by a small amount (in any case $E_1 < E < E_2$). Such an assumption is, however, at variance with the experiments on electron impact such as have been made by J. Franck and G. Hertz, if one takes into account Millikan's demonstration of the discrete nature of electricity. As a matter of fact, these experiments lead to the conclusion that energy values lying between the quantum values do not exist. From this it follows that our function Ψ does not in any way describe a homogeneous state of the system, but represents rather a statistical description in which the c_r represent probabilities of the individual energy values. It seems to be clear, therefore, that Born's statistical interpretation of quantum theory is the only possible one. The Ψ function does not in any way describe a state which could be that of a single system; it relates rather to many systems, to "an ensemble of systems" in the sense of statistical mechanics. If, except for certain special cases, the Ψ function furnishes only statistical data concerning measurable magnitudes, the reason lies not only in the fact that the operation of measuring introduces unknown elements, which can be grasped only statistically, but because of the very fact that the Ψ function does not, in any sense, describe the state of one single system. The Schrödinger equation determines the time variations which are experienced by the ensemble of systems which may exist with or without external action on the single system.

Such an interpretation eliminates also the paradox recently demonstrated by myself and two collaborators, and which relates to the following problem.

Consider a mechanical system consisting of two partial systems A and B which interact with each other only during a limited time. Let the Ψ function before their interaction be given. Then the Schrödinger equation will furnish the Ψ function after the interaction has taken place. Let us now determine the physical state of the partial system A as completely as possible by measurements. Then quantum mechanics allows us to determine the Ψ function of the partial system B from the measurements made, and from the Ψ function of the total system. This determination, however, gives a result which depends upon *which* of the physical quantities (observables) of A have been measured (for instance, coordinates or momenta). Since there can be only one physical state of B after the interaction which cannot reasonably be considered to depend on the particular measurement we perform on the system A separated from B it may be concluded that the Ψ function is not unambiguously coordinated to the physical state. This coordination of several Ψ functions to the same physical state of system B shows again that the Ψ function cannot be interpreted as a (complete) description of a physical state of a single system. Here also the coordination of the Ψ function to an ensemble of systems eliminates every difficulty.*

SUMMARY

Physics constitutes a logical system of thought which is in a state of evolution, whose basis cannot be distilled, as it were, from experience by an inductive method, but can only be arrived at by free invention. The justification (truth content) of the system rests in the verification of the derived propositions by sense experiences, whereby the relations of the latter to the former can only be comprehended intuitively. Evolution is proceeding in the direction of increasing simplicity of the logical basis. In order further to approach this goal, we must resign to the fact that the logical basis departs more and more from the facts of experience, and that the path of our thought from the fundamental basis to those derived propositions, which correlate with sense experiences, becomes continually harder and longer.

Our aim has been to sketch, as briefly as possible, the development of the fundamental concepts in their dependence upon the facts of experience and upon the endeavor to achieve internal perfection of the system. These considerations were intended to illuminate the present state of affairs, as it appears to me. (It is unavoidable that a schematic historic exposition is subjectively colored.)

I try to demonstrate how the concepts of bodily objects, space, subjective and objective time, are connected with one another and with the nature of our experience. In classical mechanics the concepts of space and time become independent. The concept of the bodily object is replaced in the foundations by the concept of the material point, by which means mechanics becomes fundamentally atomistic. Light and electricity produce insurmountable difficulties when one attempts to make mechanics the basis of all physics. We are thus led to the field theory of electricity, and, later on to the attempt to base physics entirely upon the concept of the field (after an attempted compromise with classical mechanics). This attempt leads to the theory of relativity (evolution of the notion of space and time into that of the continuum with metric structure).

I try to demonstrate, furthermore, why in my opinion quantum theory does not seem capable to furnish an adequate foundation for physics: one becomes involved in contradictions if one tries to consider the theoretical quantum description as a complete description of the individual physical system or event.

On the other hand, the field theory is as yet unable to explain the molecular structure of matter and of quantum phenomena. It is shown, however, that the conviction of the inability of field theory to solve these problems by its methods rests upon prejudice.

Appendix C

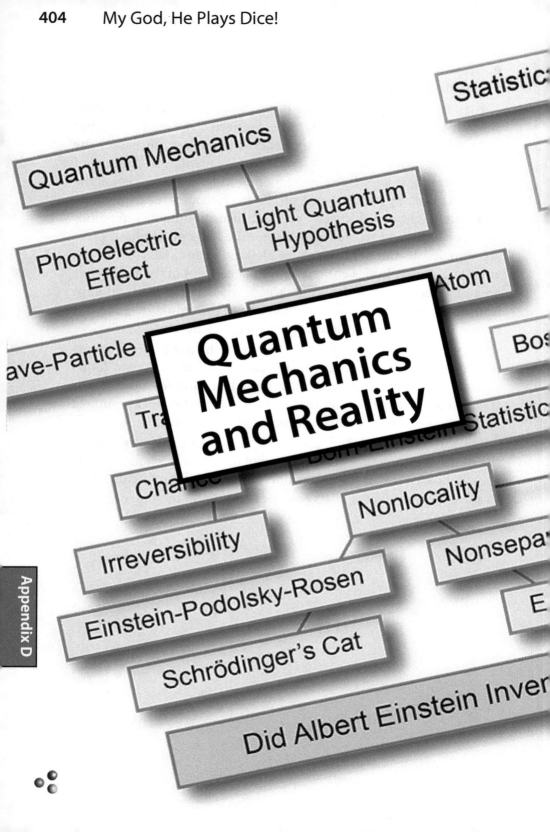

Quantum Mechanics and Reality

Dialectica, 2, issue 3-4, pp.320-324 (1948)

I. In what follows I shall explain briefly and in an elementary way why I consider the methods of quantum mechanics fundamentally unsatisfactory. I want to say straight away, however, that I will not deny that this theory represents an important, in a certain sense even final, advance in physical knowledge. I imagine that this theory may well become a part of a subsequent one, in the same way as geometrical optics is now incorporated in wave optics: the inter-relationships will remain, but the foundation will be deepened or replaced by a more comprehensive one.

I consider a free particle described at a certain time by a spatially restricted ψ-function (completely described - in the sense of quantum mechanics). According to this, the particle possesses neither a sharply defined momentum nor a sharply defined position. In which sense shall I imagine that this representation describes a real, individual state of affairs? Two possible points of view seem to me possible and obvious and we will weigh one against the other:

(a) The (free) particle really has a definite position and a definite momentum, even if they cannot both be ascertained by measurement in the same individual case. According to this point of view, the ψ-function represents an incomplete description of the real state of affairs. This point of view is not the one physicists accept. Its acceptance would lead to an attempt to obtain a complete description of the real state of affairs as well as the incomplete one, and to discover physical laws for such a description. The theoretical framework of quantum mechanics would then be exploded.

(b) In reality the particle has neither a definite momentum nor a definite position; the description by ψ-function is in principle a complete description. The sharply-defined position of the particle, obtained by measuring the position, cannot be interpreted as the position of the particle prior to the measurement. The sharp localisation which appears as a result of the measurement is brought about only as a result of the unavoidable (but not unimportant) operation of measurement. The result of the measurement depends not only on the real particle situation but also on the nature of the measuring mechanism, which in principle is incompletely known. An analogous situation arises when the momentum or any other observable relating to the particle is being measured. This is presumably the interpretation preferred by physicists at present; and one has to admit that it alone does justice in a natural way to the empirical state of affairs expressed in Heisenberg's principle within the framework of quantum mechanics.

According to this point of view, two ψ-functions which differ in more than trivialities always describe two different real situations (for example, the particle with well-defined position and one with well-defined momentum).

The above is also valid, *mutatis mutandis*, to describe systems which consist of several particles. Here, too, we assume (in the sense of interpretation Ib) that the ψ-function completely describes a real state of affairs, and that two (essentially) different ψ-functions describe two different real states of affairs, even if they could lead to identical results when a complete measurement is made. If the results of the measurement tally, it is put down to the influence, partly unknown, of the measurement arrangements.

II

If one asks what, irrespective of quantum mechanics, is characteristic of the world of ideas of physics, one is first of all struck by the following: the concepts of physics relate to a real outside world, that is, ideas are established relating to things such as bodies, fields, etc., which claim a 'real existence' that is independent of the perceiving subject - ideas which, on the other hand, have been brought into as secure a relationship as possible with the sense-data. It is further characteristic of these physical objects that they are thought of as arranged in a space-time continuum. An essential aspect of this arrangement of things in physics is that they lay claim, at a certain time, to an existence independent of one another, provided these objects 'are situated in different parts of space'. Unless one makes this kind of assumption about the independence of the existence (the 'being-thus') of objects which are far apart from one another in space which stems in the first place from everyday thinking - physical thinking in the familiar sense would not be possible. It is also hard to see any way of formulating and testing the laws of physics unless one makes a clear distinction of this kind. This principle has been carried to extremes in the field theory by localizing the elementary objects on which it is based and which exist independently of each other, as well as the elementary laws which have been postulated for it, in the infinitely small (four-dimensional) elements of space.

The following idea characterizes the relative independence of objects far apart in space (A and B): external influence on A has no direct influence on B; this is known as the 'principle of contiguity', which is used consistently only in the field theory. If this axiom were to be completely abolished, the idea of the existence of (quasi-) enclosed systems, and thereby the postulation of laws which can be checked empirically in the accepted sense, would become impossible.

III

I now make the assertion that the interpretation of quantum mechanics (according to Ib) is not consistent with principle II. Let us consider a physical system S_{12} which consists of two part-systems S_1 and S_2. These two part-systems may have been in a state of mutual physical interaction at an earlier time. We are, however, considering them at a time when this interaction is at an end.

Let the entire system be completely described in the quantum mechanical sense by a ψ-function ψ_{12} of the coordinates $q_1,\ldots$ and $q_2,\ldots$ of the two part-systems (ψ_{12} cannot be represented as a product of the form $\psi_1 \psi_2$ but only as a sum of such products). At time t let the two part-systems be separated from each other in space, in such a way that ψ_{12} only differs from 0 when $q_1,\ldots$ belong to a limited part R_1 of space and $q_2, \ldots$ belong to a part R_2 separated from R_1.

The ψ-functions of the single part-systems S_1 and S_2 are then unknown to begin with, that is, they do not exist at all. The methods of quantum mechanics, however, allow us to determine ψ_2 of S_2 from ψ_{12} if a complete measurement of the part-system S_1 in the sense of quantum mechanics is also available. Instead of the original ψ_{12} of S_{12}, one thus obtains the ψ-function ψ_2 of the part-system S_2.

But the kind of complete measurement, in the quantum theoretical sense, that is undertaken on the part system S_1, that is, which observable we are measuring, is crucial for this determination. For example, if S_1 consists of a single particle, then we have the choice of measuring either its position or its momentum components.

Any "measurement" instantaneously collapses the two-particle wave function ψ_{12}. There is no "later" collapse when measuring the "other" system S_2. The resulting ψ_2 depends on this choice, so that different kinds of (statistical) predictions regarding measurements to be carried out later on S_2 are obtained, according to the choice of measurement carried out on S_1. This means, from the point of view of the interpretations of Ib, that according to the choice of complete measurement of S_1 a different real situation is being created in regard to S_2, which can be described variously by ψ_2, $\psi_{2'}$, $\psi_{2''}$, etc.

Seen from the point of view of quantum mechanics alone, this does not present any difficulty. For, according to the choice of measurement to be carried out on S_1, a different real situation is created, and the necessity of having to attach two or more different ψ-functions ψ_2, $\psi_{2'}$, ... to one and the same system S_1 cannot arise.

It is a different matter, however, when one tries to adhere to the principles of quantum mechanics and to principle II, i.e. the independent existence of the real state of affairs existing in two separate parts of space R_1 and R_2. For in our example the complete measurement on S_1 represents a physical operation which only affects part R_1 of space.

Such an operation, however, can have no direct influence on the physical reality in a remote part R_2 of space. It follows that every statement about S_2 which we arrive at as a result of a complete measurement of S_1 has to be valid for the system S_2, even if no measurement whatsoever is carried out on S_1. This would mean that all statements which can be deduced from the settlement of ψ_2 or $\psi_{2'}$ must simultaneously be valid for S_2. This is, of course, impossible, if ψ_2, $\psi_{2'}$, etc. should represent different real states of affairs for S_2, that is, one comes into conflict with the Ib interpretation of the ψ-function.

There seems to me no doubt that those physicists who regard the descriptive methods of quantum mechanics as definitive in principle would react to this line of thought in the following way: they would drop the requirement II for the independent existence of the physical reality present in different parts of space; they would be justified in pointing out that the quantum theory nowhere makes explicit use of this requirement.

I admit this, but would point out: when I consider the physical phenomena known to me, and especially those which are being so successfully encompassed by quantum mechanics, I still cannot find any fact anywhere which would make it appear likely that requirement II will have to be abandoned.

I am therefore inclined to believe that the description of quantum mechanics in the sense of Ia has to be regarded as an incomplete and indirect description of reality, to be replaced at some later date by a more complete and direct one.

Analysis

Einstein's reality includes "bodies" and "fields." Unfortunately, continuous fields are an idealization, an abstraction, compared to material bodies. Even radiation, thought by Maxwell to be a continuous field, are in reality averages over the light quanta that Einstein himself discovered.

Einstein knows that he too is a dogmatist At all events, one should beware, in my opinion, of committing oneself too dogmatically to the present theory in searching for a unified basis [i.e., a continuous field theory] for the whole of physics.

Einstein's local reality means all properties are determined by functions in the infinitesimally small volume around a point (no "action-at-a-distance").

Einstein accepts Schrödinger's 1935 criticism of his "separation principle," now being called contiguity.

Einstein cannot accept the main fact of "entangled" systems explained to him by Schrödinger, that they cannot be separated.

Appendix D

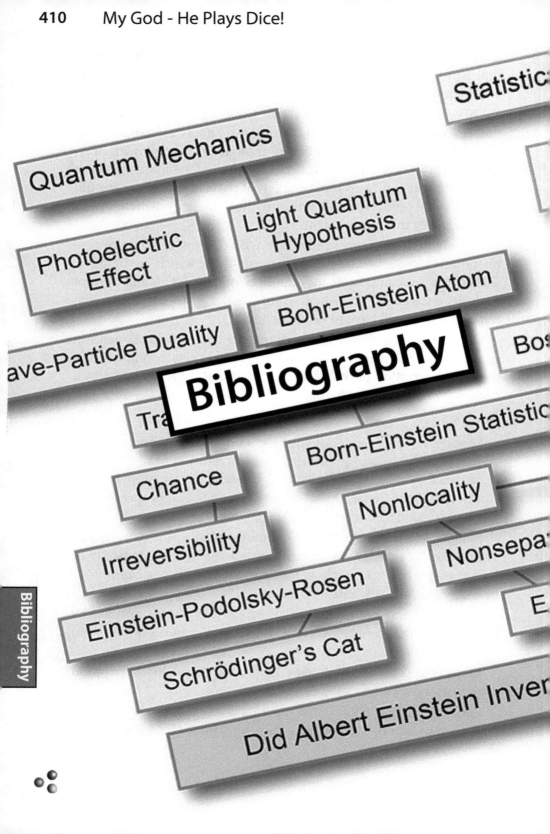

References

Afriat, A., & Selleri, F. (1998). *The Einstein, Podolsky, and Rosen Paradox in Atomic, Nuclear, And Particle Physics* New York: Plenum Press.

Albert, D. Z. (2015). *After Physics*. Harvard University Press.

Ananthaswamy, A. (2018). *Through Two Doors at Once*. Dutton.

Aspect, A. (1999). "Bell's Inequality Test: More Ideal Than Ever." *Nature*, 398(6724), 189.

Aspect, A., Grangier, P., & Roger, G. (1982). "Experimental Realization of Einstein-Podolsky-Rosen-Bohm *Gedankenexperiment*: a New Violation of Bell's Inequalities." *Physical Review Letters*, 49(2), 91.

Bacciagaluppi, G, and A. Valentini. (2009) *Quantum Theory at the Crossroads: Reconsidering the 1927 Solvay Conference*. Cambridge: Cambridge University Press.

Ball, P. (2018). *Beyond Weird*. Random House.

Ballentine, L. E. (1972). "Einstein's Interpretation of Quantum Mechanics." *American Journal of Physics*, 40(12), 1763-1771.

———. (1970). "The Statistical Interpretation of Quantum Mechanics." *Reviews of Modern Physics*, 42(4), 358.

Barrett, J. A. (1999). *The Quantum Mechanics of Minds and Worlds*. Oxford.

Barrett, J. A., & Byrne, P. (Eds.). (2012). *The Everett Interpretation Of Quantum Mechanics: Collected Works 1955-1980*. Princeton University Press.

Becker, A. (2018) *What Is Real?* Basic Books

Belinfante, F. J. (1973) *A Survey of Hidden-Variable Theories*, Pergamon Press.

Bell, J. S. (1964) "On the Einstein-Podolsky-Rosen Paradox," *Physics*, 1.3, p.195

———. (1990) "Against Measurement". In Miller (1989)..

Bell, J. S., & A. Aspect. (1987) *Speakable and Unspeakable in Quantum Mechanics: Collected Papers on Quantum Philosophy*. Cambridge University Press.

Bell, J. S., Bell, M., Gottfried, K., & Veltman, M. (2001). *John S. Bell on the Foundations of Quantum Mechanics*. World Scientific.

Beller, M, (1999) *Quantum Dialogues*, University of Chicago Press.

Bernstein, J. (1979) *Einstein*. Penguin.

———. (1991). *Quantum Profiles*. Princeton Univ. Press.

———. (2005) *Secrets of the Old One: Einstein, 1905*. New York: Copernicus.

Bitbol, M. (2007). Schrödinger Against Particles and Quantum Jumps. In Bacciagaluppi & Valentini (81-106)

Bricmont, J. (2016). *Making Sense Of Quantum Mechanics*. Springer.

———. (2017). *Quantum Sense And Nonsense*. Springer.

Brillouin, L. (2013) *Science and Information Theory*: 2nd Edition. Mineola, New York: Dover Publications.

Bohm, D. (1951) *Quantum Theory*. Prentice-Hall.

———. (1952) "A Suggested Interpretation of the Quantum Theory in Terms of 'Hidden' Variables. I," *Physical Review*, 85, p.166. "II," *Physical Review*, vol.85, p.180.

Bohm, D., & Aharonov, Y. (1957). "Discussion of Experimental Proof for the Paradox of Einstein, Rosen, and Podolsky." *Physical Review*, 108(4), 1070.

Bohr, N. (1913) "On the Constitution of Atoms and Molecules I, " *Philosophical Magazine*, v, 26, p.1

———. (1958) *Atomic Physics and Human Knowledge*. New York, Wiley

Bohr, N., & Rosenfeld, L. (1972). *Collected Works*: 13 Volumes1. North-Holland Publishing Company.

Bohr, N., French, A. P., & Kennedy, P. J. (1985). *Niels Bohr, A Centenary Volume*. Cambridge, MA: Harvard University Press.

Boltzmann, L. (2011) *Lectures on Gas Theory*. New York: Dover..

Bolles, E. B. (2004). *Einstein Defiant: Genius Versus Genius in the Quantum Revolution*. Joseph Henry Press.

Born, M. (1926) "Quantum Mechanics Of Collision Processes," *Zeitschrift für Physik*, 38, 803-827.

———. (1936) Atomic Physics.

———. (1964) *Natural Philosophy of Cause and Chance*. New York: Dover

———. Born, M. (1971). *The Born-Einstein Letters*. Macmillan, New York.

———. (1978). *My Life: Recollections of a Nobel Laureate*. Taylor & Francis.

Bricmont, J. (2016). *Making Sense Of Quantum Mechanics*. Berlin: Springer.

Broda, E., & Gay, L. (1983). *Ludwig Boltzmann Man, Physicist, Philosopher*. Ox Bow Press.

Bub, J. (1999). *Interpreting the Quantum World*. Cambridge University Press.

Byrne, P. (2010). *The Many Worlds Of Hugh Everett Iii: Multiple Universes, Mutual Assured Destruction, And The Meltdown Of A Nuclear Family*. Oxford University Press.

Cassidy, D. C. (1993) *Uncertainty: The Life and Science of Werner Heisenberg*. W. H. Freeman.

Cassirer, E. (1956) *Determinism and Indeterminism in Modern Physics*. Yale.

Cercignani, C. (2006) *Ludwig Boltzmann: The Man Who Trusted Atoms*. Oxford University Press.

Clarke, N. (Ed.). (1960). *A Physics Anthology*. Chapman and Hall.

Clauser, J. F., Horne, M. A., Shimony, A., & Holt, R. A. (1969). "Proposed Experiment To Test Local Hidden-Variable Theories." *Physical Review Letters*, 23(15), 880.

Compton, A. H., & Johnston, M. (1915). *The Cosmos of Arthur Holly Compton*. Knopf.

Darrigol, O. (2014). "The Quantum Enigma," in Janssen & Lehner. 2014, 117.

Davies, P. C. W., and Julian R. Brown, eds. (1993) *The Ghost in the Atom: A Discussion of the Mysteries of Quantum Physics*. Cambridge.

De Broglie, L. (1929) *Wave Nature of the Electron*, Nobel lecture.

Dedekind, R. (1901) "The Nature and Meaning of Numbers," in *Essays on the Theory of Numbers*, Dover (1963)

d'Espagnat, B. (1979). "The Quantum Theory And Reality." *Scientific American*, 241(5), 158-181.

DeWitt, B. S., & Graham, N. (Eds.). (1973). *The Many Worlds Interpretation of Quantum Mechanics*. Princeton University Press.

Dirac, P. A. M. (1930) *Principles of Quantum Mechanics*. 1st edition. Oxford.

Doyle, B. (2011). *Free Will: The Scandal in Philosophy*. I-Phi Press.

———. (2016a) *Great Problems in Philosophy and Physics Solved?* I-Phi Press.

———. (2016b) *Metaphysics: Problems, Paradoxes, and Puzzles Solved?*: I-Phi Press.

Dresden, M. (1987) *H.A.Kramers Between Tradition and Revolution*, Springer.

Dürr, D., & Teufel, S. (2009). *Bohmian Mechanics*. Berlin: Springer

Eddington, A. S. (1927) *The Nature of the Physical World*. Cambridge University Press.

———. (1936) *New Pathways In Science*. Cambridge University Press.

Ehrenfest, P., & Ehrenfest, T. (1959). *The Conceptual Foundations Of The Statistical Approach In Mechanics*. Cornell University Press.

Einstein, A. *The Collected Papers of Albert Einstein*, vols 1-15. Online at https://einsteinpapers.press.princeton.edu/

———. (1905a) "On a Heuristic Point of View Concerning the Production and Transformation of Light," *CPAE vol. 2*, Doc.14.

———. (1905b) "On the Movement of Small Particles Suspended in Statioary Liquids Required by the Molecular Theory of Heat," *CPAE vol. 2*, Doc.16.

———. (1905c) "On the Electrodnamics of Moving Bodies" *CPAE vol. 2*, Doc.23.

———. (1906a) "On the Theory of Light Production and Light Absorption" *CPAE vol. 2*, Doc.34.

———. (1907) "Planck's Theory of Radiation and the Theory of Specific Heat" *CPAE vol. 2*, Doc.38.

———. (1909) "On the Present Status of the Radiation Problem," *CPAE vol. 2*, Doc.56.

———. (1909) "On the Development of Our Views Concerning the Nature and Constitution of Radiation," *CPAE vol. 2*, Doc.60.

———. (1916) "Emission and Absorption of Rsdiation in Quantum Theory," *CPAE vol. 6*, Doc.34.

———. (1917) "On the Quantum Theory of Radiation," *CPAE vol. 6*, Doc.38.

———. (1922) *The Meaning of Relativity*, 5th edition. Princeton University Press

———. (1931) "Maxwell's Influence on the Evolution of the Idea of Physical Reality," in *James Clerk Maxwell: A Commemoration Volume*, Cambridge University Press.

———. (1934) *Ideas And Opinions*, New York: Bonanza Books, 1954.

———. (1936) "Physics and Reality," *Journal of the Franklin Institute*, Vol.221, No.3, March.

———. (1948) "Quantum Mechanics and Reality," *Dialectica*, 2, issue 3-4, pp.320-324.

———. (1949a) "Autobiography," in *Albert Einstein, Philosopher-Scientist*,

Library of Living Philosophers, Ed. Paul Arthur Schilpp, pp.81-89

———. (1949b) "Reply to Criticisms," in Schilpp, pp.665-688

Einstein, ., B. Podolsky, and N. Rosen. (1935) "Can Quantum-mechanical Description of Physical Reality Be Considered Complete?," *Physical Review*, 47, 777-80

Einstein, A., and M.Born. (2005) *The Born-Einstein Letters: Friendship, Politics and Physics in Uncertain Times*. Macmillan.

Einstein, A., and L. Infeld. (1961). *The Evolution of Physics: The Growths of Ideas from Early Concepts to Relativity and Quanta*. Cambridge University Press.

Einstein, A. and R. Penrose. (2005) *Einstein's Miraculous Year: Five Papers That Changed the Face of Physics*. Edited by John Stachel. Princeton, NJ: Princeton University Press.

Ellis, J., & Amati, D. (Eds.) (2000). *Quantum Reflections*. Cambridge.

Enz, C. P. (2010). *No Time To Be Brief: A Scientific Biography Of Wolfgang Pauli*. Oxford University Press

Farmelo, G. (2009). *The Strangest Man: The Hidden Life Of Paul Dirac, Quantum Genius*. Basic Books.

Feynman, R. P., & Brown, L. M. (2005). *Feynman's Thesis: A New Approach to Quantum Theory*. World Scientific.

Feynman, R. (1967). *The Character Of Physical Law*. MIT press.

Fine, A. (1996) *The Shaky Game. Einstein Realism and the Quantum Theory*. 2nd ed., University of Chicago Press.

Fölsing, A. (1997). *Albert Einstein: a Biography*. Viking.

Frank, P,. (2002) *Einstein: His Life And Times*. Cambridge, Mass.: Da Capo Press: Da Capo Press.

French, A. P. (1979). *Einstein. A Centenary Volume*. Harvard University Press.

Galison, P. (2004). *Einstein's Clocks and Poincaré's Maps: Empires of Time*. WW Norton

Galison, P., Holton, G. J., & Schweber, S. S. (2008). *Einstein for the 21st Century: His Legacy in Science, Art, and Modern Culture*. Princeton.

Gamow, G. (1970). *My World Line* (Viking, New York).

Ghirardi, G. (2005). *Sneaking a Look At God's Cards: Unraveling the Mysteries of Quantum Mechanics*. Princeton University Press.

Gilder, L. (2008) *The Age of Entanglement: When Quantum Physics Was Reborn*. Knopf: New York.

Gisin, N. (2014). *Quantum Chance: Nonlocality, Teleportation and Other Quantum Marvels*. Springer.

Gottfried, K., & Yan, T. M. (2013). *Quantum Mechanics: Fundamentals*. Springer

Greenspan, N. T. (2005) *The End of the Certain World: The Life and Science of Max Born*.

Gribbin, J. R., Gribbin, M., & Einstein, A. (2005). *Annus Mirabilis: 1905, Albert Einstein, and the Theory of Relativity*. Chamberlain Bros., Penguin.

Hacking, I. (2006) *The Emergence of Probability: A Philosophical Study of Early Ideas about Probability, Induction and Statistical Inference*. 2nd edition. Cambridge ; New York: Cambridge University Press.

Halpern, P. (2015). *Einstein's Dice and Schrödinger's* Cat. Basic Books.

Heisenberg, W. (1927) *The Physical Content of Quantum Kinematics and Mechanics, English translation in Wheeler and Zurek (1984)*

———, (1930). *The Physical Principles Of Quantum Mechanics*. U. Chicago Press.

———, (1955) "The Copenhagen Interpretation of Quantum Mechanics," in *Physics and Philosophy*.

———, (1958). *Physics and Philosophy. The Revolution in Modern Science.* Harper and Row.

———, (1971). *Physics and Beyond*. London: Allen & Unwin.

———, (1989). *Encounters with Einstein: And Other Essays on People, Places, and Particles*. Princeton University Press.

Hermann, A. (1973) *The Genesis of the Quantum Theory*, MIT Press.

Holt, J. (2018) *When Einstein Walked with Gödel,* Farrar. Straus, Giroux

Holton, G. J. (1988). *Thematic Origins Of Scientific Thought: Kepler To Einstein.* Harvard University Press.

———, (2000). *Einstein, History, and Other Passions: The Rebellion Against Science at the End of the Twentieth Century*. Harvard University Press.

Holton, G. and Y. Elkana. (1982) *Albert Einstein: Historical and Cultural Perspectives*. Princeton University Press.

Howard, D. (1985) "Einstein on Locality and Separability." *Studies in History and Philosophy of Science* 16, 171-201.

———, (1990) "'Nicht sein kann was nicht sein darf,' or the Prehistory of EPR, 1909-1935: Einstein's Early Worries about the Quantum Mechanics of Composite Systems." In *Sixty-Two Years of Uncertainty*, ed. Arthur Miller.

———, (2007) "Revisiting the Einstein-Bohr Dialogue." *Iyyun: The Jerusalem Philosophical Quarterly* 56, 57-90. Special issue dedicated to the memory of Mara Beller.

———, (2014). "Einstein and The Development of Twentieth-Century Philosophy of Science," in Janssen & Lehner. 2014, 354-376.

Howard, D. and J. Stachel (Eds.) (2000). Einstein: The Formative Years, 1879-1909. Springer Science & Business Media.

Isaacson, W. (2008) *Einstein: His Life and Universe*. New York, NY: Simon & Schuster.

Jaeger, G. (2009) *Entanglement, Information, and the Interpretation of Quantum Mechanics*. 2009 edition. Berlin: Springer.

Jaeger, L. (2019) *The Second Quantum Revolution: From Entanglement to Quantum Computing and Other Super-Technologies.* Springer

Jammer, M. (1966) *The Conceptual Development of Quantum Mechanics.* Mc-Graw Hill.

———. (1974) *The Philosophy of Quantum Mechanics: The Interpretations of Quantum Mechanics in Historical Perspective*. New York: Wiley.

———. (2000). *Einstein and Religion: Physics and Theology*. Princeton.

Janssen, M., & Lehner, C. (Eds.). (2014). *The Cambridge Companion to Einstein.* Cambridge University Press.

Jauch, J. M. (1989). *Are Quanta Real?: a Galilean Dialogue*. Indiana.

Jauch, J. M., & Baron, J. G. (1990). Entropy, Information and Szilard's Paradox. in Leff & Rex, 160-172.

Joos, E, H. D. Zeh, C. Kiefer, D. J. W. Giulini, J. Kupsch, and I-O. Stamatescu. (2013) *Decoherence and the Appearance of a Classical World in Quantum Theory*. 2nd ed. Berlin, Heidelberg: Springer.

Kaiser, D. (2011). *How The Hippies Saved Physics: Science, Counterculture, and the Quantum Revival*. WW Norton & Company.

Kastner, R. E. (2012). *The Transactional Interpretation of Quantum Mechanics: The Reality of Possibility*. Cambridge University Press.

———. (2015). *Understanding Our Unseen Reality: Solving Quantum Riddles*. Imperial College Press.

Klein, M. J. (1964) "Einstein and the Wave-Particle Duality," *The Natural Philosopher*, vol.3, p.1-49

———. (1965). "Einstein, Specific Heats, and the Early Quantum Theory." *Science*, 148 (3667), 173-180.

———. (1967). "Thermodynamics in Einstein's Thought." *Science*, 157(3788), 509-516.

———. (1970). "The First Phase of the Bohr-Einstein Dialogue." *Historical Studies in the Physical Sciences*, 2, iv-39.

———. (1979). "Einstein and the Development of Quantum Physics." Einstein: A *Centenary Volume*, 133-151.

Kox, A.J. (2014) "Einstein on Statistical Physics. Fluctuations and Atomism," in Janssen & Lehner. 2014,

Kramers, H. A. (1923) *The Atom and the Bohr Theory of Its Structure*. London: Gyldendal.

Krauss, L. M., and R. Dawkins. (2013) *A Universe from Nothing: Why There Is Something Rather than Nothing*. New York: Atria Books.

Kuhn, T. S. (1978) *Black-Body Theory and the Quantum Discontinuity, 1894-1912*. Oxford University Press.

Kwiat, P. G., Mattle, K., Weinfurter, H., Zeilinger, A., Sergienko, A. V., & Shih, Y. (1995). "New High-Intensity Source of Polarization-Entangled Photon Pairs." *Physical Review Letters*, 75(24), 4337.

Lahti, P, and P. Mittelstaedt. (1985) *Symposum on the Foundations of Modern Physics: 30 Years of the Einstein-Podolsky-Rosen Gedankenexperiment*, World Scientific Publishing.

Lanczos, C.. (1974) *The Einstein Decade, 1905-1915*. New York: Academic Press.

Layzer, D. (1975). "The Arrow of Time." *Scientific American*, 233(6), 56-69.

———. (1991) *Cosmogenesis: The Growth of Order in the Universe*. New York: Oxford University Press.

Leff, H., & Rex, A. F. (2002). *Maxwell's Demon 2 Entropy, Classical and Quantum Information, Computing*. CRC Press.

Lehner, C. (2014). "Einstein's Realism snd His Critique of Quantum Mechanics." in Janssen & Lehner. 2014, 306-353..

Lestienne, Remy, and E C Neher. (1998) *The Creative Power of Chance.* University of Illinois Press.

Levenson, T. (2017). *Einstein in Berlin.* Random House.

Lifshitz, L. D, and E. M. Landau (1958) *Quantum Mechanics: Non-Relativistic Theory.* Addison-Wesley Publishing Company.

Lindley, D.. (1996) *Where Does the Weirdness Go?* Basic Books

———. (2001) *Boltzmann's Atom: The Great Debate That Launched a Revolution in Physics.* 1st ed. Free Press.

———. (2007) *Uncertainty: Einstein Heisenberg Bohr and the Struggle for the Soul of Science.* New York; Anchor Books:, Random House.

Ludwig, G. (1968). *Wave Mechanics.* Pergamon.

Mahon, B. (2004). The *Man Who Changed Everything: The Life of James Clerk Maxwell.* John Wiley & Sons.

Maudlin, T. (2011). *Quantum Non-Locality and Relativity: Metaphysical Intimations of Modern Physics.* John Wiley & Sons.

McEvoy, J. P., & Zarate, O. (2014). *Introducing Quantum Theory: A Graphic Guide.* Icon Books Ltd.

Mehra, J.. (1975) *The Solvay Conference in Physics,* D. Reidel Publishing.

———. (1999) *Einstein, Physics and Reality.* World Scientific Publishing.

Mehra, J., and H. Rechenberg. (2001) *The Historical Development of Quantum Theory* Volumes 1-6. New York: Springer.

Mermin, N. D. (2018). *"Hidden Variables and the Two Theorems of John Bell."* arXiv:1802.10119v1 [quant-ph] 27 Feb 2018

Messiah, A. (1961) *Quantum Mechanics,* North-Holland, John Wiley & Sons.

Miller, A. I. (Ed.) (1989) *Sixty-Two Years of Uncertainty,* Springer.

———. (2002). *Einstein, Picasso: Space, Time and the Beauty That Causes Havoc:* Basic Books, Perseus.

Monod, J.. (1972) *Chance and Necessity: An Essay on the Natural Philosophy of Modern Biology.* Translated by Austryn Wainhouse. New York: Vintage Books.

Moore, R. E. (1966). *Niels Bohr: The Man, His Science, And The World They Changed.* MIT Press.

Moore, W. J. (1992) *Schrödinger: Life and Thought.* Cambridge University Press.

Musser, G.. (2015) *Spooky Action At A Distance.* Scientific American/ Farrar, Straus, Giroux.

Myrvold, W. C., & Christian, J. (Eds.). (2009). *Quantum Reality, Relativistic Causality, and Closing the Epistemic Circle: Essays in Honour of Abner Shimony.* Springe

Ne'Eman, Y. (1981). To *Fulfill a Vision: Jerusalem Einstein Centennial Symposium on Gauge Theories and Unification of Physical Force*s. Addison Wesley.

Neumann, J. von. (1955) *Mathematical Foundations of Quantum Mechanics.* Princeton: Princeton University Press.

Nielsen, M. and I. Chuang. (2010) *Quantum Computtion and Quantum Information.* Cambridge Universuty Press.

Pais, A.. (1982) *Subtle Is the Lord: The Science and the Life of Albert Einstein.* Oxford University Press.

———. (1991) *Niels Bohr's Times,: In Physics, Philosophy, and Polity.* Oxford University Press.

———. (1994). *Einstein Lived Here.* Clarendon Press.

Pais, A., M. Jacob, D. I. Olive, and M. F. Atiyah. (2005) *Paul Dirac: The Man and His Work.* Cambridge University Press.

Pauli, W., Rosenfeld, L., & Weisskopf, V. (1957). *Niels Bohr And The Development Of Physics.* McGraw-Hill.

Pauli, W.. (1980) *General Principles of Quantum Mechanics,* Springer-Verlag, Berlin

Pauli, W., L. Rosenfeld, and V. Weisskopf. (1955) eds. *Niels Bohr and the Development of Physics; Essays Dedicated to Niels Bohr on the Occasion of His Seventieth Birthday.* McGraw-Hill.

Penrose, R. (1989). *The Emperor's New Mind: Concerning Minds and the Laws of Physics.* Oxford University Press.

Planck, M. (1949) *Scientific Autobiography.* Philosophical Library,

———. (1959). *The New Science.* Meridian Books.

———. (1981). *Where Is Science Going?* Ox Bow Press.

———. (1991). *The Theory Of Heat Radiation.* Dover.

———. (1993) *A Survey Of Physical Theory.* Dover.

Price, H. (1997). *Time's Arrow & Archimedes' Point: New Directions For The Physics Of Time.* Oxford University Press.

Prigogine, Il... (1984) *Order Out of Chaos.* Shambhala.

Poincaré, H. (1952). *Science And Hypothesis.* Dover.

Porter, T. M. (1988) *The Rise of Statistical Thinking, 1820-1900.* Princeton University Press.

Price, W., & Chissick, S. (1979). *The Uncertainty Principle and Foundations of Quantum Mechanics: A Fifty Years' Survey.* John Wiley & Sons.

Reif, F.. *Fundamentals of Statistical and Thermal Physics.* (1965) McGraw-Hill Science/Engineering/Math.

Rigden, J. S. (2005). *Einstein 1905.* Harvard University Press.

Rukeyser, M., & Gibbs, J. W. (1942). *Willard Gibbs.* Ox Bow Press.

Scarani, V. (2006). *Quantum Physics: A First Encounter: Interference, Entanglement, and Reality.* Oxford University Press.

Schilpp, P. A. (1949). *Albert Einstein: Philosopher-Scientist,* Library of Living Philosophers. Evanston, Illinois.

Schlosshauer, M.A. (2008) *Decoherence and the Quantum-to-Classical Transition.* Berlin; London: Springer.

Schrödinger, E. (1935) "Discussion of Probability between Separated Systems", *Proceedings of the Cambridge Physical Society* 31, issue 4, 32 issue 1

———. (1936) "Probability Relations between Separated Systems," *Proceedings of the Cambridge Physical Society.* 32 issue 2.

Bibliography

———. (1952). "Are There Quantum Jumps?" Part I. *The British Journal for the Philosophy of Science*, 3(10), 109-123. Part II 3(11) 233-242.

———. (1989). *Statistical Thermodynamics*. Dover

———. (1995). *The Interpretation Of Quantum Mechanics: Dublin Seminars (1949-1955) And Other Unpublished Essays*. Ox Bow Press.

Schrödinger, E., & Murphy, J. (1935). *Science and the Human Temperament*. Norton and Company

Selleri, F. (Ed.). (1998). *Quantum Mechanics Versus Local Realism: The Einstein-Podolsky-Rosen Paradox*. Plenum Press

Shannon, C. E., and W. Weaver. (1948) *The Mathematical Theory of Communication*. University of Illinois Press.

Sommerfeld, A.. (1923) *Atomic Structure and Spectral Lines*. 3rd ed. London: Methuen & Co.

Stachel, J.. (1986) "Einstein and the Quantum: Fifty Years of Struggle," in *From Quarks to Quasars: Philosophical Problems of Modern Physics*," R.G. Colodny, ed.

———. (2002) *Einstein from "B" to "Z."* Birkhäuser Boston.

———. (2005). *Einstein's Miraculous Year: Five Papers That Changed the Face of Physics*. Princeton University Press.

———. (2009) "Bohr and the Photon," In: *Quantum Reality, Relativistic Causality, and Closing the Epistemic Circle*. Springer, Dordrecht.

Stuewer, R. H. (1975). *The Compton Effect. Turning Points In Physics*. Science History Publications (Neale Watson)

Stone, A. D.. (2013) *Einstein and the Quantum*. Princeton University Press.

Ter Haar, D. (1967). *The Old Quantum Theory*. Pergamon.

Van der Waerden, B. L. (1968) ed. *Sources of Quantum Mechanics*. New York, N.Y: Dover Publications.

Vedral, V. (2018). *Decoding Reality: The Universe As Quantum Information*. Oxford University Press.

Weinberg, S. (1993) *The First Three Minutes: A Modern View Of The Origin Of The Universe*. New York: Basic Books.

———. (2008) *Cosmology*. Oxford University Press.

Wheeler, J. A., and W. H.Zurek. (1984) *Quantum Theory and Measurement*. Princeton University Press,

Whitrow, G. J. (1973). *Einstein, the Man and his Achievement*. Dover.

Wigner, E. P.. (1967) *Symmetries and Reflections*. Indiana.

———. (1970). "On Hidden Variables and Quantum Mechanical Probabilities." *American Journal of Physics*, 38(8), 1005-1009.

Woolf, H. (1980). *Some Strangeness In The Proportion. A Centennial Symposium To Celebrate The Achievements Of Albert Einstein*. Addison-Wesley

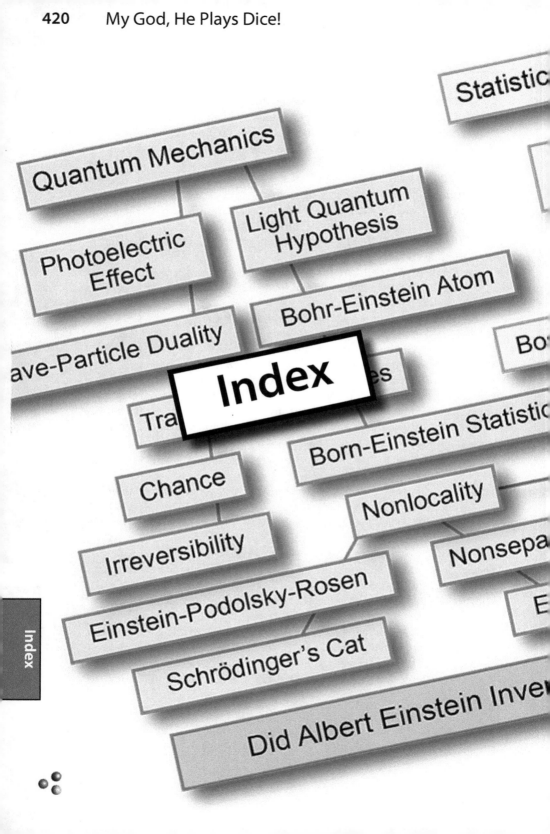

Index

Index

Books by Bob Doyle

Free Will:

The Scandal in Philosophy (2011)

Great Problems in Philosophy and Physics Solved? (2016)

Metaphysics:

Problems, Puzzles and Paradoxes, Solved? (2016)

My God, He Plays Dice!

How Albert Einstein Invented Most of Quantum Mechanics (2019)

PDFs of all of Bob's books are available for free on the I-Phi website, both complete books and as individual chapter PDFs for convenient assignment to students.

Image Credits

Some images are from websites with Creative Commons licenses or explicit permissions for non-profit and educational uses of their material, such as all the content of informationphilosopher.com and metaphysicist.com websites.

Colophon

This book was created on the *Apple Mac Pro* using the desktop publishing program *Adobe InDesign CC 2019,* with Myriad Pro and Minion Pro fonts. The original illustrations were created in *Adobe Illustrator* and *Adobe Photoshop.*

The author developed the first desktop publishing program, *Mac-Publisher,* for the Apple Macintosh, in 1984, the year of the Mac, intending to write some books on philosophy and physics. After many years of delay and a great deal of further research, books are finally in production, completing work, in his eighties, on ideas that emerged in his twenties.

Information Philosopher books are *bridges* from the information architecture of the printed page, from well before Gutenberg and his movable-type revolution, to the information architecture of the world-wide web, to a future of knowledge instantly available on demand anywhere it is needed in the world.

Information wants to be free. Information *can make you free.*

I-Phi printed books are still material, with their traditional costs of production and distribution. But they are physical pointers and travel guides to help you navigate the virtual world of information online, which of course still requires energy for its communication, and material devices for its storage and retrieval to displays.

But the online information itself is, like the knowledge in our collective minds, neither material nor energy, but pure information, pure ideas, the stuff of thought. It is as close as physical science comes to the notion of spirit, the ghost in the machine, the soul in the body.

Google It is this spirit that information philosophy wants to set free, with the help of Google and Wikipedia, Facebook and YouTube.

At a time when one in three living persons have a presence on the web, when the work of past intellects has been captured by Google Scholar, we have entered the age of *Information Immortality.*

WIKIPEDIA
The Free Encyclopedia

When you Google one of the concepts of information philosophy, the search results page will retrieve links to the latest versions of Information Philosopher pages online, and of course links to related pages in the Wikipedia, in the Stanford Encyclopedia of Philosophy, and links to YouTube lectures.

Thank you for purchasing this physical embodiment of our work. I-Phi Press hopes to put the means of intellectual production in the hands of the people.